THE ART OF ABDUCTION

THE ART OF ABDUCTION

Igor Douven

The MIT Press
Cambridge, Massachusetts
London, England

The MIT Press would like to thank the anonymous peer reviewers who provided comments on drafts of this book. The generous work of academic experts is essential for establishing the authority and quality of our publications. We acknowledge with gratitude the contributions of these otherwise uncredited readers.

This book was set in EB Garamond by the author in LuaLaTeX. Printed and bound in the United States of America.

Library of Congress Cataloging-in-Publication Data

Names: Douven, Igor, author.
Title: The art of abduction / Igor Douven.
Description: [Cambridge, Massachusetts] : Massachusetts Institute of Technology, [2022] | Includes bibliographical references and index.
Identifiers: LCCN 2021031324 | ISBN 9780262046701 (paperback)
Subjects: LCSH: Abduction (Logic) | Reasoning. | Practical reason. | Bayesian statistical decision theory.
Classification: LCC BC199.A26 D68 2022 | DDC 160—dc23
LC record available at https://lccn.loc.gov/2021031324

10 9 8 7 6 5 4 3 2 1

In memory of my father, Sjef Douven (1922–2018)

Contents

List of Figures

List of Tables

Preface

I first learned about abduction when, as a graduate student working on the scientific realism debate, I read Bas van Fraassen's *Laws and Symmetry* (van Fraassen, 1989). The thought that explanatory considerations guided our beliefs, in some form and to some extent, immediately seemed natural to me. Once I started paying attention, it occurred to me that I relied on such considerations in my reasoning on a daily basis. Not only that; it also appeared to me that I was right to do so, given how often things I had become convinced of on explanatory grounds turned out to be *true*.

There was a problem, though. Van Fraassen had given some seemingly inescapable arguments to the effect that abductive reasoning was to be avoided, on pain of irrationality. Perhaps I and other people tend to rely on this form of reasoning, but that does not make it right. There is a whole catalog of biases—systematic deviations from normative systems—that experimental studies have shown to be widespread. Letting explanatory considerations impact our beliefs might also belong there.

Van Fraassen's critique of abductive reasoning came as part of a defense of Bayesian epistemology, which at the time of his writing was still an "underground epistemology" (van Fraassen, 1989, p. 151). However, that epistemology was soon to become "all the rage" (Bovens & Hartmann, 2003, p. iii), and it has strongly dominated our thinking about confirmation, rationality, and belief revision for the last two decades. With the increasing popularity of Bayesian epistemology came further arguments purporting to show that any form of reasoning straying from the Bayesian path of wisdom—like abduction—betokened epistemically irresponsible behavior.

Since reading van Fraassen's critique, I have been trying to figure out whether my frequent, often almost automatic, reliance on abduction is something that should concern me and whether I should seek counseling. In this book, I aim to show that I am fine. That is probably good news for you, too, because, in general, people do tend to rely on abduction to some extent and in some form. This book is meant to help you understand that relying on abduction is nothing to be ashamed of and that, when done judiciously, reasoning abductively is perfectly in order.

You may have heard that abductive reasoning makes one vulnerable to Dutch bookies. Not so. Not only have I never encountered a Dutch bookie (and I am Dutch); even if they existed, it would take little effort to protect yourself against them while you continued to reason abductively. Also, there is no need to be concerned about the accuracy of your beliefs if you reason abductively. Perhaps—perhaps!—your beliefs are in this or that respect less accurate than those of your Bayesian twin (someone who is exactly like you, except that the twin uses Bayes's rule where you rely on some form of abduction), but in other respects your beliefs may well be *more* accurate. Indeed, Bayesians have been so busy pointing out possible downsides of abductive reasoning that they have completely forgotten to examine whether this form of reasoning might bring benefits that their favored machinery fails to offer and that may outweigh whatever costs abduction could be said to incur.

In thinking through the nature and status of abduction, I have been greatly helped by discussions with many people over the years. I would like to thank in particular the late Jonathan Adler, Daniel Andler, Martijn Blaauw, Luc Bovens, Peter Brössel, Filip Buekens, Otávio Bueno, the late Werner Callebaut, Alessandro Capone, Jake Chandler, Rafael De Clercq, Paul Cortois, Lieven Decock, Helen De Cruz, Kevin Demiddele, Ton Derksen, Dennis Dieks, Richard Dietz, Jacques Dubucs, Anna-Maria Eder, Paul Égré, Shira Elqayam, Pascal Engel, David Etlin, Martin Fischer, Branden Fitelson, the late Ronald Giere, Sanford Goldberg, Mario Günther, Stephan Hartmann, Rainer Hegselmann, Jan Heylen, Frank Hindriks, Pieter Hofstra, Leon Horsten, Christoph Kelp, Jakob Koscholke, Karolina Krzyżanowska, Theo Kuipers, James Ladyman, Hannes Leitgeb, Gert-Jan Lokhorst, Uskali Mäki, James McAllister, the late Ernan McMullin, Wouter Meijs, Patricia Mirabile, Fred Muller, Jennifer Nagel, Mike Oaksford, Diederik Olders, Erik Olsson, David Over, Herman Philipse, Hans Plets, Stathis Psillos, Eric Raidl, Henk de Regt, Herman de Regt, Hans Rott, Michael Schippers, Jonah Schupbach, Gerhard Schurz, Sebastian Sequoiah-Grayson, Mark Siebel, Henrik Singmann, Jan Sprenger, Jos Uffink, Mark van Atten, Jaap van Brakel, Bas van Fraassen, René van Woudenberg, Susan Vineberg, Jonathan Vogel, Jack Vromen, Verena Wagner, Jonathan Weisberg, Paul Weirich, Sylvia Wenmackers, Timothy Williamson, and Frank Zenker.

Over the last ten years, I have had the good fortune to be involved in various collaborative projects with a number of brilliant cognitive scientists and psychologists. I have learned a tremendous amount from Shira Elqayam, Peter Gärdenfors, Yasmina Jraissati, Mike Oaksford, David Over, Henrik Singmann, and Andrew Stewart. More than anything else, discussions with these friends and colleagues have let me realize that much of my own past research on rationality contributed little to a theory of rationality for real people. They encouraged me to get over my germaphobia, dive into the mud, and get my hands dirty, that is, to run my own experiments and obtain *real* data, which I found to be not nearly as spic and span as the made-up data we philosophers are used to work with. (In philosophy, we fit data to our models.) Furthermore, their lessons have had a profound influence on the present work more generally. In particular, they made me realize that while we will not be able to do completely without idealizations, and from time to time may want to do some house cleaning (what philosophers call "rational reconstruction"), we should resist the temptation to engage in robot epistemology, in which the elegance of our models counts as the main criterion of success, even if that prevents us from doing justice to a sometimes messy human psychology.

The most direct inspiration for the book came from two joint projects, one with Sylvia Wenmackers, the other with Jonah Schupbach. The first, which is described in Chapter 7, brought computational methods to bear on abduction and led to Douven and Wenmackers (2017), the second looked at abduction from an empirical perspective and led to Douven and Schupbach (2015a, 2015b), the main results of which are reported in Chapter 3. I learned as much from the collaboration with these friends as I enjoyed the interactions with them over the years.

Youness Ayaita, Brett Coppenger, Stephan Hartmann, Jon Matheson, Kevin McCain, Richard Pettigrew, Rolf Pfister, Ted Poston, William Roche, Josh Smith, Jan Sprenger, Paul Thorn, Mike Veber, and Christopher von Bülow read the entire manuscript and provided many valuable comments and suggestions, for which I thank them. Thanks also to the students in Stephan Hartmann's 2020 philosophy of science class at Ludwig Maximilian University of Munich for stimulating discussion of some of the materials presented here. Special thanks to Philip Laughlin at the Press for his encouragement and editorial guidance, and to Mykel Kochenderfer and Tim Wheeler, whose

Algorithms for Optimization (Kochenderfer & Wheeler, 2019) inspired the layout of appendix E and who generously helped me set up the software needed for making typeset Julia code look beautiful.

I am grateful to Daniel Andler, Anouk Barberousse, Jean-Baptiste Rauzy, and Pierre Wagner for their help with various matters administrative and institutional. In recent years, the friendship and collaboration with Lieven Decock, Shira Elqayam, Rainer Hegselmann, Karolina Krzyżanowska, Patricia Mirabile, and Hans Rott have meant much to me, as have the friendship and continuing discussions about virtually everything with Mark van Atten. Finally, I am immensely grateful for the love and support of my wife, Jacqueline.

Acknowledgments

In this volume, I use parts of previously published materials. Chapter 2 uses parts of my entry on abduction in *The Stanford Encyclopedia of Philosophy* (Douven, 2017a) and of my chapter on underdetermination in *The Routledge Companion to Philosophy of Science* (Douven, 2008). Section 3.4 is based on "Can the evidence for explanatory reasoning be explained away?," which appeared in *IEEE Transactions on Cognitive and Developmental Systems* (Douven, 2020a), and chapter 4 reuses parts of "Inference to the best explanation made coherent," published in *Philosophy of Science* (Douven, 1999), and parts of "Inference to the best explanation, Dutch books, and inaccuracy minimisation," published in *Philosophical Quarterly* (Douven, 2013). Chapter 5 relies heavily on "Scoring in context," which was published in *Synthese* (Douven, 2020b), and parts of chapter 6 come from "The ecological rationality of explanatory reasoning," which appeared in *Studies in History and Philosophy of Science* (Douven, 2020c). Chapter 7 extends "Optimizing group learning: An evolutionary computing approach," published in *Artificial Intelligence* (Douven, 2019a), and chapter 8 is based on "Can the skepticism debate be resolved?," which appeared in *Synthese* (Douven, 2007). I gratefully acknowledge permission to reuse these materials.

1
Introduction

> [T]he wisdom we mortals possess does not merely consist of remembering things past and apprehending the present, but on the basis of these two activities being able to predict the future, which is considered by serious men to be the highest form of human intelligence.
>
> —Giovanni Boccaccio, *The Decameron*

1.1 Why You Should Want to Know About Abduction

If you are consulting this book with malicious intent, looking for ways to expand your criminal pursuits, I encourage you to stop reading and to better your life. However, you may instead be a researcher—a cognitive scientist, a psychologist, a linguist, or a philosopher—who is interested in a type of inference that gives pride of place to explanatory considerations and that makes judgments on the quality of explanations and ultimately on explanatory superiority, which are crucial to what to believe and how to change our beliefs. You may have heard some good things about this type of inference but also, more likely, some bad things, and you may be wondering what to make of those things or how to reconcile them. In that case, I encourage you to read on.

If you are a psychologist, I encourage you to read on because there is accumulated evidence that abduction and, more generally, explanatory considerations play a key role in a great number of high-level cognitive processes, including categorization (e.g., Williams & Lombrozo, 2010, 2013; Edwards et al., 2019; Vasilyeva & Lombrozo, 2020), generalization (e.g., Lombrozo & Gwynne, 2014), learning (e.g., Chi et al., 1994; Baillargeon & DeJong, 2017;

Rittle-Johnson & Loehr, 2017; Walker et al., 2017; Ruggeri, Xu, & Lombrozo, 2019), understanding (Keil, 2006; Legare & Lombrozo, 2014; Walker & Lombrozo, 2017), interpreting behavior (Jern, Derrow-Pinion, & Piergiovanni, 2021), semantic processing (Krzyżanowska, 2015; Douven, 2016a; Douven et al., 2018, 2020; Mirabile & Douven, 2020; Douven, Elqayam, Mirabile, in press; Stewart et al., 2021; Rostworowski, Pietrulewicz, & Będkowski, in press), and reasoning and belief change (Pennington & Hastie, 1988, 1992, 1993; Douven & Schupbach, 2015a, 2015b; Johnston et al., 2017; ch. 3 examines in detail experimental evidence showing how pivotal explanation is in the realm of reasoning and belief change).

It has been suggested that abduction is also operative at a more basic level of cognitive functioning. Hermann von Helmholtz (1867, section 26) famously argued that perception is not the passive reception of sensory input that it might naïvely be thought to be but that it involves active, albeit typically unconscious, inference: A tower that we see in the distance projects only a tiny image on our retinas. If we perceive the tower as tall, that is the result of an inference that we make on the basis of the retinal image and an estimate (based on previous experience) of the distance between us and the tower. In this connection, Helmholtz (1867, p. 447ff) refers to John Stuart Mill's work on inductive inference. That may already be an implicit appeal to abduction, if Gilbert Harman (1965, 1968) is right that induction is a special kind of abduction (see also Thagard, 1978; Lycan, 1988; Weintraub, 2013). Leaving induction and its relation to abduction to the side for the moment, note passages in Helmholtz (1867) that suggest that he saw a clear connection between perception and explanatory reasoning. For instance, he attributes the occurrence of perceptual ambiguities—"when various . . . interpretations of an impression are possible"—to the fact that a person may vacillate between different explanations of the impression, or that different persons may settle for different explanations (Helmholtz, 1867, p. 440).

Helmholtz strongly influenced Irvin Rock's work on the relation between perception and inference, which is more explicit in relating perception directly to explanation. Rock (1983) spends several chapters discussing how explanatory reasoning helps to resolve cases of perceptual ambiguity. In chapter 6 of that book, he advances the hypothesis that what we perceive is guided by explanatory reasoning. More exactly, his claim—buttressed by references to the outcomes of a great number of experiments that Rock and others conducted

Figure 1.1: Illusion of a transparent square overlaying a tiled square (*left*), which according to Rock supports the hypothesis that explanation-guided inference is involved in perception; the illusion is disrupted when no easy explanation of the stimulus in terms of transparency is available (*right*). (*Source*: Rock, 1983, p. 140.)

over the years—is that the percept that we settle on is the one that offers the best explanation of the stimulus.

Consider the left panel of figure 1.1, which is reproduced from Rock (1983, p. 140). People tend to perceive this as a tiled square overlaid by a smaller square that is transparent, because—Rock argues—this offers the most elegant explanation of the stimulus. An alternative is to see the square as consisting of a number of unrelated figural fragments, but that explanation is not nearly as elegant. Note also how easily the illusion is disrupted by complicating the explanation made in terms of an overlaying transparent square, as in the right panel of figure 1.1. When we look at the left panel, we imagine that the transparent square is tinted and that the tint has a *consistent* effect on how we perceive the different colors of the tiles as seen through the overlying square. For the right panel, to maintain the explanation of looking through a transparent square for the right panel, we would have to assume *either* that the square is itself divided into four parts, each part tinted differently, and that the overlying square is rotated relative to the larger square exactly so that the different tints are aligned with the tiles, *or* that the underlying square does not consist of four tiles, two of which are black and two of which are light gray, but either is tiled in a more complicated way or has four tiles that are colored in more complicated ways, *or* make still more convoluted assumptions. That is obviously too much, which is why the transparency illusion does not, or not easily, arise for the right panel.

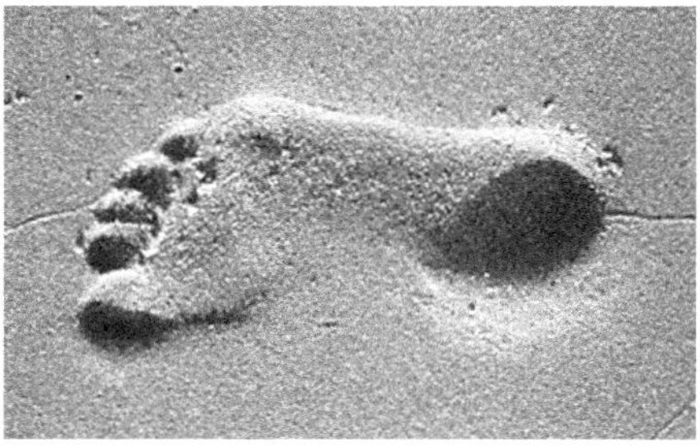

Figure 1.2: We tend to see this as a bas-relief footprint illuminated from above, but in reality it is an indented footprint illuminated from below. (Photo courtesy of Manuel Cazzaniga.)

Rock's book describes many more well-documented effects concerning perception, which he takes to indicate that in general we perceive what best explains the stimuli. These effects include stroboscopic motion, in which a stationary object *seems* to move, for instance, because different parts of it are illuminated shortly after each other (Rock & Ebenholtz, 1962; Sigman & Rock, 1974), anorthoscopic form perception, in which we perceive an object by viewing only small parts of it at a time (Rock & Halper, 1969; Rock & Gilchrist, 1975), and the kinetic depth effect, which (among other things) lets us interpret projections of three-dimensional objects onto a screen as shadows of three-dimensional objects rather than as transformations of two-dimensional objects (Rock & Smith, 1981). Rock (1983, ch. 6) canvasses various other potential explanations for these and other effects (e.g., empiricism, according to which interpretation is a mere matter of frequency of previous exposure, and the constancy view, which lets us prefer the interpretation that keeps the percept constant through changes of viewing angle or location), but he argues that his proposal offers a better explanation for those effects than do the competing accounts. Thus, he argues abductively for an abductive theory of perception.

It is also not hard to understand, abductively, why we tend to perceive figure 1.2 as a bas-relief (i.e., convex) footprint, although in reality the photo shows a normal concave footprint lit from below instead of, as is normally the case, from above. Judgments of explanation quality are typically relative to background assumptions, and precisely because objects are normally illuminated from above, the stimulus presented by figure 1.2 is best explained by assuming that the footprint rises from the sand. But that judgment is easily overruled: if people are given clues that there might be a light source *below* the footprint, they no longer perceive the footprint as rising out of the sand but instead see it as indented (Morgenstern, Murray, & Harris, 2011). It is readily appreciated how, with this kind of clue, the assumption that we see a bas-relief footprint will probably no longer be the best explanation of the stimulus.

This explanation of the bas-relief illusion is particularly interesting because the illusion has often been alleged to provide evidence in favor of the so-called Bayesian brain hypothesis, according to which the human nervous system operates largely as if it obeyed Bayesian prescriptions. From this perspective, the illusion highlights that we always start from prior probabilities or simply priors: when the brain computes the percept from the stimulus, it assigns a high prior to the light-from-above hypothesis and a low one to the light-from-below hypothesis. Jakob Hohwy (2013) makes a scientifically and philosophically informed case for this Bayesian approach to perception (though see Colombo, Elkin, & Hartmann, 2021). Although he does stress the importance of abduction for his case, he sees it as an essentially Bayesian principle (p. 25). He does not support this by any arguments, however, and as we explain later in this chapter, the claim is not as easy to maintain as he may think.

Jerry Fodor (2001) also recognizes the key role that abduction plays in cognitive processing. As he puts it, "by all the signs, the cognitive mind is up to its ghostly ears in abduction" (p. 78). In his opinion, that makes the prospects for real progress in cognitive science dim, because, in his opinion, "we do not know how abduction works [and so] we do not know how the cognitive mind works . . . " (Fodor, p. 78).[1] I am inclined to agree that at a minimum we do not know *much* about abduction and also that this hampers progress in cognitive science—and not only in that field. That was one important motivation for

1. Fodor (2001) thinks that, in particular, the computational theory of mind, which he otherwise favors, cannot account for the reliability of abduction (which he takes as a given). As Rellihan (2009) notes, however, there are independent reasons for holding that an adequate theory of mind must be hybrid, acknowledging the importance of both computational and associative mechanisms, and abduction is most plausibly accounted for in terms of the latter. Moreover, in view of the recent successes of deep learning (LeCun, Bengio, & Hinton, 2015; Goodfellow, Bengio, & Courville, 2016), Fodor's critique of associative theories of mind has come to appear much less compelling than it did twenty years back.

writing this book. Later in this chapter, I present some ideas about what may have stood in the way of promulgating a better understanding of abduction.

If you are a linguist, I encourage you to read on because abduction has been said to be fundamental to determining what a speaker means by an utterance. Specifically, it has been argued that decoding utterances is a matter of figuring out the best explanation of why someone said what she said when she said it. Even more specifically, authors working in the field of pragmatics have suggested that hearers invoke the Gricean maxims of conversation (Grice, 1989) to help them work out the best explanation of a speaker's utterance whenever the semantic content of the utterance is insufficiently informative for the purposes of the conversation, or is *too* informative, or is off topic, or is implausible, or is otherwise odd or inappropriate. As Yan Huang (1994, p. 2) puts it, "What pragmatics does is to provide a set of . . . *explanatory* principles which constrains the interpretation or production of an utterance whose linguistic representation has already been antecedently cognized" (emphasis mine).[2]

To go into a little more detail, pragmatics starts from the observation that true utterances can still be misleading. The semantic content of a speaker's utterance can be φ, but if the best explanation of why she made the utterance on the given occasion is, or entails, that she (also) wanted to convey ψ, then her utterance can be misleading even if φ is true—because ψ may be false. For example, we will typically interpret an utterance of

(1) Some of the guests are bringing flowers

as implicating (to use the technical term for what is conveyed over and beyond semantic content) that not *all* guests are bringing flowers. After all, if that were the case, it would be difficult to explain why the speaker did not simply *say* that all guests are bringing flowers (assuming a context in which the speaker is in a position to see whether all, or only some, guests are bringing flowers). So, if all guests are bringing flowers, then an utterance of (1) is misleading, even if the speaker is not *saying* anything false.

Or suppose a graduate student tells his supervisor

(2) You have published some papers that I really like.

2. In the same vein, see Bach and Harnish (1979), Dascal (1979), Hobbs (2004), Gabbay and Woods (2005, ch. 4), Chien (2008), and Douven (2012a); see also various papers in Bunt and Black (2000).

The supervisor can see two possible explanations of why the student uttered this sentence. One is that the student wanted to convey that he read some (but not all) of her papers and liked all the ones he read; the other is that he read some or all of her papers and liked some of those he read and some not so much. Depending on context, one or the other explanation may appear better to the supervisor. For instance, the first explanation may top the second if the supervisor is a prolific writer and the student is known to be somewhat lazy. In that case, the supervisor may infer that the student did not read all of her papers. If, on the other hand, the supervisor has so far only published a handful of papers, and the student is known for his critical attitude and for not being intimidated in the least by any of the professors, that may favor the second explanation. Naturally, the two explanations may also be in close competition, in which case the inference of an implicature can be guarded at best. This explains why implicatures can vary in strength (Douven, 2012a; Douven & Krzyżanowska, 2019). Furthermore, the example highlights a phenomenon that is discussed in chapter 3, to wit, that the extent to which the best explanation stands out as being the best can vary and can have a significant impact on people's willingness to infer the truth of that explanation.

If you are a philosopher, you may not need to be convinced of the importance of abduction, given how common appeals to this form of reasoning are in your field. For instance, social epistemologists have argued that our trust in other people's testimony rests on abductive reasoning (e.g., Harman, 1965; Adler, 1994; Fricker, 1994, 2017; Lipton, 1998). According to Jonathan Adler (1994, pp. 274–275), "[t]he best explanation for why the informant asserts that *P* is normally that . . . he believes it for duly responsible reasons and . . . he intends that I shall believe it too." That is why we are normally justified in trusting an informant's testimony.

And a number of philosophers of science hold that abduction is a cornerstone of scientific methodology (e.g., Boyd, 1984, 1985, 1990; Harré, 1986, 1988; Psillos, 1996, 1999; Lipton, 2004; Schupbach, 2017). According to Timothy Williamson (2017, pp. 334–335), "[t]he abductive methodology is the best science provides," and Ernan McMullin (1992) even goes so far as to call abduction "the inference that makes science." Relatedly, it has been said that abduction is the predominant mode of reasoning in medical diagnosis: physicians tend to go for the hypothesis that best explains the patient's symptoms (Josephson & Josephson, 1994; Dragulinescu, 2016; Stilgenbauer & Baratgin,

2018).[3] We look at some examples of the use of abduction in science in chapter 2 and consider abduction specifically in the context of medical decision making in chapters 6 and 7.[4]

In philosophy, abduction is frequently invoked in objections to so-called underdetermination arguments. We consider the issue of underdetermination at some length in chapter 2, and chapter 8 is devoted in its entirety to arguably the most prominent underdetermination argument in philosophy or anywhere: the argument for Cartesian skepticism. For now, suffice it to say that underdetermination arguments proceed by arguing that the data, or whatever counts as such in a given debate, are insufficient to warrant the adoption of this or that hypothesis, where the hypothesis at stake often appears commonsensical to us (that there is an external world, that others are conscious beings like us, that thanks to science we know a good deal about the mechanisms and processes causing observable events, and so on). While perhaps most prominent in epistemology and the philosophy of science, such arguments have either been advanced or are implicit in other areas of philosophy as well. Concomitantly, we find appeals to abduction in all those areas.

For instance, in metaphysics David Lewis (1973, 1986) is famous for defending modal realism, according to which other possible worlds are as real as our own world. Although typically met with an "incredulous stare"—as Lewis (1973, p. 86) acknowledges—he advocates the position because, he thinks, it offers the best explanation for a vast array of phenomena, including how we use modal expressions like "necessary" and "possibly" and how we talk about causes and effects. Similarly, Williamson (2016) gives an abductive argument for the hypothesis that there are objective modalities (which does not entail modal realism but still goes against, for instance, standard empiricist thinking according to which all modalities have a merely verbal status; see, e.g., van Fraassen, 1977). Philosophers of language have invoked abduction in debates

3. Interestingly, Patel and Groen (1986, 1991) and Patel, Groen, and Arocha (1990) found that the *degree* to which physicians rely on abductive reasoning may vary with their level of experience. In particular, they report a number of experiments showing that less experienced physicians rely more heavily on abductive reasoning than their more experienced colleagues.

4. Although they are not strictly speaking scientists, judges are truth-seekers whom we expect to reason as carefully as scientists. Michelon (2019) shows that they frequently rely on abduction, not only when assessing the evidence that is brought before the court but also in determining which legal principles have been tacitly assumed in relevant jurisprudence. On the role of abduction in the law, see also Walton (2005).

about the metaphysical status of truth and its supposed bearers. It has been argued, for instance, that minimalist conceptions of truth (see, e.g., Horwich, 1990) are metaphysically too lightweight to carry the explanatory workload required of truth by our current best theories of meaning and interpretation (for discussion, see Devitt, 1991, pp. 31–32; Davidson, 1996; Horsten, 2011, ch. 5). It has further been argued that we are warranted to believe in the existence of propositions because of the various explanatory roles that they play (e.g., van Elswyk, 2019). Possibly the oldest use of abduction in metaphysics is found in the philosophy of religion, in which authors through the ages have tried to argue that the existence of god best explains this or that part of reality, like the beauty or order of the world, human morality, or that people have religious beliefs (e.g., Hildebrand & Metcalf, in press; for discussion, see Schupbach, 2005, and De Cruz & De Smedt, 2015).[5]

Outside of metaphysics, Williamson (2017) argues that the case for classical logic over nonclassical logics is abductive: classical logic gives the best *overall* account of our various deductive practices, even if perhaps this or that nonclassical logic offers a more elegant solution of the semantic paradoxes.[6] And in his *Doing Philosophy* (Williamson, 2018), he notes that platonists defend their position "because they find it crucial for the best explanation of what mathematicians are doing" (p. 44). Indeed, in the same book (ch. 6) he argues extensively that abduction is central to philosophy in general, even as central as it is to science.

Finally, at the risk of sounding overly ambitious, I encourage you to read on for the sake of scientific progress. There are different ways in which science can progress. There is the kind of progress that makes headlines: the possibility of quantum entanglement over hundreds of miles established, a new disease discovered, the oldest yet humanoid skull found, a new treatment for cystic fibrosis developed, the first close-up pictures of the moons of Jupiter, artificial neural networks able to accurately predict protein structures from sequences of amino acids, and so on. But there is much scientific progress of a more prosaic variety about which the general public will not necessarily

5. On the role of abduction in metaphysics, see also Hawley (2006), Sider (2009), Biggs and Wilson (2017), and Schurz (2021).

6. On the role of abduction in logic, see also Priest (2005, p. 217), Russell (2015, 2019), and Hjortland (2019).

hear but that may be as important and consequential as the more newsworthy findings.[7]

For example, in recent decades much progress has been made in the field of data visualization. Thanks to the development of heat maps in the 1990s, we can more easily recognize patterns in the expression of genes in different tissues than was previously the case (Weinstein, 2008). Similarly, the development of so-called actograms—a kind of diagrams for representing the activity of organisms throughout the day—has contributed much to our understanding of the circadian rhythm (Sheredos et al., 2013).

More to the present point is Ian Hacking's (1984, p. 70) observation that "[t]he quiet statisticians have changed our world—not by discovering new facts or technical developments but by changing the ways that we reason, experiment and form our opinions about it." Progress is to an important part due to the fact that we have collected more data, which in turn is to an important part due to our having improved and expanded our *techniques* for collecting data. But progress is no less due to the fact that we have learned more about what we can and cannot conclude from the data in our possession and that we have learned more about how to determine what the data tells us about our theories and ultimately the mechanisms and processes that generated the data. In other words, we have learned more about what can be learned from the data we have. We owe much of this to Hacking's "quiet statisticians."

The quote from Hacking might be understood to suggest that the development of statistics has been completed. The opposite is the case, however: the development of statistics, and more generally of research methods, is an ongoing process that is accompanied by discussion and disagreement but that from time to time produces results that virtually all in the scientific community regard as progress. This concerns, for example, the development of statistical techniques that allow us to make more accurate estimates or to determine more precisely how certain we can be of a prediction.

Here is an illustration that is also relevant to subsequent discussion, in which we look in detail at some empirical data about explanatory reasoning. Regression analysis is one of the most commonly used statistical techniques for determining the influence (if any) of one or several variables (the "independent variables" or "predictor variables") on another variable (the "dependent

7. In fact, if Stegenga (2018) is right, medical progress, which is arguably the kind of progress most consistently covered by the media, tends to be overpraised.

variable" or "outcome variable"). The result of such an analysis can help us predict the value that the dependent variable will take on the basis of measured values of the independent variables. Suppose, for instance, that you are interested in the relationship between font size and reading speed. To investigate this, you let people read texts that are set in the same font but in different sizes, and you measure the speed at which they read the text. No doubt there will be variation in your participants' reading times, and if you are lucky, a regression analysis will find a significant relationship between reading time and font size.

Regression analysis has proven to be a more than highly valuable statistical technique. Not so long ago, however, a number of statisticians noted that in applications such as our fictional example, the method is not optimal. In that example, we assume that you will find variation in reading times that you can then relate to variation in font size. But it is to be noticed that part of the variation that you find may have to do with the fact that you have used different test subjects: not everyone is sensitive to variations in font size in the same way. This variation among test subjects can influence the outcome of your analysis in a way and to an extent that you do not want. In this experiment, you are not interested in individual differences among people; you want to establish the relationship between font size and reading speed *in general*.[8]

In response to this problem, so-called mixed-effects models have been developed (sometimes called "hierarchical linear models"). These models are intended to accommodate that there can be different sources of variance in the data, some of which may not be of interest to us. In this example, some variance will come from the fact that you manipulated the font size—the variance that interests you—but some will also come from the fact that the participants will vary in important respects. Some will be faster readers in general than others, some will have better eyesight than others, some will be more motivated to participate than others, some will be better rested at the

8. Stated more technically, the problem is that in standard regression analysis, the data are assumed to be independent and identically distributed. In particular, the independence part of this assumption—the probability of a data point taking a specific value is independent of which values are taken by the other data points—is typically violated when each participant contributed more than one data point: data points coming from the same participant will often be correlated. Violations of assumptions of statistical procedures are not always harmful, but violating the independence assumption in the case of a regression analysis *is* harmful (Judd, Westfall, & Kenny, 2012).

time of the experiment than others, and so on. Sometimes we are interested in precisely such differences. But in our fictional example we assume that you are *not* interested in those. Mixed-effects models help us to separate interesting from uninteresting variance in the data and thereby help us avoid discrediting the variable or variables of interest for failing to explain variance that is due merely to elements in our experimental design that we consider to be random (like the participants we happened to enroll, or the particular items we used as materials). Think of the way in which this is achieved as first creating a regression model for each participant separately and then constructing from those individual models a kind of aggregate model. It has been shown that this type of analysis tends to yield more accurate estimates of the particular parameters of interest than traditional regression models do (Baayen, 2008; Zuur et al., 2009; Singmann & Kellen, 2020).

My hope is that a more thorough study of abduction than we have seen so far will lead to a similar kind of scientific progress—that, more exactly, and with certain qualifications made subsequently, we will be able to get more mileage out of our evidence by paying special attention to explanatory connections between that evidence and whatever hypotheses that we are considering.[9] According to the current mainstream, however, there is no way to assign a role to such connections that would not make us deeply irrational; or at least there is no *interesting, substantive* way to do it. So, if we are to make the kind of progress that, I believe, a closer attention to abduction would allow us to make, we must first remove a number of persistent misunderstandings about it that prevent the mainstream from deeming abduction even worthy of serious attention.

In short, I encourage you to read on if you are interested in a mode of reasoning that appears central to numerous cognitive processes, that people tend to find entirely natural after you have told them about it, that may be so routine and automatic that often we do not even notice when we engage in

9. Here, I am in the company of Haig (2005a, 2005b), who expects much from the "abductive theory of method" that he proposed for the behavioral sciences. However, although he is not concerned to articulate a specific version of abduction, from the brief statement of abduction that he gives (e.g., Haig, 2005a, p. 381) it appears that he endorses the version labeled ABD1 in section 2.2.2, which is shown to be inadequate. He also does not address any of the concerns that have been raised about the tenability of abduction, and takes the reliability of abduction for granted (Haig, 2005a).

it, but that is also explicitly appealed to in various domains of research, and whose study finally raises the prospect of making some real scientific progress.

1.2 Two Common Misconceptions

If abduction is so ubiquitous, plays all these roles in the cognitive lives of people, and has the other features mentioned previously, why do we know so little about it and perhaps even, if Fodor is right, simply have *no* idea of how it works? When pressed, we may have difficulty saying what exactly abduction is. Moreover, many who feel attracted to the idea of abduction do not seem to know what to make of the standard objections to abduction. Finally, even if psychologists may be interested in various *descriptive* issues surrounding abduction, the increasing popularity of Bayesian statistics raises the question of why we should still care about the *normative* status of abduction. Indeed, in light of the successes about which Bayesian statistics can boast (Kruschke, 2011; Gelman et al., 2013; McElreath, 2020), any claim to the effect that the study of abduction could contribute to the progress of science would appear overblown, possibly even ludicrous.

To be fair, as is clear from the citations given at the outset of the previous section, psychological research into how people rely on abduction has been picking up steam in recent years. However, I firmly believe that abduction would have received still more attention were it not for some persistent misunderstandings about its normative status. I believe this even if Shira Elqayam and Jonathan Evans (2011) are right that psychologists should concern themselves with descriptive issues only and should leave normative issues to others.[10] Most psychologists working on reasoning appear to know the philosophical literature fairly well. If all they read there about abduction are deprecating remarks, they may not receive much incentive to prioritize empirical work on abduction. Rather, they may come to think of abductive reasoning as a further bias that may deserve no *special* attention, given that we already know quite a bit about biases. (See section 3.4 for more on this.) In philosophy, misunderstandings about abduction have hampered research into it to an even greater extent. It is an overarching aim of this book to rectify those misunderstandings. I begin by describing what I see as the two main ones.

10. This claim has been contested; see the responses to these authors in the same issue of *Behavioral and Brain Sciences* in which their target paper appeared.

1.2.1 *Nothing but a Slogan*

So far, we have given only the sketchiest characterization of abduction, as a mode of inference that makes explanatory considerations relevant to what we are licensed to believe. That may be excusable for an introductory chapter, but Bayesians and others have complained that it is difficult to find *anywhere* a statement of abduction that is more specific than such slogan-like characterizations. That alone can appear a reason to prefer Bayes's rule over abduction: rather than a slogan, Bayes's rule is a mathematically precise formula. It would thus seem that those of us who are still attracted to the idea of abduction have to provide a formulation of it that is nearly as precise as the formulation of Bayes's rule.

First, it is simply not true that there are no precise versions of abduction. In various chapters, we look at versions that are *as* precise as Bayes's rule and are also completely cast in the language of probability theory. Admittedly, while these versions serve important dialectical roles in the book, I am not vouching for any of them as adequate update rules. That is at least partly because I am not presently willing to commit to any specific theory of explanation,[11] nor am I willing even to commit to the claim that explanation quality can be measured in probabilistic or more general formal terms, as some authors have claimed or at least assumed (e.g., McGrew, 2003; Glass, 2007, 2012a, 2018, 2021a, 2021b; Schupbach, 2011a; Schupbach & Sprenger, 2011; Crupi & Tentori, 2012; Brössel, 2015; Cohen, 2016; Sprenger & Hartmann, 2019, ch. 7).[12]

11. Nor need I do so; see Prasetya (in press). Prasetya distinguishes the approach to explanationism taken in this book ("ampliative explanationism," as he calls) from the heuristic and objective approaches to explanationism (as defended by, e.g., Lipton, 2004 and Dellsén, 2018, and, respectively, Weisberg, 2009, Poston, 2014, and Climenhaga, 2017) and argues that those other approaches do commit one to particular theories of explanation.

12. That may in fact be unlikely if, as some have argued (e.g., de Regt & Dieks, 2005; de Regt, 2017; but cf. Strevens, 2013), explanation is conceptually tied to understanding or kindred notions (e.g., articulated awareness; see Woody, 2004). The reason is that understanding may be too complex a notion to be captured in strictly probabilistic or formal terms. On the other hand, in a major approach to causality (Pearl, 1988, 2000; Sloman, 2005; Ali et al., 2010; Ali, Chater, & Oaksford, 2011; Fernbach & Erb, 2013; Oaksford & Chater, 2013, 2014, 2017, 2020a; Sprenger, 2018; Hartmann, in press), this notion is analyzed by means of graphical models, which are at the same time statistical models, encoding relations of probabilistic dependence and independence. Insofar as explanation is *causal* explanation, work on these so-called causal Bayes nets may also help to capture the notion of explanation and to quantify explanation quality. To be sure, not every explanation is a causal explanation (Wouters, 1995, 2007; Bueno

Nevertheless, in the following we often *pretend*—again for dialectical purposes—that explanation can be captured in probabilistic terms. But even if explanation quality is adequately captured by one or more formal measures, is it not still a problem that, as I said, we encounter a *variety* of precise versions of abduction? How in that case are we supposed to pick the right one? The answer is that, no, it is not necessarily a problem if there is not a best mathematically precise version of abduction, just as it is not necessarily a problem if there is *no* such version. The suggestion that I make is that the advocates of abduction should rather embrace the previous critique instead of trying to counter it. The suggestion, to be more exact, is that abduction is best thought of as a slogan such as that explanatory power is of confirmation-theoretic significance or similar circumscriptions previously discussed, and in each situation this slogan has to be fleshed out in the best possible way for that situation. Finding the best elaboration for a given situation may be a matter of experience or talent. Some people may be better at this than others, maybe because they have had more practice, or have better insights, or have a combination of the two. That may make them better scientists or generally more successful in life. Finding the elaboration of the slogan that best fits a given situation may even sometimes be a matter of luck. And we should reckon with the possibility that sometimes the elaboration can be made formally precise, and sometimes not.[13]

To clarify the suggestion, I start by offering a couple of analogies. At various junctures, we support claims by using computer simulations. The simulations were all written in Julia, a language for scientific computing (Bezanson et al., 2017). One of the main features of this language (which is not unique, though many languages do *not* have it) is called "multiple dispatch." This

& Colyvan, 2011; Mancosu, 2015; Lange, 2016; Kasirzadeh, in press; Reutlinger, Colyvan, & Krzyżanowska, in press; Saatsi, in press), but if Kitcher (1981) and others are right, then at least some types of noncausal explanation can be understood in terms of coherence (internal coherence, coherence with the data, or coherence with background knowledge). Inasmuch as the currently best accounts of coherence are all of a probabilistic variety (e.g., Shogenji, 1999; Olsson, 2002, 2005; Bovens & Hartmann, 2003; Fitelson, 2003; Meijs, 2005; Douven & Meijs, 2007; Meijs & Douven, 2007; Siebel & Wolff, 2008; Schupbach, 2011b; Schippers, 2014; Koscholke, 2016; Koscholke & Jekel, 2017; Schippers & Schurz, 2017; Koscholke & Schippers, 2019; Koscholke, Schippers, & Stegmann, 2019), such types of explanation can still be analyzed in strictly probabilistic terms.

13. This is also why I am not hopeful that we can have a general logic of abduction, parallel to deductive logic. For some interesting attempts, see Aliseda (2006) and Meheus and Batens (2006).

feature offers the capability of defining different so-called methods for a given function; the methods pertain to different argument types or combinations of such types. For instance, in Julia although the + operator always performs the same operation at some abstract level of description, at the most basic level what happens when that operator is called on a number of arguments is different for different types of arguments or combinations of types of arguments. One can see this clearly by comparing the assembly code (a very basic representation of the Julia code, close to the "metal") for two operations, for example, +(5, 6) and +(5.0, 6.0). In the first case, + operates on integers, and in the second, it operates on *floating point numbers*, or *floats*, which are your computer's way of representing as best it can the real numbers.[14] As a result, the assembly code is quite different in the two cases, even though the higher-level code—the code that we write—contains nothing to suggest this.[15] This is quite similar to what I claim about abduction, namely, that at a general level it can be described as a mode of reasoning that assigns special weight to explanatory factors but that this broad description needs further filling in, where the filling-in may vary per application context.

Here is another analogy: psychologists studying problem solving have identified a number of general strategies that people use to solve problems. Among them is a strategy often called "divide and conquer," which comes down to dividing a problem into more manageable subproblems, solving the subproblems, and obtaining a solution to the initial problem by collating the solutions to the subproblems. This strategy will appear familiar to everyone because we all frequently rely on it. It is a strategy that parents recommend to their children and teachers to their students. For instance, mathematics teachers tell their students that although there are several ways to calculate the area of a pentagon, the simplest one is to cut the pentagon into three triangles,

---

14. How well it can represent them depends on the computer's architecture, but no computer can represent them perfectly because computers are finite things (see Goldberg, 1991, or Overton, 2001, for details). To see the difference between how your computer stores, for instance, the number 1 as an integer and how it stores the same number as a float, you can use Julia's `bitstring` function, which yields the bit representation of a number, and compare the output of `bitstring(1)` with that of `bitstring(1.0)`. See appendix E on how to use Julia.

15. To inspect assembly code in Julia, one can use the `@code_native` macro; for our example, compare the output of `@code_native +(5, 6)` with that of `@code_native +(5.0, 6.0)` To see all the methods associated with a function, one can call the `methods` function on it. For instance, executing `methods(+)` in Julia will show the 184 methods for the + function that are currently implemented in the language.

calculate the areas of those triangles, and then add the outcomes. Similarly, but at a more advanced level, they instruct students to integrate a function over a complicated region by trying to divide the region into smaller regions such that one obtains nice bounds on one's integrals and then summing up the integrals over those smaller regions. In a very different example, piano teachers recommend that their students study difficult passages by grouping the notes in certain ways, practicing the groups separately, and after each group has been mastered, playing the passage in its entirety.

But note that the divide-and-conquer strategy, too, needs filling in. For what are the subproblems? And how should one collate the solutions to the subproblems to obtain the solution to the overarching problem? This is difficult, if not impossible, to say in general. Discovering the best way to divide a larger problem and the best way to put the partial solutions together is sometimes far from straightforward, and some people may be better at these things than others. In piano practice, it is known that one way of grouping the notes in a passage can be much more efficient than others in reducing the difficulty of that passage, and some piano teachers are famous for having a special knack for finding clever groupings for tricky passages; the best of them will even tell you that different groupings may work for different students (no two hands are the same, as the saying goes, though the more relevant observation is probably that no two brains are the same). Successful mathematical modeling is also often a matter of identifying the most expedient way to split a problem into smaller ones.

Again, one of the claims argued for in this book is that it is no different with abduction: different situations may call for different ways of letting explanatory considerations guide one's belief changes. Picking the right way for the given situation is a skill that may come more easily to some people than to others but that most—hopefully, all—of us possess to some degree and may be able to hone, with enough dedicated practice.[16] Various authors (e.g., Day & Kincaid, 1994; Lipton, 2004) have argued that abduction is context sensitive in that its use heavily relies on background knowledge, most notably,

16. This is not to deny that in the hands of *some* people abduction is a risky tool. As Mirabile and Horne (2019) show, people prone to "conspiracist ideation" (Swami et al., 2011), who have a general tendency to adopt conspiracy theories, often come to hold such theories on explanatory grounds. There are at least parts of the practice of abductive reasoning that these people have not fully mastered or even not mastered at all.

for making judgments of explanatory goodness.[17] The context sensitivity that flows from the current proposal is more extreme, in that it makes the exact shape that abductive reasoning should take dependent on context as well (while also acknowledging the kind of context sensitivity discussed by previous authors).

The idea advanced here can also be related to Rudolf Carnap's work on concept explication. Concept explication, as Carnap conceived it, is "[t]he task of making more exact a vague or not quite exact concept used in everyday life or in an earlier stage of scientific or logical development, or rather of replacing it by a newly constructed, more exact concept" (Carnap, 1947a, pp. 7–8). He saw this task as central to much of what philosophers do, and he was clear about when—in his view—it has been carried out successfully. Next to being more exact, the replacing concept should also bear some similarity to the replaced concept, and it should meet some adequacy conditions, like being simple and fruitful (Carnap, 1950, p. 5). Think of the slogan-like formulation of abduction as the not quite exact concept. Then, as subsequently shown, that can be given more exact explications that are arguably adequate, albeit that an explication that is adequate in one situation need not be so in another situation. In that second situation, however, a different explication of the slogan may be adequate.[18]

My proposal is meant to be in accord with a more general claim that has been forcefully advocated in cognitive science and psychology, to wit, that philosophers have been wrong in their "universalist" approach to the question of the rationality of belief change. Gerd Gigerenzer and various of his collaborators, and more recently also Elqayam, have argued that the idea—prominent among philosophers—that rationality can be captured by a number of universally valid principles has been an illusion all along. According to these researchers, we can understand the notion of rationality only if we take into account (1) the various ways—biological and cognitive—in which humans are limited; (2) the goals that people have; and (3) the environment or environments in which humans operate and interact in pursuit of their goals. Biological, cognitive, and environmental constraints, as well as goals, can all vary greatly from one individual to another and from one moment to another. This led Elqayam, Gigerenzer, and others to conclude that

17. On page 5, we discussed an example of the context sensitivity of abduction even as it functions in lower-level cognitive processes.

18. Thanks to Youness Ayaita for helpful discussion here.

whether someone reasons or acts rationally depends on whether the reasoning or behavior helps the person achieve whatever goal or goals she pursues in whichever environment she is in, given the cognitive resources available to her in that environment. Abduction, being a broadly circumscribed rule that can, however, be tailored to one's needs, depending on the circumstances, fits much more naturally with this so-called ecological conception of rationality than the supposedly one-size-fits-all rule that Bayesians pretend to offer. This is one of the key points argued in chapters 6 and 7.

The foregoing is not to say that the current popularity of Bayesianism is completely undeserved, nor is that something argued in this book. As explicitly pointed out, adopting Bayesian norms may be the right thing to do in certain contexts. But Bayesian *imperialism*, according to which those norms should govern our behavior, cognitive as well as practical, in each and every context, is misguided. This imperialism has gone at the expense of proposals that assign a guiding role to explanatory considerations. It is shown here how taking into account such considerations can offer important benefits, which should be reason to give the said proposals a fairer hearing than philosophers and also many psychologists have so far been willing to do.

To be clearer still about the position that I advance in this book, I refer to William Lycan's (2002, p. 417) useful typology of versions of "explanationism" (as he and some others refer to the broad idea that explanation has confirmation-theoretic significance). In this typology, "Weak Explanationism" is the view that we can rationally reach a conclusion via abductive reasoning; however, it is left open whether abduction is a fundamental form of reasoning or parasitic on, say, Bayesian reasoning. "Sturdy Explanationism" adds to Weak Explanationism the claim that abduction is fundamental indeed. Whereas both Weak and Sturdy Explanationism are compatible with there being other forms of rational nondeductive reasoning, "Ferocious Explanationism" is *not*: according to this position, abduction is the only such form.[19] Finally, the strongest version of explanationism—which Lycan calls (rather infelicitously, in my opinion) "Holocaust Explanationism"—is the view that abduction is the only rational form of reasoning *tout court*, meaning that, for instance, even our reliance on deductive reasoning is to be justified on explanatory grounds. In terms of this typology, the position to be defended in the following is closest to Sturdy Explanationism, the main qualification

19. Harman qualifies as a ferocious explanationist in this typology; see page 2.

being that, in my view but not in Lycan's (as far as I can tell), it would be wrong to think of abduction as a single specific rule of inference; "abduction," as explained, is to be conceived as a blanket term denoting a broad idea to be filled in differently in different contexts.

1.2.2 Right but Boring, or Interesting but Wrong

Over the last two decades, Bayesian confirmation theory has firmly established itself as the dominant view on confirmation; currently, one cannot effectively discuss a confirmation-theoretic issue without making clear whether and, if so, why one's position on that issue deviates from standard Bayesian thinking. Abduction assigns a confirmation-theoretic role to explanation. In the probabilistic versions previously hinted at, and subsequently examined in great detail, explanatory considerations help to make some hypotheses more credible and others less so. By contrast, Bayesian confirmation theory makes no reference at all to the concept of explanation. Does this imply that abduction is at loggerheads with the prevailing doctrine in confirmation theory? Answer *yes*, and many people are likely to tell you that although your answer suggests that abduction might be *interesting* for offering some genuine alternative to Bayes's rule, it also shows that you are on the wrong track. Those people will claim that any systematic way for changing your beliefs that deviates from Bayes's rule will make you irrational. If, on the other hand, you answer *no*, then the mainstream will think that you may be right but not that your idea presents a significant challenge. If you are not offering a real *alternative* to Bayes's rule (or some generalization thereof), then how is your proposal not boring?

A number of authors have argued that abduction is compatible with Bayesianism indeed, but that this does not make abduction boring. To the contrary, it should be considered a much-needed supplement. At this time, the fullest defense of this view has been given by Peter Lipton (2004, ch. 7); as he puts it, Bayesians should also be explanationists.[20]

This requires some clarification. What could it mean for a Bayesian to be an explanationist? In order to apply Bayes's rule and determine the probability for hypothesis H after learning evidence E, the Bayesian agent has to determine the probability of H conditional on E. For that, she needs to

20. For similar defenses, see Okasha (2000), McGrew (2003), Poston (2014, ch. 7), and Niiniluoto (2018, ch. 6).

assign unconditional probabilities—*prior probabilities*, or *priors*—to H and E as well as a probability to E given H, called "the likelihood of H on E." How is the Bayesian to determine these values? Probability theory gives us *more* probabilities once we have *some*; it does not give us probabilities from scratch. When H implies E or the negation of E, or when H is a statistical hypothesis that bestows a certain chance on E, then the likelihood follows "analytically."[21] However, this is not always the case, and even if it were, there would remain the question of how to determine the priors. This is where, according to Lipton, abduction comes in. In his proposal, Bayesians ought to determine their prior probabilities and, if applicable, likelihoods on the basis of explanatory considerations.

Exactly how are explanatory considerations to guide one's choice of priors? The answer to this question is not as simple as one might initially think. Suppose that you are considering what priors to assign to a collection of rival hypotheses and you wish to follow Lipton's suggestion. How are you to do this? A seemingly obvious—although somewhat vague—answer may go like this: Whatever exact priors you are going to assign, you should assign a higher prior to the hypothesis that best explains the available data than to any of its rivals (provided there is a single best explanation).[22] Note, though, that your neighbor, who is a Bayesian but thinks confirmation has nothing to do with explanation, may well assign a prior to the best explanation that is even higher than the one you assign to that hypothesis. In fact, his priors for best explanations may even be consistently higher than yours, not because in his view explanation is somehow related to confirmation—it is not, he thinks—but just because. . . . In this context, "just because" is a perfectly legitimate reason, because any reason for fixing one's priors counts as legitimate by Bayesian standards. According to mainstream Bayesian epistemology, priors (and sometimes likelihoods) are up for grabs, meaning that one assignment of priors is as good as another, provided that both are coherent (that is, they obey the axioms of probability theory). Lipton's recommendation to the Bayesian to be an explanationist is meant to be entirely general. But what should your

21. This claim assumes some version of Lewis's (1980) Principal Principle (see p. 78), and it is controversial whether or not this principle is analytic; hence the scare quotes.

22. To forestall misunderstanding, your prior for a hypothesis is not the probability you assign to that hypothesis before *any* data are in. The terms "prior" and "posterior" are to be understood relative to whichever data are newly obtained and subsequently to be accommodated. In Lindley's (1972, p. 2) oft-cited phrase, "Today's posterior is tomorrow's prior."

neighbor do differently if he wants to follow the recommendation? Should he give the same prior to any best explanation that you, his explanationist neighbor, give to it, that is, *lower* his priors for best explanations? Or should he rather give even *higher* priors to best explanations than he already does?

Perhaps Lipton's proposal is not intended to address those who already assign highest priors to best explanations, even if they do so on grounds that have nothing to do with explanation. The idea might be that as long as one does assign highest priors to those hypotheses, everything is fine, or at least finer than if one does not do so, regardless of one's reasons for assigning those priors. The answer to the question of how explanatory considerations are to guide one's choice of priors would then presumably be that one ought to assign a higher prior to the best explanation than to its rivals, *if* this is not what one already does. If it is, one should just keep doing what one is doing.

A more interesting answer to the question of how explanation is to guide one's choice of priors has been given by Jonathan Weisberg (2009). We said that mainstream Bayesians regard one assignment of prior probabilities to be as good as any other. So-called objective Bayesians do not do so, however. These Bayesians think priors must obey principles beyond the probability axioms in order to be admissible. Objective Bayesians are divided among themselves over exactly which further principles are to be obeyed, but at least for a while they agreed that the Principle of Indifference (Keynes, 1921) is among them. Roughly stated, this principle counsels that, absent a reason to the contrary, we give equal priors to competing hypotheses. However, in its original form the Principle of Indifference may lead to inconsistent assignments of probabilities and so can hardly be advertised as a principle of rationality (Gillies, 2000, ch. 3). The problem is that there are typically different ways to partition logical space (the space of all logically possible worlds) that all appear equally plausible given the problem at hand and that not all of them lead to the same prior probability assignment, even assuming the Principle of Indifference (see section 3.2.1 for details). Weisberg's proposal amounts to the claim that explanatory considerations may favor some of those partitionings over others. Perhaps we will not always end up with a unique partition to which the Principle of Indifference is to be applied, but it would already be progress if we ended up with only a handful of partitions. In that situation we could arrive in a motivated way at our prior probabilities by proceeding in two steps, namely, by first applying the Principle of Indifference to the partitions separately, thereby possibly obtaining different assignments of priors, and by then taking a weighted average of the priors thus obtained, where the weights, too, are to depend on explanatory considerations. The result would again be a probability function—the uniquely correct prior probability function, according to Weisberg.

The proposal is intriguing as far as it goes but, as Weisberg admits, in its current form, it does not go very far. For one thing, it has nothing to say about how explanatory considerations are to determine the weights required for the second step of the proposal. For another—and this seems the bigger problem—it may be idle to hope that taking explanatory considerations into account will in general leave us with a manageable set of partitions, or that even if it does, this will not be due merely to the fact that we are overlooking a great many prima facie plausible partitionings of logical space to begin with.

Another interesting suggestion about the connection between abduction and Bayesian reasoning is that the explanatory considerations may serve as a heuristic to determine, even if only roughly, priors and likelihoods in cases in which we would otherwise be clueless and could do no better than guessing (Niiniluoto, 1999a; Okasha, 2000; McGrew, 2003; Lipton, 2004, ch. 7; Cabrera, 2017; Dellsén, 2018). This suggestion is sensitive to the well-recognized fact that we are not always—indeed are quite often *not*—able to assign a prior to every hypothesis of interest or to say how probable a given piece of evidence is conditional on a given hypothesis. Consideration of that hypothesis's explanatory power might then assist us in gauging the prior to assign to it or the likelihood to assign to it on the given evidence.

Bayesians, especially the more modest ones, might want to retort that the Bayesian procedure is to be followed only if (1) priors and likelihoods can be determined with some precision and objectivity, or (2) likelihoods can be determined with some precision and priors can be expected to "wash out" as more and more evidence accumulates, or (3) priors and likelihoods can both be expected to wash out. In the remaining cases, Bayesians might say, we should simply refrain from applying Bayesian reasoning. A fortiori, then, there is no need for an abduction-enhanced Bayesianism in these cases. Further, some incontrovertible mathematical results indicate that, in the cases that fall under (1), (2), or (3), our probabilities will converge to the truth anyway (Gaifman & Snir, 1982). Consequently, in those cases there is no need for the kind of abductive heuristics that the previously mentioned authors suggest.[23]

23. For similar concerns, see Weisberg (2009, section 3.2).

Finally, Stathis Psillos (2000) proposes yet another way in which abduction might supplement Bayesian confirmation theory, a way very much in the spirit of Peirce's conception of abduction, which is briefly discussed in section 2.2.1. The idea is that abduction may assist us in selecting plausible candidates for testing, where the actual testing will follow Bayesian lines. However, Psillos (2004) concedes that this proposal assigns a role to abduction that will strike committed explanationists as being too limited.

If the explanationists' attempt to cozy up with Bayesians makes their proposal boring, attempts to sell abduction as *deviating* from Bayes's rule are nonstarters, or so a number of influential authors hold. Why is that? We are typically given one of two arguments: the dynamic Dutch book argument and the argument from inaccuracy minimization. These arguments are too intricate and too important—they have been the main obstacles to a wider acceptance of abduction—to be dealt with in a single section. We devote a whole chapter to them. To briefly address them here: the dynamic Dutch book argument is alleged to show that using any rule for belief change other than Bayes's rule may lead one to assess as fair a number of bets that together ensure a financial loss, come what may, and thus in the end that using any rule for belief change other than Bayes's rule makes one irrational; the argument from inaccuracy minimization purports to show that if one uses some non-Bayesian rule for belief change, then one's mental representation of the world is bound to be less sharp than it would be were one to use Bayes's rule instead.

As we shall see, neither argument is ultimately compelling, and each fails for more than one reason. Perhaps the simplest but also most important reason for buying into neither is that, even if they were sound (which they are not), absolutely nothing follows from the mere claim that vulnerability to dynamic Dutch books and / or failure to minimize inaccuracy is *really, really bad*. That could be a reason to abandon abduction if it had at the same time been shown that there is nothing *really, really good* to be had by reasoning abductively, something that is out of reach for people sticking to Bayes's rule, or any other rule. Furthermore, none of the critics of abduction has even bothered to consider that possibility. In fact, as shown subsequently, it is more than a possibility; there is positive reason to hold that abduction offers benefits that we would forego by consistently adhering to Bayes's rule. Put more generally, if you believe we should judge epistemic norms (such as rules for belief change) by their consequences, then make sure that you are not too narrowly focused on only certain *kinds* of consequences, possibly leaving others that might be just as relevant, or even more relevant, out of consideration. In my view, many Bayesians criticizing abduction have erred in this regard.

My hope is that, if you do read on, I will ultimately be able to convince you that abduction is interestingly different from Bayes's rule and that there is nothing intrinsically untoward about it. As mentioned, abductive reasoning may involve a bit of art, of imagination, and of creativity; and although—as with all matters artistic—some of us may have a greater inherent talent for abductive reasoning than others, I firmly believe that we can all improve our reasoning skills by getting a better understanding of how abduction works (see, in the same vein, Mirabile, 2020).

1.3 Overview

The following are the main goals I hope to achieve in this book:

(1) To clarify what abduction is and to explain why we should care about it. I argue that abduction is the general idea that explanation has a role to play in confirmation, where however this idea needs to be specialized per context of use; I further argue that we should study abductive reasoning both for normative and for descriptive reasons.

(2) To convince the reader that, in assessing the rationality of abductive reasoning, our standards should attend to known facts about human psychology: the study of abduction should be part of an effort to develop a comprehensive theory of rationality for *real* people, not for a future make of computationally unconstrained robots.

(3) To show that the bad things you may have heard about abduction are vastly overblown and, at a minimum, are misinterpreted: the standard arguments against abduction make assumptions the advocates of abduction are by no means forced to commit to, and even if those arguments are granted, they only show that there can be costs attached to reasoning abductively, not that this mode of reason is cost-ineffective.

(4) To show that there are actually quite a few good things to report about abduction, in particular that, in the right hands, abduction is a powerful tool: computer simulations are used to show that, in various types of contexts, we may be better off using abduction than the currently more popular Bayes's rule; it will also be shown that abduction can do

the sort of heavy lifting that philosophers have always hoped it would do for them.

Next is an overview of how each chapter is meant to help realize these goals.

Chapter 2 states the idea of abduction in greater detail by examining abduction from a number of different angles. Abduction is distinguished from other modes of inference; we briefly trace its historical roots; some applications are described; we consider some common definitions of abduction in the philosophical literature; and we discuss the problem of underdetermination, which is the problem that most often leads philosophers to invoke abduction.

Chapter 3 examines reasoning research concerning how judgments of explanation quality impact people's beliefs and especially changes of belief. Special attention is given to research showing that people deviate from Bayesian principles in ways one would expect to occur if they were influenced by explanatory considerations. The aim is to raise psychologists' interest in explanatory reasoning. An equally important goal of the chapter is to make clear that if one wants to maintain that any violation of Bayesian principles betokens irrationality, then people are massively irrational (which some philosophers may not care about, but some certainly will). Moreover, the chapter provides some inspiration for formulating probabilistic versions of abduction that are compared with Bayes's rule along dimensions of performance that arguably matter to us.

Chapter 4 canvasses the two main objections that have been brought against abduction, both already mentioned: the dynamic Dutch book argument and the inaccuracy-minimization argument. The chapter first argues that it is a mistake to think that abduction, or any other rule for belief change, can be judged in isolation from other rationality principles, including decision-theoretic principles. Instead, it is to be viewed as part of a package of rules, and there are packages that include abduction and that allow one to use abduction without having to live in fear of Dutch bookies. The chapter then argues that critics who level the charge that abduction fails to minimize inaccuracy overlook that there are different senses of inaccuracy that are relevant in the context of the present debate, and that while abduction may fail to minimize inaccuracy in one sense, there is a different, arguably more important, sense of inaccuracy that is not touched at all by these critics' arguments.

Chapter 5 argues that one need not even grant that abduction fails to minimize inaccuracy in *any* sense! The same chapter also takes issue with the claim that, in contrast to the dynamic Dutch book argument, the inaccuracy-minimization argument is a distinctively epistemic (as opposed to pragmatic) argument.

That abduction does not actually fall prey to the dynamic Dutch book and inaccuracy-minimization arguments does not automatically mean that it has anything to recommend it. Chapter 6 gives abduction a positive grounding. It uses computer simulations to show that versions of abduction can vastly outperform Bayes's rule in terms of obtaining a best balance between the speed with which our probabilities converge to the truth and the overall accuracy of those probabilities. That is precisely because abduction is more flexible and can be fine tuned to meet the specific needs of the situation in which we happen to be. The chapter also demonstrates how evolution may have predisposed us to reason abductively.

Chapter 7 adds a social perspective to the foregoing by studying artificial agents who update, either by Bayes's rule or by some version of abduction, on evidence they receive directly from the world but whose belief states are also impacted by those of others in their community. In such a social setting, updating via abduction appears to be particularly advantageous. Here, too, we find grounds for believing that evolution may have favored groups of epistemically interacting agents who reason abductively.

Chapter 8 highlights another benefit of abduction by putting it to work in probably the best-known underdetermination argument, namely, the argument for skepticism regarding the external world. Various authors have argued that we are warranted in believing in the existence of an external world because that belief best explains our sensory evidence. I explain why the argument for that claim is not as straightforward as it has appeared to many, in particular by pointing out that previous authors arguing abductively against the skeptic have done insufficient justice to skeptical concerns about admitting as evidence anything that goes beyond the purely phenomenal and also about rules of reasoning—most notably, abduction—that might enable one to move from premises about the phenomenal alone to a conclusion about the external world. The chapter argues that we can acknowledge those concerns and still be able to resolve the skepticism debate.

2

What Is Abduction?

In this chapter, I introduce explanatory reasoning from three different angles. First, I give some real-world examples that are offered as clear and uncontroversial instances of explanatory reasoning. To forestall misunderstanding, I stress that I do not present these examples by themselves to show or even to suggest that explanatory reasoning is a form of reasoning in its own right that cannot be adequately modeled, for example, in Bayesian terms. The book as a whole *is* intended to show that the kind of reasoning that these examples exhibit is not, or at least not in all cases, best modeled in those terms; that we can model them by explicitly factoring in explanatory considerations, even in ways inconsistent with the tenets of Bayesianism; and that there is nothing necessarily wrong with the resulting model. The examples to be considered shortly are merely meant to give the reader—especially the reader new to the topic of abduction—an impression of what abductive reasoning actually is.

Next, I look at some of the extant theory concerning this mode of reasoning, in particular what philosophers have been saying about it. There is a welter of proposals concerning the shape that explanatory reasoning might take. Section 1.2.1 has made it clear that, in my opinion, such different proposals can peacefully coexist.

Finally, I describe the main source of interest in abduction, at least in philosophy, which is the belief that, as previously mentioned, abduction may contribute to resolving cases of underdetermination, that is, cases in which observation alone, or observation plus logic, is not enough to settle a question or even to arrive at a reasonably grounded answer to the question, because a number of rival answers appear equally reasonable. Various philosophers have argued that abduction can be invoked to break such deadlocks. Abduction is used to this effect in chapter 8, where I respond to the Cartesian skeptic.

2.1 Abduction in the Wild

2.1.1 Newton and the Raid on the Medway

In 1667, under the leadership of Michiel Adriaanszoon de Ruyter the Dutch fleet invaded the Thames and inflicted what is still the greatest ever defeat on the British navy.[1] At the time of this Raid on the Medway, as the invasion is also called, Isaac Newton was a Fellow at Trinity College, Cambridge, where the rumble of the battle could be heard. In his *Memoirs of Sir Isaac Newton's Life*, William Stukeley reports about it as follows:

> Their guns were heard as far as Cambridg, and the cause was well known; but the event was only cognisable to Sir Isaac's sagacity, who boldly pronounc'd that they had beaten us. The news soon confirm'd it, and the curious would not be easy whilst Sir Isaac satisfy'd them of the mode of his intelligence, which was this; by carefully attending to the sound, he found it grew louder and louder, consequently came nearer; from whence he rightly infer'd that the Dutch were victors. (Stukeley, cited in Westfall, 1980, p. 194)

According to Richard Westfall (1980), the "mode of his intelligence" referred to here struck the other Fellows in Cambridge with awe. How *did* Newton arrive at the conclusion that the Dutch had won? What made it *right*—if it *was*—for him to infer from the sound of the gun fire getting louder that the Dutch had beaten the British fleet?

2.1.2 Discovering Neptune

Uranus was discovered in 1781, and by the mid-nineteenth century it had become clear that its orbit deviated from the orbit predicted by Newton's theory of gravity. Astronomers realized that this finding could betoken a straightforward falsification of the theory. But that thought was anathema to most at the time, given the theory's otherwise hugely impressive track record. Another possibility was that the predictions derived from Newton's theory were based on faulty calculations. New calculations by the French mathematician Urbain Le Verrier showed that earlier ones indeed contained some mistakes. On the other hand, those new calculations still did not result

1. I have found that this fact is virtually unknown outside the Netherlands. In the Netherlands, by contrast, almost everyone knows the story of the invasion, probably because it is the only success in the whole military history of our country.

in an orbit that accorded with the available data. Le Verrier considered various possible explanations of why Uranus's orbit was inconsistent with predictions based on Newton's theory; for instance, that Uranus was itself orbited by a giant moon that steered it off course, or that it was hit by a comet, or that it was hampered by something like an ether. In the end, he came to think that the wobbles in Uranus's orbit were attributable to the gravitational pull of another, so far unknown, planet (Levenson, 2015, p. 36). He also thought that if it existed, he could calculate the position of the planet at that time. He first sent his calculations to the astronomers of the Parisian observatory. When his countrymen were slow to act, however, and had after several weeks not made any attempt to verify Le Verrier's prediction, he turned for help to the German astronomer Johann Galle of the observatory in Berlin. Galle wasted no time. As soon as he had received Le Verrier's message, he directed his telescope to the location in the sky where according to Le Verrier's calculations the planet should be visible, and within hours he discovered what soon became known as "Neptune."

2.1.3 Non-discovering Vulcan

Not long after the discovery of Neptune, Le Verrier concerned himself with an at the time unexplained irregularity (from the viewpoint of Newtonian gravitation) in the orbit of yet another planet, Mercury. The discrepancy was much smaller than in the case of Uranus's orbit but large enough to worry astronomers of the time. In 1859, Le Verrier proposed that this discrepancy was to be explained by the presence of a small, as yet unknown, planet, a proposal that seemed entirely reasonable in light of how Neptune had been discovered. Soon reports arrived of observations of this supposed new planet—which had been dubbed "Vulcan"—and examining records of past observations revealed many that could also be classified as sightings of such a planet. On the other hand, those observations were used to calculate the orbit of Vulcan, from which predictions were derived about where and when one should be able to observe it again, and although there were sporadic reports—typically by amateur enthusiasts—of sightings that seemed to confirm those predictions, more rigorously organized attempts to spot Vulcan remained unsuccessful. This did not, however, seal the fate of Vulcan. Le Verrier considered the possibility that Vulcan was so small that it would be hard to observe from earth or that it could be observed only under especially favorable conditions.

However, doubts kept lingering, precisely because if the planet was so small, its mass would probably also be too small to completely explain the phenomenon that got the whole inquiry started. Also, failed attempts to observe Vulcan kept piling up, and over time, absence of evidence came to be regarded as evidence of absence (Levenson, 2015, p. 108). The solution to the puzzle of why Mercury's orbit deviated from Newtonian predictions came only in 1915, when Albert Einstein presented his General Theory of Relativity, which explained the wobble in Mercury's orbit in terms of the effect mass has on the curvature of spacetime (Einstein, 1915).[2]

2.1.4 On the Origin of Species

Charles Darwin's *Origin of Species* (1859) is one of the most important works of science of all time, according to some *the* most important (see, e.g., Wilson, 2009, p. xv). In the work, Darwin argued famously for a theory of the diversity of life forms and of the evolution of species by postulating mechanisms of individual variation and selective retention, and against the hypothesis that species are the immutable results of a divine act of creation. A revolutionary theory at the time of its inception, Darwin's proposal nevertheless gained rapid acceptance in the scientific community. That may have occurred, at least in part, because the theory was able to explain a great variety of facts, some of which Darwin did not even contemplate at the time he wrote *Origin*. As he put it in the sixth edition of the work,

> It can hardly be supposed that a false theory would explain, in so satisfactory a manner as does the theory of natural selection, the several large classes of facts above specified. It has recently been objected that this is an unsafe method of arguing; but it is a method used in judging of the common events of life, and has often been used by the greatest natural philosophers. The undulatory theory of light has thus been arrived at; and the belief in the revolution of the earth on its own axis was until lately supported by hardly any direct evidence. (Darwin, 1876, p. 421)

The facts that Darwin refers to as having been specified previously come from a variety of domains, including embryology, morphology, paleontology, and biogeography, and most of those facts were seen as wholly unconnected, prior

2. Levenson (2015) gives a fascinating and highly detailed account of the fruitless hunt for Vulcan.

to the advancement of Darwin's theory. As various authors have pointed out (e.g., Ruse, 1975, 1995; Thagard, 1977), Darwin was heavily influenced by the philosopher of science and polymath William Whewell, who had made a strong case for the confirmatory power of what he called "consilience of inductions," which occurs if a theory is successfully proposed to explain a given body of data and is then found to also explain, without requiring any further adjustment, one or more other bodies of data, none of which had played a role in the conception of the theory. Such a consilience of inductions brings about a "unification of two or more hitherto disparate areas of understanding beneath one or a few high-level hypotheses or established laws" (Ruse, 1995, pp. 2–3) and thereby provides a theory with a "stamp of truth" (Whewell, 1847, II, p. 66). In the preceding passage, Darwin appears to concur, as did and continue to do many of the advocates of the theory of natural selection: that this theory can explain such a great variety of facts in such a satisfactory fashion seems to rule out its falsity.

2.1.5 The Discovery of the Electron

The electron was the first known subatomic particle. It was discovered in 1897 by the English physicist Joseph John Thomson. Thomson had conducted experiments on cathode rays in order to determine whether they are streams of charged particles. Improving an experimental setup that had earlier been used (unsuccessfully) by Heinrich Hertz, Thomson let cathode rays travel between two parallel metal plates. He then showed that when those plates were oppositely electrified, the cathode rays were deflected. (That Hertz's experiment had shown no effect was due, Thomson argued, to the fact that Hertz had failed to sufficiently evacuate the cathode tube.) From this result, Thomson concluded that cathode rays are indeed charged particles, reasoning as follows:

> As the cathode rays carry a charge of negative electricity, are deflected by an electrostatic force as if they were negatively electrified, and are acted on by a magnetic force in just the way in which this force would act on a negatively electrified body moving along the path of these rays, I can see no escape from the conclusion that they are charges of negative electricity carried by particles of matter. (Thomson, 1897, p. 302)

Today's physicists are still convinced that Thomson drew the right conclusion.

2.1.6 *Gödel's Demons*

Kurt Gödel was the greatest logician of the twentieth century and possibly the greatest logician ever. He proved the completeness of first-order logic (logical truth coincides with finite tinkering in a system of axioms and rules that most freshmen are comfortable working with in just a couple of weeks) and the incompleteness of arithmetic (no consistent system of axioms and rules captures the notion of arithmetic truth). He was a genius, but he also suffered from mental illness all his life (Wang, 1996, ch. 1). Whether or not due to his illness, he was a strong believer in ghosts and demons. In a conversation with Georg Kreisel, another brilliant logician, he asked whether Kreisel had noticed the peculiar asymmetry that there are many more unpleasant events taking place in the universe than pleasant ones. According to Gödel, this was best explained by assuming the existence of demons. Kreisel responded that he was not so sure and that it might be in our nature to notice unpleasant things more than we do pleasant things and that, more generally, we may have an inborn tendency to be unhappy with what surrounds us. Gödel agreed that we have this tendency but also thought that it, too, was best explained by the existence of demons (Kreisel, 1980).

2.1.7 *A Head in Jackson*

On June 11, 2017, CNN reported on its website that in Jackson, Mississippi, a head had been found on the front porch of a home and that a headless body had been found less than a mile away. At that point in time, according to CNN, a medical examiner was still trying to determine whether the head and body were from the same person. Two days later, another CNN report confirmed that they were indeed from the same person (and the victim was identified). That was what I had believed, without the shadow of a doubt, from the moment I first read the story on June 11. Perhaps you also read this story when it broke and drew the same inference I drew. Even if not, you will still see why the conclusion appeared entirely reasonable to me, immediately on June 11, before the medical examiner's report was released.

2.2 Deduction, Induction, Abduction

The foregoing cases should give you some idea of what abduction is and how it works in practice. To acquire a more systematic understanding of abduction, we first need to make some basic distinctions concerning inference. Abduction is normally thought of as one of three major types of inference, the other two being deduction and induction. The distinction between deduction, on one hand, and induction and abduction, on the other hand, corresponds to the distinction between necessary and non-necessary inferences. In deductive inferences, what is inferred *must* be true if the premises from which it is inferred are true; that is, the truth of the premises *guarantees* the truth of the conclusion. A familiar type of example is inferences instantiating the schema

All Φs are Ψs.
Object a is a Φ.
Hence, a is a Ψ.

If all Φs are Ψs, and a is a Φ, then there is simply no way for a not to be a Ψ. Any argument of the above form is said to be (deductively) *valid*.[3]

Not all inferences are of this variety. Consider, for instance, the inference of "Alice is rich" from "Alice lives in Chelsea" and "Ninety-seven percent of the people who live in Chelsea are rich." Here, the truth of the first sentence is not guaranteed (but only made likely) by the joint truth of the second and third sentences. Differently put, it is not necessarily the case that if the premises are true, then so is the conclusion: it is logically compatible with the truth of the premises that Alice be a member of the small minority of nonrich inhabitants of Chelsea. Hence, the said inference is *invalid*.

It is standard practice to group non-necessary inferences into *inductive* and *abductive* ones. Inductive inferences form a somewhat heterogeneous class, but for present purposes they may be characterized as those inferences that are based purely on frequency information, whether provided numerically, as in the preceding example about Alice (which falls into the inductive category), or in a more qualitative fashion (e.g., "Most people living in Chelsea are rich").

The mere fact that an inference is based on statistical data is not enough to classify it as an inductive one. You may have observed many gray elephants and no nongray ones, and infer from this that all elephants are gray because that would provide the *best explanation* for why you have observed so many

3. Some authors include in the class of deductively valid arguments those that hold in virtue of their form *and* certain meaning postulates, such as the argument with premise "Marble a is red" and conclusion "Marble a is colored" (see, e.g., Carnap, 1952).

gray elephants and no nongray ones. You would thereby make an abductive inference.

The best way to distinguish between induction and abduction may be the following: while both are *ampliative*, meaning that the conclusion has more information content than is in the premises, in abduction there is an implicit or explicit appeal to explanatory considerations, whereas in induction there is not; in induction, there is *only* an appeal to observed frequencies or statistics.

Not all philosophers agree that there is a distinction to be made between inductive and abductive inference. As intimated previously, Harman believes that inductive inference is really a form of abduction. For instance, he would argue that we infer that Alice is rich in the above example because, if true, that would best explain why she lives in Chelsea, given the background knowledge that 97 percent of the Chelsea population are rich. There is research suggesting that people actually tend to prefer explanatory information over statistical information in making inferences (e.g., Bes et al., 2012). Hence, maybe people *would* reach the conclusion that Alice is rich on the basis of explanatory considerations rather than by merely drawing on the provided frequency information. To me, however, the appeal to explanation seems strained in this case.

This book concerns abduction, so if inductive inferences are really just a particular type of abductive inferences—which I do not believe—then so much the better: it would expand the book's market. It *would* be problematic if Richard Fumerton (1980, 2017) were right: that what philosophers tend to regard as abductive inferences are in fact merely inductive ones, albeit topped with a reference to explanation that is not doing any real work.

Fumerton may certainly be right that *some* cases advertised as paradigm cases of abduction actually only exploit frequency information, or at any rate that their appeals to explanation are gratuitous. For instance, elsewhere (Douven, 2017a) I used a little story from the opening pages of Hilary Putnam's *Reason, Truth and History* (Putnam, 1981) to illustrate the use of abductive reasoning. In this story, you are walking along the beach when you see what looks like a picture of Winston Churchill in the sand. You realize that what you see could be the trace of an ant crawling on the beach. But I suggested that a by far better explanation in this case is that someone intentionally drew a picture of Churchill in the sand—which is what you would supposedly be justified to believe. Here, one might claim that your inference would be justified merely by extrapolating from previous experiences. It is not so much that you will have seen many portraits of known persons in the sand and that all those were known to have been drawn by humans. To the best of my recollection, the closest to a portrait in the sand that I have ever seen was a smiley face. However, your reference class need not be confined to portraits in the sand. Most likely, you will never have seen any portrait—on paper, on canvas, on a T-shirt, or on the sidewalk—that was not drawn by someone but instead was a random configuration of, for example, dust specks that just *happened* to look like a portrait of a known person.

Even if this is granted, however, it is not nearly enough to establish the general claim that abductive inferences are just misunderstood inductive inferences. In particular, I take the previous section to consist of cases that are much more naturally understood as cases of abductive reasoning than as cases of inductive or, for that matter, deductive reasoning. The appeal to explanation in those cases appears anything but gratuitous.

To start with the head-in-Jackson case, it is clear that the relevant conclusion could not have been reached on the basis of a deductive argument: that a head and a headless body found in each other's proximity belong together does not follow from anything we know about heads, bodies, decapitations, or anything else. If it did, what was the medical examiner needed for? In that case, the Jackson police should have a put a logician on the case.

As for the possibility that the conclusion was really reached on the basis of frequency information, note that there was no such information on the CNN website. There was nothing about murder rates in Jackson (or anywhere else) nor was there information about findings of heads and headless bodies and how often, in such findings, the head and headless body were from the same person. Rather, I suspect that such findings are so rare that there are simply no reliable statistics that one could consult. And even if reliable statistics *were* available in this case, they were not available to *me*, and so they did not play a role in the reasoning that led *me* to arrive at the conclusion that head and body were from the same person.

Nor could we easily have reasoned by analogy, which is sometimes treated under the heading of inductive inference (Carnap, 1945, 1980; Skyrms, 1993a; Hill & Paris, 2013; Paris & Vencovská, 2017).[4] It was not as if last year there

4. Lindenbaum-Hosiasson (1941) sees reasoning by analogy as being distinct from inductive reasoning, the latter being concerned with increasing our confidence in laws or general hypotheses, the former with increasing our confidence that a certain observable but as yet unobserved fact obtains. A number of later authors (e.g., Thagard, 1989, 2000; Schurz, 2008a)

had been the finding of a leg and a legless body near each other, which then turned out to belong together, and the year before the finding of a shoulder and a shoulderless body near to each other, and so on. Such situations had not occurred, at least not to my knowledge, so that again, even if they had occurred, I could not justifiably reason by analogy from those situations to the conclusion that the head and body that were found in Jackson, Mississippi, were from the same person.

But then what did make me, and presumably many others who read the same story on June 11, so certain that the head and body belonged together even before the medical examiner's conclusion was reported in the media? It is hard to see how the conclusion, on June 11, 2017, could *not* have been the outcome of an abductive inference. At a minimum, that the head and body belonged together was, at the time, clearly the best explanation of the evidence known then. To be sure, it was not the only explanation: there might have occurred two killings, say, followed by two decapitations, with the head of one victim being dumped close to where the body of the other victim was dumped, whether coincidentally or as planned by the perpetrators. But, at least in light of the information on the CNN website on June 11, no other candidate explanation appeared nearly as plausible as the conclusion I drew: that the head and the body were from one and the same person. It was not a conscious decision on my part to make an abductive inference in this case, but to the extent that I have introspective access to my own reasoning, I would say that this is exactly the inference I did make then.

Similarly in Newton's case, the "mode of intelligence" that stunned his colleagues is most plausibly thought of as involving abductive reasoning. It is exceedingly difficult to think of a reasonable set of premises—reasonable from Newton's perspective at the time—from which the conclusion that the Dutch had won follows deductively. Nor did the Dutch—or any other nation that possessed a sizable fleet in the second half of the seventeenth century—invade England frequently enough for Newton's reasoning to be naturally construed as inductive. Instead, it seems that what led Newton to his conclusion is that a Dutch victory was the *best explanation* for his evidence. There are various potential explanations of why the sound of the cannon fire grew louder and louder that do *not* involve a Dutch victory. For instance,

have presented analogical reasoning as being essential to abductive reasoning, or at least as a mode of reasoning involved in some abductions.

the British fleet might have defeated the Dutch, but then that victory might have been followed by a mutiny in which the British seamen turned against their own headquarters. However, this and other potential explanations are topped, in terms of explanatory goodness, by the hypothesis that the Dutch fleet had beaten the British.

Or again, in the case of Thomson's reasoning about what his experimental results showed, the conclusion that cathode rays consist of negatively charged particles does not follow logically from those results, nor could Thomson draw on any relevant frequency information. More generally, as Gerhard Schurz (2008a; 2019, section 11.2.5) and Ruth Weintraub (2017) argue, it is hard to imagine how we could ever arrive at postulating an unobservable entity merely on the basis of inductive reasoning. It is much more reasonable to suppose that Thomson could "see no escape from the conclusion" about cathode rays because that conclusion was the best—in this case presumably the only plausible—explanation of his results that he could think of. I take this and the other cases just discussed to offer strong support for the claim that abduction is a type of inference genuinely distinct from induction.

If great minds like Newton, Le Verrier, and Thomson rely on abduction, that certainly speaks in favor of this mode of reasoning. But great minds can go wrong and can, like the rest of us, be subject to biases. Thus we should still ask whether abduction is any good. That question takes up much of the rest of the book, but the preceding stories already allow us to make a couple of pertinent observations.

The first is that abduction is certainly not fail-safe, as the cases of Vulcan and Gödel's demons show. The Gödel case is special, of course, precisely because it involves someone whose mind was at least in some ways not functioning properly (even if in other ways it was functioning brilliantly). Given how background knowledge informs our judgments of explanation quality, we cannot expect abduction to be any good when applied by someone with deranged ideas about reality, as Gödel's certainly were. If one's picture of reality is already distorted, then one may well come to regard the existence of demons to best explain why there is so much misery in the world.[5]

Le Verrier's view of the world was *not* distorted, as far as is known. Not only was it reasonable for him to judge that the anomalous rate of precession of the perihelion of Mercury's orbit was best explained by the existence of

5. See also chapter 1, footnote 16.

an as yet undiscovered planet, but one could argue that the later judgment was even more reasonable than the earlier one, which had led to the discovery of Nepture. After all, in the case of Mercury, the judgment was additionally backed by the known success of the analogous earlier proposal. If a proposed hypothesis explaining a given phenomenon is similar to one known to explain a similar phenomenon, that should increase the degree of acceptability of the former (Lindenbaum-Hosiasson, 1941; Carnap, 1980; Thagard, 1989, 2000). Nevertheless, Le Verrier's later judgment turned out to be false.

That abduction sometimes leads to false conclusions is what you would expect, given that it is an ampliative type of inference. In itself, that is not a reason to dismiss it. It would of course be wonderful if we came by purely with deductive reasoning, such that our reasoning would never introduce falsehoods, provided that we took some care in checking the validity of the steps in our arguments. But as Gerhard Schurz and Ralph Hertwig (2019, p. 19) rightly remark, while deduction has "maximum ecological validity"—meaning that it is entirely general and can be relied on in every possible situation—it also has "very low applicability," in that "the prevalence of deductive inferences with *nontrivial* conclusions is low." In other words, we are practically bound to use ampliative types of reasoning, but then we must reckon with the fact that, at times, our conclusions will be false even if the premises from which we start are not.

Even though abductive reasoning sometimes produces falsehoods, it may still be *reliable*: it may be that on *most* occasions when we reason abductively we will reach a true conclusion, provided that we start from true premises. A quick count shows that of the six cases discussed in the previous section at least five exhibit abductive reasoning on the basis of true premises (the Gödel case may be the exception, although we do not know exactly what his premises were), and in four of those five the reasoning also led to a true conclusion. But the sample that we are considering is not only small, it is also highly nonrandom and hence not a good basis for an assessment. For all that has been shown, my being correct in the case of the head in Jackson was a lucky hit for a mode of reasoning that should still be advised against.

Nonetheless, the foregoing suggests that we might be able to show abduction reliable simply by counting the successful applications. We merely have to widen our sample and make sure that it is truly representative, and then the proportion of cases in which abduction worked should give us an indication of its reliability (or otherwise). That would amount to an inductive justification of abduction. There have also been attempts to justify abduction abductively. I am sympathetic to all such attempts, at least to the extent that I believe the justification for abduction must proceed empirically on the basis of its success rate. Trying to justify abduction a priori, by deducing it from first principles, is a hopeless endeavor, I believe (see also section 6.2), just as we cannot hope to justify induction by deducing it from first principles (Schurz, 2019). In fact, even a deductive justification of deduction may be out of reach (Haack, 1976). So, I maintain that abduction should be justified empirically. However, doing so is not nearly as straightforward as one might initially think, at least not if the aim is to justify it *broadly*, as opposed to justifying it for a limited class of cases.

Note also that a justification of abduction must consist of more than a demonstration of its reliability. Abduction might be reliable but possibly not as reliable as some other ways of forming or changing our beliefs. In that case, it could be foolhardy to rely on abduction instead of some more reliable rule. It *could* be but need not be, because those other rules might have other defects that make them unsuited. However, the same may hold for abduction: perhaps it generally leads to true conclusions when applied on the basis of true premises, but actually relying on it could still get one into serious trouble (e.g., because it exposes one to Dutch bookies).

But the justificatory status of abduction is a topic for later chapters. We are still in the process of clarifying what abduction is. In the preceding discussion, we have positioned abduction relative to the other main types of inference. Next, we have a look at how abduction has been characterized by various philosophers. Philosophers have actually used the term “abduction” in two somewhat different senses. In the sense in which it is used most frequently in modern literature, it refers to the place of explanatory reasoning in *justifying* hypotheses; abduction in this sense is also known as “Inference to the Best Explanation.” By contrast, in the older sense, “abduction” refers to the place of explanatory reasoning in *generating* hypotheses. This book concerns abduction strictly in the modern sense.[6] Nevertheless, we start with a historical digression by briefly going into how the term “abduction” was introduced in the philosophical literature.

---

6. However, I agree with Niiniluoto (1999a, p. S442) that it is not always possible to distinguish between the two roles of abduction. For a book-length treatment of how the two roles relate to one another, see Niiniluoto (2018).

2.2.1 *Historical Digression: Peirce on Abduction*

The term "abduction" was coined by Charles Sanders Peirce in his work on the logic of science. He introduced it to denote a type of nondeductive inference that was different from the already familiar inductive type. It is, however, a common complaint that no coherent picture emerges from Peirce's writings on abduction.[7] Yet it is clear that, as Peirce understood the term, "abduction" did not quite mean what it is currently taken to mean (Campos, 2011; McAuliffe, 2015; Yu & Zenker, 2018). One main difference between his conception and the modern one is that whereas according to the modern conception, abduction belongs to what the logical empiricists called the "context of justification"—the stage of scientific inquiry in which we are concerned with the assessment of theories—for Peirce abduction had its proper place in the context of *discovery*, the stage of inquiry in which we in the first place try to generate theories that may later be assessed. As he writes, "[a]bduction is the process of forming explanatory hypotheses. It is the only logical operation which introduces any new idea" (1934, section 172); elsewhere he writes that abduction encompasses "all the operations by which theories and conceptions are engendered" (1934, section 590). Deduction and induction, then, come into play at the later stage of theory assessment: deduction helps to derive testable consequences from the explanatory hypotheses that abduction has helped us to conceive, and induction finally helps us to reach a verdict on the hypotheses, where the nature of the verdict is dependent on the number of testable consequences that have been verified.[8]

7. Perhaps this is not surprising, given that he worked on abduction throughout his career, which spanned a period of more than fifty years. For a concise yet thorough account of the development of Peirce's thoughts on abduction, see Fann (1970); in addition, Niiniluoto (2018) contains valuable historical background information.

8. Schurz (2008a) defends a view of abduction that is very much in the Peircean spirit. In this view, "the crucial function of a pattern of abduction . . . consists in its function as a search strategy which leads us, for a given kind of scenario, in a reasonable time to a most promising explanatory conjecture which is then subject to further test" (Schurz, 2008a, p. 205). Note, however, that Schurz does acknowledge a role for abduction in the justification of hypotheses, though that role is, in his view, a relatively minor one. Schurz's paper is also of interest because of the rich typology of patterns of abduction that it puts forth. Most notably, he distinguishes between *selective* and *creative* abduction. The first is invoked when we face a large space of possibilities from which to select a hypothesis, and the second serves to introduce new concepts or entities. For instance, the abductive reasoning that led Newton to conclude

As Harry Frankfurt (1958) notes, however, the foregoing view is not as easy to make sense of as might at first appear. Abduction is supposed to be part of the logic of science, but what exactly is *logical* about inventing explanatory hypotheses? According to Peirce (CP 5.189), abduction belongs to logic because it can be given a schematic characterization, to wit, the following:

> The surprising fact, *C*, is observed.
> But if *A* were true, *C* would be a matter of course.
> Hence, there is reason to suspect that *A* is true.

But Frankfurt rightly remarks that this is not an inference leading to any new idea. After all, the new idea—the explanatory hypothesis *A*—must have occurred to one *before* one infers that there is reason to suspect that *A* is true, for *A* already figures in the second premise.

Frankfurt then goes on to argue that a number of passages in Peirce's work suggest an understanding of abduction, not so much as a process of *inventing* hypotheses but rather as one of *adopting* hypotheses, where the adoption of the hypothesis is not as being true or verified or confirmed but as being a worthy candidate for further investigation. On this understanding, abduction could still be thought of as part of the context of discovery. It would work as a kind of selection function or filter, determining which of the hypotheses that have been conceived in the stage of discovery are to pass to the next stage and be subjected to empirical testing. The selection criterion is that there must be a reason to suspect that the hypothesis is true, and we will have such a reason if the hypothesis makes the relevant observed facts a matter of course. This would indeed make better sense of Peirce's claim that abduction is a logical operation.

Nevertheless, Frankfurt ultimately rejects this proposal as well. Given, he states, that there may be infinitely many hypotheses that account for a given fact or set of facts (which Peirce acknowledged), it can hardly be a sufficient condition for the adoption of a hypothesis (in the preceding sense) that its truth would make that fact or set of facts a matter of course. At a minimum, abduction would not seem to be of much use as a selection function. One may doubt whether this is a valid objection, however. Previewing some of the discussion forthcoming in this chapter in connection with underdetermination arguments, I note that it is by no means clear that "accounting for

---

that the Dutch had won was of the former kind, whereas the abductive reasoning that led Thomson to postulate the existence of electrons was of the latter.

a given fact" is to be identified with "making that fact a matter of course." For all Frankfurt says, for a hypothesis to account for a fact, it is enough if it entails that fact. It is dubitable that entailment is sufficient for explanation, or at least for what we would regard as a satisfactory explanation. Given that it seems reasonable to read the phrase "making a given fact a matter of course" as "giving a satisfactory explanation of that fact," one could argue in response to Frankfurt's objection that even if there are an infinity of hypotheses that account for a given fact, there may still be only a handful that could be said to explain it satisfactorily. It is for Peirce scholars to decide whether this proposed interpretation is plausible in the light of Peirce's further writings.

Even if "making a given fact a matter of course" can be read as "giving a satisfactory explanation of that fact," it is remarkable that there is in Peirce's writings on abduction no reference to the notion of *best* explanation. Some satisfactory explanations might still be better than others, and there might even be a unique best one. This idea is crucial in all recent thinking about abduction. Therein lies another main difference between Peirce's conception of abduction and the modern one, which is the only one with which we are concerned from here on.

2.2.2 *How Contemporary Philosophers Characterize Abduction*

The core idea of abduction is often said to be that explanatory considerations have confirmation-theoretic import, or that explanatory success is a (possibly fallible) mark of truth, or something similar. As mentioned previously, critics have pointed out that formulations like these are slogans at best and that they can be cashed out in a great variety of ways. "Which is it?", the critics want to know. As also noted, I think it is fine to conceive of abduction as a slogan that can be filled in per context of use. Nevertheless, a number of philosophers have proposed general statements of abduction that are meant to go beyond the known slogans. We look at these not just for the sake of completeness but also because some of them might be, if not universal explications, at least candidate fill-ins for certain contexts. Moreover, it is important to point out that a number of well-known objections against abduction that some might take to militate against the idea of abduction in general really pertain only to specific proposed precisifications.

That is for instance true of the following version, which is common in textbooks:

ABD1 Given evidence E and candidate explanations $H_1, \ldots, H_n$ of E, infer the truth of *that* H_i that best explains E.

This is often immediately followed by the comment that although hopefully headed in the right direction, it may be slightly too strong, and that the inference that abduction warrants is only to the *probable* truth of the best explanation, or to the *approximate* truth of the best explanation, or to the *probable approximate* truth.

The main problem with ABD1 is not solved by any of these suggestions, however. Because abduction is ampliative, as explained earlier, it will not be a sound rule of inference in the strict logical sense, however exactly it is explicated. It can still be *reliable* in that it mostly leads to true conclusions whenever all premises are true. An obvious necessary condition for ABD1 to be reliable in this sense is that *most of the time* when E is true and H best explains E, then H is true as well (or H is approximately true, or probably true, or probably approximately true). But this would not be *enough* for ABD1 to be reliable, because ABD1 takes as its premise only that some hypothesis is the best explanation of the evidence *as compared with other hypotheses in a given set*. Thus, if the rule is to be reliable, it must hold that, at least typically, the best explanation relative to the set of hypotheses that we consider would also come out as best in comparison with any other hypotheses that we might have conceived (but for lack of time or ingenuity or for some other reason did not conceive). In other words, it must hold that at least typically the *absolutely* best explanation of the evidence is to be found among the candidate explanations that we have come up with, because otherwise ABD1 may well lead us to believe "the best of a bad lot" (van Fraassen, 1989, p. 143).

How reasonable is it to suppose that this extra requirement is usually fulfilled? Not at all, presumably. To believe otherwise, we would have to assume some sort of privilege on our part, to the effect that when we consider possible explanations of the data, we are somehow predisposed to hit upon, inter alia, the absolutely best explanation of this data. After all, hardly ever will we have considered, or will it even be possible to consider, *all* potential explanations. As Bas van Fraassen (1989, p. 144) compellingly shows, it is a priori rather implausible to hold that we are thus privileged.

In response to this "argument of the bad lot," one might argue that the challenge to show that the best explanation is always or most often among the hypotheses considered can be met without having to assume some form of privilege. Given the hypotheses that we have managed to come up with, we can always generate a set of hypotheses that jointly exhaust logical space. Suppose that $H_1, \ldots, H_n$ are the candidate explanations that we have so far been able to conceive. Then simply define $H_{n+1} := \neg H_1 \wedge \cdots \wedge \neg H_n$ and add this new hypothesis as a further candidate explanation to the ones that we already have. Obviously, the set $\{H_1, \ldots, H_{n+1}\}$ is exhaustive, in that one of its elements must be true. Following this simple procedure would seem enough to make sure that we never miss out on the absolutely best explanation.[9]

Alas, there is a catch. Even though there may be many hypotheses H_j that imply H_{n+1} and, had they been formulated, would have been evaluated as better explanations for the data than the best explanation among the candidate explanations that we started out with, H_{n+1} itself will in general be hardly informative; in fact, in general it will not even be clear what its empirical consequences are. Suppose that we have as competing explanations the Special Theory of Relativity and Lorentz's version of the ether theory. Then, following the preceding proposal, we may add to our candidate explanations the further hypothesis that neither of these two theories is true. But surely this new hypothesis will be ranked quite low qua explanation—if it will be ranked at all, which seems doubtful because it is wholly unclear what its empirical consequences are. This is not to say that the suggested procedure may never work. The point is that in general it will give little assurance that the best explanation is among the candidate explanations that we consider.[10]

Van Fraassen (1989) took the argument of the bad lot to show that if abduction can be made to work at all, it must be in the form of a probabilistic rule akin to Bayes's rule, but different from it in that it explicitly factors in explanatory considerations. And such rules, he argued, can be shown to be incoherent by Lewis's dynamic Dutch book argument. Before we turn to probabilistic versions of abduction and the issue of their coherence, it is to be pointed out that the argument of the bad lot is not nearly as general as van Fraassen appears to suppose.

9. See Lipton (1993) for a proposal along these lines. See Schupbach (2014) for a different response, and see Dellsén (2017) for discussion. McCain and Poston (2019) dispel a seeming objection to abduction related to the argument of the bad lot.

10. The argument of the bad lot, the response to it summarized in the preceding paragraph, and the problem with that response, pointed out here, were all foreshadowed in a rudimentary form in van Fraassen (1980, pp. 21–22).

A promising response to the argument begins with the observation that it capitalizes on a peculiar asymmetry or incongruence in ABD1. The rule gives license to an *absolute* conclusion—that a given hypothesis is true—on the basis of a *comparative* premise, namely, that that particular hypothesis is a better explanation of the evidence than the other hypotheses available (see Kuipers, 2000, p. 171). This asymmetry is not avoided by replacing "truth" with "probable truth" or "approximate truth." In order to avoid it, one has two general options.

The first option is to modify the rule so that it requires an absolute premise. For instance, following Alan Musgrave (1988) or Lipton (1993), one may require the hypothesis whose truth is inferred to be not only the best of the available potential explanations but also to be *satisfactory* (Musgrave) or *good enough* (Lipton), yielding the following variant of ABD1:

ABD2 Given evidence E and candidate explanations $H_1, \ldots, H_n$ of E, infer the truth of *that* H_i that explains E best, *provided* H_i is satisfactory / good enough qua explanation.

Even apart from the asymmetry problem, ABD2 immediately strikes us as more plausible than ABD1. Who, after all, would want to infer the truth of a poor explanation of the evidence, even if the competing explanations were poorer still? But what counts as satisfactory or good enough? Lipton and Musgrave make no attempt to make these notions precise. An interesting proposal has been made by Finnur Dellsén (2021). In a nutshell, his proposal is that for an explanation to be good enough it must have survived sufficiently many attempts to replace it with a better alternative. One imagines that what counts as sufficiently many attempts and even what counts as a better alternative may vary somewhat from one context to another, which would be consistent with our idea that abduction is a slogan that needs to be filled in per context of use.

We might want to add a further qualification to ABD2. For as Alexander Bird (2010) notes, even if there is a best explanation for the evidence at hand and that best explanation is also good enough, it is still an open question how much better an explanation it is than its closest competitor(s). The answer to this question, he thinks, may make a difference in what we can infer from the evidence. Specifically, he suggests that given a best and good-enough explanation, we are entitled to infer its truth only if it is significantly better than its closest competitor. If the second-best explanation of the evidence is perfectly good as well and is just barely topped by the best one, then according to Bird, "our faith in that slightly better one must be slim" (Bird, 2010, p. 346).[11]

A second way to formulate a symmetric or congruous version of abduction is by having it sanction, given a comparative premise, only a comparative conclusion; this option, too, can in turn be realized in more than one way. Here is one way to do it, which has been proposed and defended in the work of Theo Kuipers (e.g., 1984, 1992, 2000; see also Niiniluoto, 2018, p. 56):

ABD3 Given evidence E and candidate explanations $H_1, \ldots, H_n$ of E, if H_i for some i with $1 \leqslant i \leqslant n$ explains E better than H_j for all j: $1 \leqslant j \leqslant n$ such that $j \neq i$, infer that H_i is closer to the truth than any of those other hypotheses.

Clearly, ABD3 requires an account of closeness to the truth, but many such accounts are on offer today (see ch. 5), some of which would certainly be suitable for our purposes. Still, while it circumvents the best-of-a-bad-lot objection, ABD3 could be criticized for being too weak. After all, it gives no indication whatsoever of how far from the truth the best explaining hypothesis is.

There is a more general concern that one might have about replacing ABD1 by either ABD2 or ABD3. Abduction has been touted as a powerful weapon by scientific realists. It is supposed to assist us in choosing between (or among) empirically equivalent theories, thereby breaking the deadlock of underdetermination, as discussed in the next section, but it has also been wielded more directly by realists, who proclaimed that their position yields the best explanation for the instrumental successes of modern science and that this gives good reason for holding realism to be true (see section 6.2 for more on this).

Van Fraassen has been careful to state the realism–antirealism debate as being, ultimately, about the goal or goals of science. According to him, realists deem the goal of science to consist of uncovering the hidden processes and mechanisms behind the phenomena and thereby to arrive at theories that are true not just as far as they pertain to the observable part of the world but also as far as they pertain to the unobservable part. For antirealists, or at least the sort of antirealists to whom van Fraassen belongs, the goal is only to arrive at empirically adequate theories, which give a true description of the observable

11. In the same vein, see Josephson and Josephson (1994, ch. 1).

part of the world but not necessarily of the unobservable part. Note that this official statement of the opposition between realism and antirealism is entirely silent on the question of whether science is *able* to arrive at theories that are also true of the unobservable. However, antirealism would not be of much interest if it conceded that science does typically give us theories going beyond mere empirical adequacy but that nevertheless it is not *essential* to science to do so; the former is what we really care about. And indeed, in much of his work, van Fraassen has sought to undercut realist arguments to the effect that science is in the process of uncovering what lies beneath the phenomena. His attacks on abduction belong to this part of his work.

But if we reject ABD1 and adopt ABD2 or ABD3 in its stead, can abduction still serve realist purposes? With regard to ABD2, van Fraassen could ask why it should matter to the truth of a theory that it explains the evidence to *our* satisfaction. We might be way too undemanding. How to tell? And ABD3 leaves open the possibility that even our best-explaining theories are *very* far removed from the truth, which is incompatible with realism, at least in spirit. In response, it could be said that if ABD3 does not support full-blown realism, it still allows us to hold on to a position interestingly stronger than van Fraassen's brand of antirealism. For even if aiming at truth is too ambitious a goal, if ABD3 is a reliable rule of inference, we would be definitely underplaying our hand if we aimed no higher than at empirical adequacy. For then the goal that is minimally within the reach of science is to bring us ever closer to the truth, also on a theoretical level. (I am assuming here that in general scientists are able to think up hypotheses explanatorily superior to the ones already available.)

That is a weaker position than scientific realism as commonly conceived, which maintains that we have good reason to believe that contemporary scientific theories are approximately true and not only that they are closer to the truth than their predecessors were.[12] However, those realists not satisfied with the just-mentioned convergent realism should not despair if no stronger explication of the idea of abduction could be coherently maintained. As a moment's reflection shows, ABD3 licenses an inference to the unqualified truth of the *absolutely best* or *perfect* explanation. That is to say, if one is sure that however many other potential explanations for the data may have

12. This is rough. For a more careful statement of scientific realism, see Hoefer and Martí (2020).

gone unconsidered, none can equal the best of those that have been considered, then it makes no difference whether one applies ABD1 or ABD3. Consequently, if one is able to show that realism, as commonly conceived, is a perfect explanation for the observed predictive accuracy of science (i.e., that no unborn hypothesis can even equal realism as an explanation of that fact), then it suffices to show that ABD3 is a reliable rule of inference in order to defend the strong version of realism. I am skeptical that this can be accomplished (Douven, 1995). However, I would be content with a convergent realism as suggested here, as are others who consider themselves to be in the realist camp (e.g., Niiniluoto, 1999b; Kuipers, 2000). What is more, in chapter 8, I argue that we can probably do better than that, specifically that we are in a position or can move to a position from which we can ascertain the existence of many unobservables.

A more general criticism, which pertains to all these above explications of abduction as well as to their variants, is that they are stated in *categorical* terms. All these versions issue a yes-or-no verdict: they either do or do not license an inference, which for most authors means that one either is or is not warranted in accepting or believing a hypothesis, or believing that it is closer to the truth than its rivals, or believing that it is probably true, and so on.[13] That may suggest that abduction belongs to, in Richard Foley's (1992) terminology, the epistemology of (categorical) belief rather than the epistemology of degrees of belief. For many Bayesians, this alone would disqualify abduction because as they see it, the notion of categorical belief has no place in scientific philosophy; categorical belief talk is loose talk. And those Bayesians who *are* willing to condone the notion of categorical belief and so might be open to considering seriously a version of abduction cast in categorical terms do so only on the condition that the connection between categorical beliefs and probabilities or degrees of belief can be made formally precise—and the prospects of satisfying

13. For instance, according to Musgrave (1988, p. 239, italics omitted), "[i]t is reasonable to accept a satisfactory explanation of any fact, which is the best available explanation of that fact, as true." And Psillos holds that abduction "authorises the acceptance of a hypothesis H, on the basis that it is the best explanation of the evidence" (2004, p. 83, italics omitted), and that the guiding idea behind abduction is that "a hypothesis is accepted on the basis of a judgement that it best explains the available evidence" (2007, p. 442). Similarly, according to Greco (2015, p. 510), in abductive reasoning "one concludes that something is the case on the grounds that this best explains something else one believes to be the case."

that condition are bleak at best (Douven & Williamson, 2006; Douven & Rott, 2018; but cf. Chandler, 2013; Wenmackers, 2013; Leitgeb, 2017).[14]

I have two comments on this. First, I am not aware of an argument to the effect that advocates of abduction are committed to the epistemology of belief and so should reject out of hand any explication of the core idea of abduction in probabilistic terms. While such explications are sparse, they are not altogether absent from the literature. It was previously mentioned that van Fraassen, in the context of criticizing abduction, claimed that it had a chance of being tenable only if it could take the form of a probabilistic update rule, similar to Bayes's rule. He then went on to suggest the following version of abduction, which is of precisely that kind:[15]

EXPL Let $\mathcal{H} = \{H_i\}_{1 \leqslant i \leqslant n}$ be a set of mutually exclusive and jointly exhaustive hypotheses, and let Pr and Pr′ designate a person's degrees-of-belief function before and after learning evidence E, where it is assumed that $\Pr(E) > 0$. Then that person updates her degrees of belief abductively precisely if it holds for all j that

$$\Pr'(H_j) = \frac{\Pr(H_j)\Pr(E \mid H_j) + f(H_j, E, \mathcal{H})}{\sum_{k=1}^{n} \left(\Pr(H_k)\Pr(E \mid H_k) + f(H_k, E, \mathcal{H})\right)},$$

with

$$f(H_j, E, \mathcal{H}) = \begin{cases} c & \text{if } H_j \text{ best explains } E, \\ 0 & \text{otherwise,} \end{cases}$$

for some c: $0 < c < 1$.

14. What to make then of the widely shared intuition that categorical and graded belief do hang together closely? As Douven and Elqayam (2021) argue, it is possible to account for this intuition even if there is no *formal* connection between the two. Drawing on recent work on dual processing (e.g., Evans, 2007a, 2008, 2010a, 2010b), they show how categorical and graded belief can be *psychologically* connected in a way that explains the said intuition but that almost certainly cannot be captured in logico-mathematical terms. For related takes on the psychological dimension of the categorical-versus-graded-belief contrast, see Weisberg (2020) and Nagel (2021).

15. In the present context, it would make sense to call this rule "ABD4," but I have been calling it "EXPL" in various papers for about twenty years, and I have become somewhat attached to this label, which I is why I am using it here. Stephan Hartmann suggested, jokingly, that I call the principle "van Fraassen's principle," in analogy of van Fraassen's (1989) calling the Principal Principle "Miller's principle" to "honor" Miller's (1966) attempt (unsuccessful, as others soon discovered; e.g., Howson & Oddie, 1979) to refute the Principal Principle.

This looks more complicated than it is. It is helpful to think of updating via EXPL as consisting of three simple steps: in the first step, we update on the evidence via Bayes's rule; in the second, we add a bonus c to the post-Bayesian-update degree of belief assigned to the best-explaining hypothesis; and finally, we renormalize so that the degrees of belief assigned to the members of $\mathcal{H}$ sum to one in the end.

As stated here, EXPL needs some qualifications. What if two or more hypotheses are tied for explanatory bestness? The obvious solution in that case is to split the bonus evenly between or among those hypotheses. Furthermore, it could be said that, in most instances, we will judge explanation quality not just in light of the most recently acquired piece of evidence, but in light of the totality of our evidence; indeed, that might be what we *should* do (Carnap, 1947b; Good, 1967). If so, the bonus should go to the hypothesis (hypotheses) that best explains (explain) our new evidence E together with whatever evidence we already possessed. And finally, it is noted that EXPL really provides a continuum of abductive rules, one for each value of c. Van Fraassen uses the specific instance of EXPL with $c = .1$ to argue that abduction is incoherent—a claim with which we take issue in chapter 4—though the exact value of c is inessential to his argument; that is to say, any other value would have done just as well for that argument (except the value 0, which, if it were permitted, would effectively reduce EXPL to Bayes's rule). We shall encounter instances of EXPL on numerous later occasions, and comments on van Fraassen's proposal are postponed to those occasions. For now, the important point is that abduction is not tied to the notion of categorical belief, but can take a probabilistic form that allows giving some weight to judgments of explanation quality in the process of adjusting one's degrees of belief in the light of new evidence.[16]

The second comment is that although most (but not all) attention in the remainder of the book goes to probabilistic versions of abduction, this should not be taken to indicate that I believe that there is something wrong with categorical versions or that such versions have only secondary importance. To the contrary (and to repeat), a central message of the book is that it is fine to regard abduction as a slogan that, upon use, is to be explicated in a way that

16. Note, incidentally, that using an instance of EXPL would not be inconsistent with bringing explanatory considerations to bear on the assignment of priors and / or likelihoods, in the way suggested by Lipton and others. For reasons given previously, it may be hard to make this suggestion precise, but that is another matter.

depends on the context. Given that we sometimes think and talk about what we believe (categorically) and sometimes think and talk about how confident we are in this or that, it seems entirely reasonable to suppose that we will sometimes want to have at hand a categorical explication of abduction and sometimes a probabilistic one (if we are open to reasoning abductively at all). Nonetheless we will focus mainly on probabilistic versions because the main criticisms of abduction have come from proponents of Bayesianism, and I see no other way to counter those criticisms effectively than by relying on versions of abduction that are somewhat similar to Bayes's rule.

I end this section with two more general remarks. First, it merits emphasis that there is no need to choose between abduction in the Peircean sense and abduction in the more modern sense. One can consistently commit to a role for explanatory considerations in the process of generating theories (Peircean abduction) and a role for such considerations in the process of trying to justify a theory. Indeed, this is what Brian Haig (2005a) does in his abductive theory of method that was mentioned previously (see ch. 1, footnote 9).[17] In his view (Haig, 2005b), for instance, exploratory factor analysis is an abductive technique (abductive in the Peircean sense) that facilitates the generation of theories by suggesting a number of so-called latent variables that together might be sufficient to explain the phenomena at hand (the "manifest" variables). But he also sees a key role for explanatory reasoning later in the scientific process, when a number of potential explanations have been generated and we are to choose in which of those to invest our confidence. In particular, he sees a key role here for the concept of explanatory coherence, as proposed by Paul Thagard (1989, 2000), which allows one to rank theories according to how well they do with respect to criteria of explanatory goodness.

This brings me to the second remark, which concerns those criteria in relation to the idea of explanatory bestness. In section 1.2.1, I have stated that I aim to remain noncommittal regarding explanation. For the purposes of this book, it is immaterial whether explanations are best thought of as arguments of a specific type (in the manner of Hempel, 1965), as unifying different and seemingly unrelated pieces of evidence (Friedman, 1974; Kitcher, 1981), or as revealing the cause or causes of the explanandum (Salmon, 1984).

17. Niiniluoto (2018) also sees a role for abduction both in the Peircean and in the modern sense, as does Schurz (2008a), as mentioned in footnote 8 in this chapter. However, in Schurz's view the role that abduction plays in the context of justification is only minor; its primary role is in the context of discovery.

Indeed, according to some authors, a plurality of concepts of explanation are needed to understand how that notion functions in science and in everyday life, and it is a mistake to hold that there is exactly one correct account of explanation (Woody, 2015; Lange, 2016; Colombo, 2017; Khalifa, Doble, & Millson, 2020; Schupbach and Sprenger, in their 2011 paper, are also open to this possibility). Here it suffices to note that all these positions are compatible with the widely held view that explanation quality is a matter of scoring high on the so-called theoretical virtues, which include simplicity, scope, fruitfulness, internal coherence, coherence with background knowledge, and mathematical elegance.[18]

It has often been remarked that these virtues may conflict with each other (e.g., McMullin, 1996; Niiniluoto, 1999a; Schurz, 2008a) and then will have to be balanced against one another. Exactly what relative importance they are to be given may differ per context of application, and even how they are to be interpreted may vary (Longino, 1995). For instance, there is a plethora of measures of coherence (see ch. 1, footnote 12 for references), and here, too, it has been argued that we should be open to the possibility that there is not one best such measure but that different measures may be called upon on different occasions (Schippers, 2014). Note that this view on explanation quality and its relation to abduction dovetails nicely with the general message that abduction is best seen as a slogan that needs to be precisified on a per-case

18. See, for instance, McMullin (1996), Papineau (1997, p. 9), and Niiniluoto (1999a). Different authors may weigh the individual virtues differently. For instance, Einstein (1950, p. 62) thought that especially *simplicity* should guide theory choice, because "we are justified in feeling that Nature is the realisation of what is mathematically simplest." Kornblith (1993, p. 33) highlights the importance of *scope*—the ability of a theory to account for phenomena that would seem to have nothing to do with each other—when he explains why Brownian motion and Gay-Lussac's law of combination of gases together provide such strong evidence for the existence of atoms: "Although either phenomenon, by itself, might perhaps be explained by some as yet unthought hypothesis, the fact that two entirely disparate phenomena are naturally explained by one and the same hypothesis lends dramatic support to the atomic theory." And Zahar (1973) argues that Einstein's Special Theory of Relativity superseded the Lorentz–Fitzgerald–Poincaré version of the ether theory because it was more *fruitful* in that it suggested various new theories to be tested. Also recall Whewell's (1847) idea that a *consilience of inductions*—which combines elements of simplicity and scope—can provide a theory with a "stamp of truth." For empirical work on the theoretical virtues, see Lombrozo (2007); Khemlani, Sussman, and Oppenheimer (2011); Colombo, Bucher, and Sprenger (2017); Johnson, Valenti, and Keil (2019); and Shimojo, Miwa, and Terai (2020).

basis: the precisification may pertain as well to the criteria by which we judge explanations.

2.3 Abduction and Underdetermination

Sometimes all it takes to settle a dispute is to obtain more data. Suppose that we disagree about whether Bob is at his desk. We walk over to his office, and there he is, sitting at his desk. Done! Obtaining more data is not always so easy. Newton and his friends had to wait a little before the data came in that confirmed that the Dutch had won. When Newton concluded that the Dutch had won, the data alone were not enough to settle the issue. Sometimes obtaining data that would settle an issue may be out of reach, even forever. If our dispute concerns the question of whether Bob is a conscious being, someone with an inner life, or is only *behaving* as if he had an inner life, then there appears to be no equivalent to walking over to his office or waiting for an eyewitness report; no amount of data may be enough to show who of us is right, at least not conclusively, and not on its own. Similarly if we are wondering whether we are actually *seeing* Bob at his desk or instead are *hallucinating* him to be there; or more generally still, if we cannot agree on what the data are that might help us settle our dispute.

All these situations exhibit a phenomenon known as "underdetermination," meaning that in these situations, our data is compatible with various hypotheses of interest being true. That predicament will often only be temporary, as in the case of Newton or even more so in the case in which all we need to do to settle our dispute is walk over to Bob's office. But philosophers have come up with cases in which, according to them, the predicament is permanent: no amount of data could ever warrant a choice for one or the other hypothesis.

So far in this chapter we have looked at a number of uncontroversial applications of abduction, and we have also reviewed various proposals to explicate abduction. This section discusses the most paradigmatic use of abduction in philosophy, to wit, as a tool to break the deadlock in alleged cases of underdetermination. Advocates of abduction believe that, sometimes, where the data cannot settle an issue, the data together with an appeal to explanatory considerations *can*. That is so whether the underdetermination is thought to be only temporary or permanent, although the underdetermination cases that have been most widely discussed in the philosophical literature are invariably of the permanent variety. Accordingly, the uses of abduction that have been most widely discussed also concern precisely such cases.

It will therefore be useful to describe underdetermination cases generically and to point at reasons for why we should care about them. Arguably, applications of abduction in cases of permanent underdetermination are the more interesting ones. It is thus worth also discussing why we should believe that there are such permanent cases of underdetermination. Finally, we want to be more explicit about how abduction may facilitate to resolve them.

2.3.1 *What Does It Mean to Say That One Thing Is Underdetermined by Another?*

Underdetermination is a central issue in various areas of analytic philosophy. Underdetermination claims are often adduced to argue that our epistemic position vis-à-vis a given part of reality is less impressive than we would have hoped or thought it was, and in any event there is usually a lot at stake in arguments concerning some underdetermination claim(s). Some well-known philosophical debates can be regarded as turning, at bottom, on whether or not a given underdetermination claim must be accepted and concomitantly on whether or not we must resign ourselves to some modest (typically, very modest) epistemic position concerning the part of reality at issue.

For instance, in the philosophy of science one frequently encounters claims to the effect that a particular theory is (or is not) underdetermined by the evidence, or even that all scientific theories (or at least all those belonging to a certain interesting class of theories) are underdetermined by the evidence and even by all evidence that we might ideally possess. What is typically, and roughly, meant by such claims is that having *all* the available evidence will still not allow us to determine the truth-value of the theory, respectively of any theory (or any theory belonging to some designated class). To make this both more precise and more general, we can let underdetermination be a relationship between distinct classes of propositions and hold for different combinations of "know" and "justifiedly believe."[19] We might for instance say that one class of propositions, C_1, ⟨know, know⟩-underdetermines another class of propositions, C_2, if and only if knowing every member of C_1 is not

19. To make this *entirely* general, one might even consider any combination of epistemic attitudes, though I doubt that others than the just-mentioned ones will yield philosophically interesting underdetermination claims.

enough to know any member of C_2. Similarly, C_1 ⟨know, justifiedly believe⟩-underdetermines C_2 if and only if knowing every member of C_1 is not enough even to be justified in believing any member of C_2.

Although not usually stated in this way, most underdetermination claims encountered in the philosophy of science seem to be about ⟨know, justifiedly believe⟩-underdetermination of some given class of rival scientific theories by a class of propositions expressing (relevant) evidence—and not only those in the philosophy of science. One of the most central underdetermination claims in epistemology can be rendered as follows: The class of propositions expressing all your sense data throughout your entire life, as well as those you might have had at a certain moment in your life, ⟨know, justifiedly believe⟩-underdetermines the class of propositions {You are a brain in a vat, You are an embodied brain}. And a well-known underdetermination claim from the philosophy of mind is that the class of truths about a person's behavior ⟨know, justifiedly believe⟩-underdetermines, among many others, the class of hypotheses {The person has an inner life, The person does not have an inner life}. Unless stated otherwise, by "underdetermination" is meant ⟨know, justifiedly believe⟩-underdetermination.

2.3.2 *Why Is Underdetermination Philosophically Interesting?*

Nothing philosophically really interesting follows from an underdetermination claim in itself. Suppose that C_1 underdetermines C_2, so that knowing every member of C_1 will not justify us in believing any member of C_2. That does not mean that justifiedly believing in members of C_2 is impossible. Perhaps we simply do not need to know any member of C_1 in order to come to have justified beliefs in, or even knowledge of, the members of C_2, for instance, because we have direct epistemic access to the latter or epistemic access via some third class of propositions, C_3.

In interesting underdetermination claims, there is always some allegedly important epistemic distinction between the two classes of propositions referred to in the claim. To one class we are supposed to have a fairly direct cognitive access or at least a direct access given certain idealizing assumptions which in the context of the discussion are mostly deemed innocuous (or, at any rate, permissible). To the other class we at least *prima facie* seem to have cognitive access, if at all, only via the former, that is, it seems that the propositions in the latter class can be known or at least justifiedly believed, if at all, only because we can know the propositions in the former.

Arguments involving underdetermination claims come in two main varieties, which one might call the "modus tollens type" and the "modus ponens type." Arguments of the former type are meant to establish either the existence of a class of data to which we have some kind of cognitive access—although it is not obvious that we have that access—or the existence of one or more rules of inference that we justifiably rely on, although it is not obvious that we rely on that / those rule(s), or at least it is not obvious that we are justified in doing so. The modus ponens variety is meant to establish some form of skepticism.

In arguments of the modus tollens type, it is standardly presented as a given that we know / justifiedly believe all or at least some members of a class of propositions, C_1, and that there is at least some initial plausibility to the thought that our knowledge of / justified belief in these propositions depends entirely on our knowledge of (some of) the members of another class of propositions, C_2. But, it is then claimed, C_2 underdetermines C_1. The point then typically argued for is *either* that our knowledge of / justified belief in the elements of C_1 (insofar as we have it) must, appearances to the contrary notwithstanding, depend on more than just our knowledge of (some of) the members of C_2 (if it depends on that at all) *or* that there must be other rules than those most obviously available to us (like the rules of first-order logic) by dint of which we can come to have knowledge of / justified beliefs in (some of) the elements of C_1 on the basis of our knowledge of the members of C_2. This is almost invariably accompanied by some proposal as to what the something more, or the other rule(s), could be.

We find an underdetermination claim of this type in Lewis (1986, p. 107), when he discusses the requirements for a functional theory of mental content and argues that in order to be able to assign content to functional states, we must rely on principles of fit, roughly to the effect that the assignment of contents to a person should tend to make her behavior come out as serving her desires according to her beliefs. But, writes Lewis (1986, p. 107),

> principles of fit can be expected to underdetermine the assignment of content very badly. Given a fitting assignment, we can scramble it into an equally fitting but perverse alternative assignment. Therefore a theory of content needs a second part: as well as principles of fit, we need "principles of humanity," which create a presumption in favour of some sorts of content and against others.

This is a kind of attempted transcendental deduction of the existence of principles that we use in interpreting each other, where—note—it is taken as a given that interpretations (or assignments of contents) are in fact not generally underdetermined.

The other type of argument involving underdetermination claims is more common in philosophy. The basic structure of arguments of this type is the following: We can know / justifiedly believe the members of a class of propositions, C_1, if at all, only by knowing the propositions in another class, C_2; but C_2 underdetermines C_1; hence we cannot know any member of C_1. Well-known examples of this type are the Cartesian argument for external-world skepticism and various arguments for more restrictive forms of skepticism, such as skepticism about other minds. To this type also belongs one of the main arguments—if not *the* main argument—for scientific antirealism, the position in the philosophy of science that counsels agnosticism as the proper epistemic attitude vis-à-vis scientific theories, because, it is claimed, scientific theories are underdetermined by the evidence.

2.3.3 *Why Believe Underdetermination Claims?*

While underdetermination claims are found in many areas of philosophy, the most detailed discussion of underdetermination has taken place in the philosophy of science, specifically in the debate between scientific realists and scientific antirealists. Central to this debate is whether science is in the business of uncovering the hidden structure of reality, laying bare the unobservable goings-on responsible for the observable goings-on, as scientific realists hold, or whether it can give us, at best, theories that are empirically adequate, meaning that they correctly represent the observable part of the world but not necessarily the unobservable part, as scientific antirealists and in particular so-called constructive empiricists (van Fraassen, 1980) hold.[20] A main motivation for scientific antirealists is that they believe theory choice in science to be underdetermined.[21] Precisely because underdetermination has received so much detailed attention in the philosophy of science literature,

20. At least this is the interesting take on what is at stake in this debate. The boring take on it is that the debate is strictly about the goal of science and not about what science may be able to achieve (see p. 49).

21. Underdetermination is not the only motivation for scientific antirealism. For a systematic discussion of the various reasons scientific antirealists have advanced for their position, see Psillos (1999).

we focus on underdetermination in science for a while. As subsequently discussed, much of that discussion carries over to underdetermination arguments in other areas.

The standard antirealist argument for the thesis that scientific theories are underdetermined by the evidence involves two premises. The first is the following:

EE For each scientific theory there are empirically equivalent rivals,

where an empirically equivalent rival to a theory is a contrary theory (a theory inconsistent with it) that at least in light of the data alone—any possible data—will necessarily be accorded the same confirmation-theoretic status. Naturally one could consider weaker versions of EE ("for most theories . . . ," "for many theories . . . ," "for some theories . . . ," "for all theories with such-and-such features . . . ," and so on), which in combination with the subsequent premise KE, would all seem to yield somewhat different versions of scientific antirealism, but for simplicity we stick to EE here.

The important thing to note is that if EE is correct, then no matter how many empirical tests a theory has already passed, this success cannot on its own be taken as an indication that the theory is true, for each of its empirically equivalent rivals will or would pass the same tests just as successfully. Thus, unless the data refute a theory, no amount of them suffices to determine its truth-value.

Although EE does not by itself yield any antirealist conclusions, it does do so together with the following premise sometimes called "Knowledge Empiricism":

KE If the data alone do not suffice to determine a theory's truth-value, then nothing does.

Indeed, from EE and KE it follows directly that the truth-value of any scientific theory must forever remain beyond our ken. Crucially, in the context of our present discussion, notice that KE says, in effect, that explanatory power is at most of pragmatic value and has no epistemic significance.

Arguments for EE either extrapolate from supposed historical cases of empirical equivalence or try to prove formally the existence of empirically equivalent rivals to any scientific theory. As to the former, we are often pointed to the empirical equivalence of the Special Theory of Relativity and the ether theory in the Lorentz–Fitzgerald–Poincaré version (briefly mentioned previously) and respectively that of standard quantum mechanics and Bohmian mechanics. As to the latter, John Earman (1993) has proposed various plausible formalizations of the notion of empirical equivalence and used them to prove some propositions that can all be regarded as establishing interesting versions of EE (see Douven & Horsten, 1998, for discussion). Similar results have been obtained by other authors.

Arguments for KE typically try to raise doubts about the truth-conduciveness of any *prima facie* reasonable candidate criterion for theory choice beyond conformity with the data. For instance, antirealists have argued that there is no a priori reason to believe that reality is simple rather than complex.[22] A further point they have raised is that even if it be granted that the world is simple in *some* sense of that word, it still need not be simple in the parochial sense that its nature or structure is easy to grasp for creatures with *our* cognitive capacities (van Fraassen, 1980, p. 90).

2.3.4 What Can One Say in Response to Underdetermination Claims?

Not many philosophers are happy to accept scientific antirealism. It is not surprising, then, that several responses to the preceding antirealist argument for underdetermination are to be found in the literature. It merits remark that most of the responses now to be discussed, which all concern the antirealist argument for underdetermination, have close parallels in debates about underdetermination in other parts of philosophy.

One type of response against the antirealist argument denies EE, or at least maintains that currently we have no reason to believe that there exist (interesting) empirically equivalent rivals to any scientific theory. We might regard as an early token of this the logical positivists' argument that apparently empirically indistinguishable rivals are really just notational variants of one another. But their response was based on a verificationist view of meaning that nowadays is almost universally regarded a failure.

It seems a better strategy to tackle directly, or at least to try to raise doubts about, the arguments that have been given in support of EE. For instance, it may be pointed out that historical evidence for the thesis is very sparse. Advocates of the thesis point time and again to the two historical cases mentioned in the previous section. Patently, however, having two actual cases of

22. Some realists agree. See, for instance, Niiniluoto (1999b, pp. 183–184).

empirical equivalence in science seems hardly enough to support the claim that all scientific theories have empirically equivalent rivals nor even that a substantial number of theories have such rivals. Yet that is about all the historical evidence that we have ever been given! This objection might be countered by saying that the sparseness of actual examples of empirically equivalent rivals is explained by the fact that in scientific practice it is typically hard enough to come up with even one theory that fits the data and is also consistent with accepted background theories, let alone with a number of such theories. But one can also, and perhaps with more right, draw an altogether different moral from this fact, as Gerard 't Hooft (1994, p. 27) for instance does about the fundamental laws of physics when he states that "[t]he requirement that [the fundamental laws of physics] must agree with the very restrictive postulates of both quantum mechanics and general relativity has up to now proved so difficult to realize in any physical model that one is tempted to suspect that not more than one model will exist which agrees with all this."

In an influential paper, Larry Laudan and Jarrett Leplin (1991) have argued that in fact no number of actual examples of allegedly empirically equivalent theories can support EE, because, they contend, such theories may really be only *temporarily* indistinguishable by the data.[23] They do not mean to suggest that cases of theories that happen to be indistinguishable in light of the data that we currently have manifest only the kind of underdetermination we found in the case of Newton's conclusion about the Dutch victory and so are unable to support EE. Rather their point is, first, that our conception of data may change over time and, in particular, that the line between the observable and the unobservable may shift due to new technological advances; and second, as is widely acknowledged, that theories have observational consequences only when conjoined with so-called auxiliaries, and over time we may come to hold different views about the hypotheses that we deem eligible to figure as auxiliaries in the derivation of observational consequences from theories, so that over time theories may come to have different observational consequences.[24]

Still, the antirealist would seem in her right to insist on a conception of data, or at least of the observable, that is *not* susceptible to change over time.

23. See also Leplin (1997), and see Douven (2000) for discussion.

24. Fodor and Lepore (1992) make basically the same point when they argue that "our knowledge of confirmation relations is a posteriori," so that there is no way to decide a priori which empirical beliefs are relevant to the confirmation of a given hypothesis.

Van Fraassen (1980) has argued that there is an epistemically significant distinction between claims whose truth-value can be ascertained by observation with the naked eye and ones whose truth-value cannot thus be ascertained. And, almost by definition, no technological advances are going to affect that distinction. As for the variability of auxiliaries, Richard Boyd (1984) may be right that advocates of EE can successfully respond to this point by reformulating the thesis in terms of "total sciences," which include both theories and auxiliaries.

In their paper Laudan and Leplin further complain that while many philosophers of science seem to believe that there exists some algorithm for generating empirically equivalent rivals to any given theory, such an algorithm is nowhere to be found in the literature. This seems right. At the same time one wonders why such an algorithm should be called for. It seems that a proof of EE—whether or not that shows how *effectively* to construct empirically equivalent rivals—would offer the advocates of EE all that they could wish for. And as we saw previously, such proofs have for instance been given by Earman.[25]

It must be admitted, however, that at least the existing proofs of EE appear to rely on assumptions that are open to dispute. For instance, the proofs in Earman (1993) crucially depend on the assumption that scientific theories can be formulated in first-order languages, which may well be false. It may in effect be very hard to prove EE in a way that could suit the antirealist's needs. The main stumbling block here is that there seems to be no purely logical characterization of the notion of empirical equivalence. Of course, it is not uncommon to find empirically equivalent rivals defined as contraries that have the same logical consequences in the observational part of some designated vocabulary—which *is* a logical characterization. But while this characterization of empirical equivalence may be perfectly all right if hypothetico-deductivism is assumed, that confirmation theory is not part of the current orthodoxy. Indeed, it would be a mistake to think that the notion of empirical equivalence can be defined without (at least implicit) reference to a confirmation theory.

To buttress this point, it is good to briefly consider the underdetermination problem from a Bayesian perspective. We subsequently go deeper into

25. It should be noted that Earman's paper appeared after the publication of Laudan and Leplin's paper; it was in effect at least partly meant as a response to that paper.

the details of Bayesian confirmation theory; here an informal understanding suffices to appreciate that, for Bayesians, EE would be unable to establish any (interesting) underdetermination claim if empirical equivalence were defined in the way just suggested, because that two or more theories have the same observational consequences does not imply that they bestow the same likelihoods on all evidence statements that they do not entail (but are consistent with). And the latter are, from a Bayesian point of view, just as relevant to determining a theory's confirmation-theoretic status as are its observational consequences. In fact, given a purely subjective version of Bayesian confirmation theory, which imposes no constraints on rational degrees of belief beyond the axioms of probability theory, no underdetermination claim would seem to follow from EE even if empirically equivalent theories were defined to be theories that bestow the same likelihoods on all evidence statements. That confirmation theory would for instance allow one to assign a prior probability of 0 to all empirically equivalent rivals of a given theory, so that unless the data refute it, the theory may eventually come to have probability 1 (which is typically thought to suffice for justified credibility). On the other hand, for versions of Bayesian confirmation theory that are only slightly stronger, if the existence of empirically equivalent rivals in the just-defined sense is assumed, then interesting underdetermination claims *can* be derived. Suppose for instance we add to the subjective theory the apparently still quite weak principle that, given any set of empirically equivalent theories, there is no *unique* element of that set that receives highest prior probability. Then it is a direct consequence of Bayes's theorem that at no point in time will any of the theories have a unique highest posterior probability, no matter what evidence one may come to possess.

So, whether two theories will have the same confirmation-theoretic status given any amount of evidence depends at least in part on the confirmation theory that is being assumed, and thus proofs of EE may be of limited interest at best: even if it can be shown that all theories have empirically equivalent rivals given our current best confirmation theory, there is no guarantee that they will have these rivals given some still-to-be-developed confirmation theory that we may come to prefer one day.

However, it may be questioned whether EE is really all that important for antirealist purposes. For instance, one may wonder why an argument for underdetermination must be based on the claim that there *actually* exist empirically equivalent rivals to any scientific theory. Is it not enough to observe that any scientific theory *might* have such rivals? But first, whether the mere possibility that these rivals exist really undercuts any confirmation that we might otherwise have for a given scientific theory will (again) depend on one's confirmation theory. Furthermore, most philosophers of science would regard a position based on the assumption of the mere possibility of empirically equivalent rivals as of academic interest at best and not as a live option. That may explain why no scientific antirealist has tried to argue for underdetermination along these lines. More generally, there are to the best of my knowledge simply *no* underdetermination claims, in philosophy or elsewhere, based on the mere possibility that a hypothesis of interest has an empirically equivalent rival.

There may be a better way to argue that even if EE cannot be maintained, an interesting argument for underdetermination can go through. Consider that it may be rather simple to devise—on paper, that is—an experiment that would enable us to distinguish between two theories whereas the experiment is practically impossible to carry out. This possibility is anything but academic. For instance, David Atkinson (2003, p. 216) argues that while string theory is testable in principle, in order to really test it "one would have to produce energies that are ten to the power sixteen . . . times higher than those that [the biggest particle accelerator] will produce in 2005." Indeed, he concludes that "[i]t seems safe to say that we will never be able to produce energies anywhere near this value, and that string theory can never be confronted with the crucial test of experiment" (p. 216). One does not have to agree with Atkinson's pessimism regarding the testability of string theory in order to appreciate how similar practical considerations may apply quite generally in science. And if every scientific theory should have rivals that are indistinguishable from it by any evidence we might be practically able to obtain, even if the theories *are* distinguishable by evidence that we could obtain in principle, that would seem to serve an underdetermination argument for scientific antirealism no less than does EE. Because every empirically equivalent rival to a theory is a rival that is indistinguishable from it given the evidence that could practically be obtained, but not vice versa, the claim that every theory has rivals indistinguishable from it by that sort of evidence is patently weaker than EE.

The other main type of response to the argument from underdetermination, a response that I believe to be more promising than the attack on EE, is of course to deny KE. KE is a better target anyhow if we are interested in underdetermination arguments in general and not just those concerning scientific theories. After all, whereas it may be exceedingly difficult to come up with an empirically equivalent rival to a given *scientific* hypothesis, the bar is much lower when it comes to thinking up empirically equivalent rivals to the external-world hypothesis or the hypothesis that other people are conscious beings. References to hallucinations, dreams, envatted brains, zombies, and the like, have been generally found sufficient to raise doubts concerning our epistemic standing vis-à-vis the aforementioned hypotheses. Here, too, it has been questioned whether the rival hypotheses are really empirically equivalent to their nonskeptical counterparts (see, e.g., Putnam, 1994). Again, however, we can grant that these are cases of genuine empirical equivalence, and still no skeptical conclusion follows without KE. Thus, if we find grounds for gainsaying KE, that should help us across the board, both in the scientific and in the nonscientific domain.

Denials of KE nowadays mostly take the form of an attempt to defend abduction, which is also the approach taken in this book. If we accord confirmation-theoretic import to explanatory force, then if of two theories that both conform to the data one better explains those data than the other, that speaks epistemically in favor of the former. So, can we defend abduction? This question occupies us in much of the rest of the book. But before starting the defense of abduction, I go in some detail into what is known about how people actually rely on explanatory considerations in their reasoning. Some of those results serve as inspiration, further on, for formulating alternatives to the previously stated rule EXPL, and those rules will be instrumental in my defense of abduction.

To end this chapter, I should note that there is a third type of response to underdetermination—heralded by some of the American pragmatist philosophers, though no longer advocated by anyone—that could be viewed as an independent response but also as constituting an easy defense of abduction. According to this response, KE is false because theoretical virtues such as simplicity, scope, and more generally explanatory power, are truth-conducive *by definition*: possessing those virtues is simply constitutive of being true, together of course with being in accordance with the data. This response immediately raises a question: what if two empirically equivalent theories that are in accordance with the data do equally well in light of the theoretical virtues, too? Surely there is no guarantee that this will *never* happen. Here the pragmatist answer is that even though such theories may appear to be rivals, they are not really. Rather they are different, equally legitimate, conceptualizations of reality, and may both be true ("true in their conceptual schemes," as it is often put). In this vein, W. V. O. Quine (1992, p. 100) has suggested that one may "oscillate" between the different conceptualizations "for the sake of added perspective from which to triangulate on problems." Very similar passages are to be found in the writings of Putnam after his conversion to pragmatic realism (see, for example, his 1981; see also Decock & Douven, 2012). One reason why this type of response is no longer popular today is that most believe it to share all the problems that beset pragmatist accounts of truth and language generally (see, e.g., Schmitt, 1995). Like these accounts, the pragmatic response to underdetermination seems to issue in what many think is a self-refuting relativism. Recent work on concepts, in particular on the *rationality* of concepts, may come to rescue here. It has been argued that some conceptual schemes are more rational than others (Douven, 2020d; Douven & Gärdenfors, 2020), where however—in line with a conception of rationality to be endorsed subsequently (chs. 6.2 and 7)—there is no universally rational conceptual scheme, but rather which such scheme to use depends on context. But this is not a line of argumentation pursued in the present work.

3
The Psychology of Abduction

Without data, you're just another person with an opinion.

—W. Edwards Deming

3.1 Into the Mud

Lipton, whose work on abduction has done much to shape the current discussion, viewed that work as providing "a partial answer to the descriptive problem . . . of giving a principled account of the way we actually go about making non-demonstrative inferences" (Lipton, 2004, p. 142). So far, we have only pointed at some anecdotal evidence suggesting that abduction is not some exotic form of reasoning of strictly academic interest but rather captures an important part of how "we actually go about making non-demonstrative inferences." By today's standards, even in philosophy (in which empirically informed approaches have been gaining in popularity), I do not think that there is a way of tackling the descriptive question of how exactly we reason abductively without having recourse to the experimental techniques developed in cognitive psychology. In this chapter, we go beyond anecdotal evidence by giving a close look at data that directly bears on abductive reasoning. The data reveals in some detail what form this reasoning takes in certain well-delineated contexts. Part of the data subsequently inspires the statement of precisifications of abduction akin to EXPL.

The first pages of this book briefly discuss research, some of it going back to Helmholtz, about how explanatory considerations may shape what we perceive, even if only subconsciously. Research into how explanation connects to higher-level cognitive processes, such as learning, categorization, and reasoning—research that has been only cited—is of a much more recent date. In retrospect, it is surprising that psychologists did not become interested in explanation earlier, because at least some of that research shows explanation to play a major role in cognition. For instance, Michelene Chi and colleagues (1994) show that prompting students to self-explain a text about the circulatory system led to better results on a later test than reading the study materials twice; this was so even if the self-explanations were misguided.[1] Furthermore, Joseph Williams and Tania Lombrozo (2010) have presented experimental findings showing that participants who were asked to *explain* why items belonged to a certain category were much more likely to discover the rules underlying the categorization than participants who were asked to *describe* the items (but who were also tasked to discover the categorization rule). There is, however, experimental work showing that the influence of explanation on cognition is not always for the good, in that, for example, it leads us to overestimate probabilities (Koehler, 1991) or it lets us take into account base rates that are evidentially irrelevant (Johnson, Rajeev-Kumar, & Keil, 2016).

In this chapter, we look at experimental results more directly concerned with the connection between explanation and belief formation or belief change—which is what abduction is about. Most of that research is of a very recent date. To appreciate the findings to be discussed, it is important to know about the dominant paradigm in current reasoning research, the so-called New Paradigm in the psychology of reasoning (Over, 2009; Elqayam & Over, 2012; Elqayam, 2018; Oaksford & Chater, 2020b). We start by outlining the fundamentals of that paradigm, which includes stating in greater formal detail the basics of Bayesianism, on which the paradigm builds. We then look at experimental findings concerning belief and belief change that allow us to compare a strict Bayesian view with one that leaves room for the possibility of abductive reasoning, in deviation from strict Bayesian principles. An attempt to explain away the apparent influence of explanatory reasoning on belief change as a result of systematic bias is shown to fail. In fact, it turns out that this influence can benefit our reasoning in that it makes us more accurate.

1. See, in the same vein, Sidney, Hattikudur, and Alibali (2015), who report that when they prompted students to self-explain examples of fraction division, that improved those students' conceptual learning.

3.2 The New Paradigm and Bayesian Rationality

From its beginnings in the 1960s, most notably through the work of Peter Wason, the psychology of reasoning was committed to the view that the norms of rational reasoning were given by classical logic.[2] Researchers saw it as their task to determine the extent to which people obeyed those norms, and to see whether any deviations from the norms could be attributed to systematic biases. Wason's (1968) celebrated selection task was the first of many whose results were all interpreted as showing that, in general, people were poor at logical reasoning and subject to a variety of biases. In the selection task, participants are presented with four cards, which they are informed have a letter on one side and a number on the reverse. The cards are laid out on a table, one with A showing, one with K, one with 2, and one with 7. The participants are tasked to determine whether it holds as a rule that a card has a 2 on one side if it has an A on the other side. They are asked to select the cards that must be turned in order to find out whether the rule holds for the four cards. Logic suggests turning the cards that show the A and the 7, which are the only ones that could provide falsifying instances, but that is not what people typically do; instead, they typically select the cards showing the A and the 2 or the A card alone.

At first, psychologists tried to explain these and similar apparent violations of logic in terms of cognitive limitations. The idea was that people *are* rational, at least in principle, but due notably to our limited working memory, we make performance errors when going through the necessary logical steps. Not much later, however, a number of psychologists started to question the logic-oriented approach generally. Logic was developed to facilitate mathematical reasoning, in which we use truth-preserving inference rules to derive theorems from a typically small set of supposedly self-evident axioms. Although logic has been immensely useful, it is to be acknowledged that much of our reasoning is not of that sort. Recall Schurz and Hertwig's (2019) previously cited point that the maximum ecological validity of logical

2. Wason was influenced by the work of the developmental psychologists Inhelder and Piaget (see in particular their 1958 book). But while Inhelder and Piaget held that logic was also a descriptively accurate model of how adults reason, Wason saw it only as a normative model and indeed took his work on the selection task described in the main text to have refuted the descriptive claim. In the terminology of Elqayam and Evans (2011), Wason was willing to commit only to *prescriptive* logicism, not to *descriptive* logicism; Inhelder and Piaget endorsed both.

inferences—in any context, if you use the rules of logic to derive a conclusion from a set of true premises, that conclusion is guaranteed to be true—is offset by their low applicability.

More specifically with respect to the selection task, researchers began to realize that participants may have differing degrees of confidence in the propositions involved in a reasoning experiment, and that those degrees of confidence and not just the logical form of the problem with which they are presented may impact their responses. For instance, experimenters interested in whether people reason in accordance with some argument form that is logically valid may ask their participants to accept certain premises and then ask to what degree they are willing to accept some conclusion, where the premises and conclusion together instantiate the argument form at issue. As mentioned, the suspicion was that the experimenters' request to assume the premises does not always suffice to completely shield against an impact on the participants' responses of their degrees of confidence in the premises. Numerous experiments confirmed this suspicion to be correct (Oaksford & Chater, 2001). In fact, it turned out that not only participants' degrees of confidence in the premises can tilt the results but their confidence in the conclusion can do so as well: various experiments showed participants to be more inclined to accept the conclusion of an argument if they already believed that conclusion to be likely, independent of the form of the argument; this effect is known as "belief bias" (see Evans, Barston, & Pollard, 1983).

Guided by such observations, psychologists came to regard degrees of confidence, or degrees of belief, as central to human reasoning. Concomitantly, they came to think that the norms of rationality should primarily pertain to such degrees of confidence—about which classical logic has little to say (but not nothing, as subsequently shown). Instead, taking their cue from work by Frank Ramsey (e.g., 1926) and Bruno de Finetti (e.g., 1937), and also influenced by more recent developments in philosophy and elsewhere, psychologists endorsed what is now generally known as "Bayesianism" as an encompassing account of rationality for graded beliefs.

3.2.1 Bayesian Basics

De Finetti, Ramsey, and others contributing to the development of Bayesianism took for granted that our beliefs come in degrees. That should not raise any eyebrows, as it is introspectively clear that we believe some things more strongly or to a higher degree than others; we do not need empirical evidence for that (even though by now there is a wealth of supporting evidence).[3] Then the following are presented as the core principles of rationality:

Synchronic coherence (SC) At any time, our degrees of belief should conform to the probability calculus.

Diachronic coherence (DC) We should adapt our degrees of belief to the receipt of new information by means of Bayes's rule.

SC refers to the probability calculus, and DC to Bayes's rule; let us be specific about these.

Where "$\bot$" stands for an arbitrary contradiction, "$\top$" for an arbitrary tautology, and "$\vdash$" for the relation of logical consequence, a probability function is any real-valued function Pr that satisfies the following axioms:[4]

Axiom 3.1 For all propositions φ, $\Pr(\bot) = 0 \leqslant \Pr(\varphi) \leqslant 1 = \Pr(\top)$.

Axiom 3.2 For all propositions φ and ψ, if $\varphi \vdash \neg\psi$, then $\Pr(\varphi \vee \psi) = \Pr(\varphi) + \Pr(\psi)$.

In words, probabilities lie in the interval [0, 1]; contradictions always receive probability 0; tautologies always receive probability 1; and whenever two

3. Note, however, that for de Finetti and Ramsey, who were both operationalists, introspection had no value. Instead, they identified a person's degree of belief in a given proposition with her fair betting quotient for that proposition, that is, the price at which she is willing to take either side of a bet that pays off a certain positive amount of money if the proposition turns out true, and nothing otherwise. For instance, if a person is indifferent between buying and selling a bet on φ that pays y dollars if this bet's price is x dollars, then, it is said, she believes φ to a degree of x/y. Thus, if you are willing to take either side of a bet that pays $1,000 if it rains tomorrow (and nothing otherwise) if its price is $750, then it is said that you believe the proposition that it will rain tomorrow to a degree of 0.75. Because, de Finetti and Ramsey thought, we can always make measurements in this way, it always makes sense to attribute degrees of belief to people. From a modern psychological perspective, however, it is perfectly legitimate to cite introspective evidence for the existence of degrees of belief (Douven & Uffink, 2003).

4. Standard probability theory builds on *classical* logic; so, for instance, "$\vdash$" is taken to designate the relation of *classical* logical consequence. For attempts to define probability in a way that is neutral regarding the underlying logic, see Weatherson (2003); see Dietz (2010) for critical discussion.

propositions exclude one another, the probability of their disjunction is the sum of their separate probabilities.[5]

It is common practice in presentations of probability theory to define a shorthand notation for one especially important concept, to wit, that of conditional probability:

$$\Pr(\psi \mid \varphi) \;=_{\text{df}}\; \frac{\Pr(\varphi \wedge \psi)}{\Pr(\varphi)}, \quad \text{provided } \Pr(\varphi) > 0.$$

We read $\Pr(\psi \mid \varphi)$ as the probability of ψ given (or "conditional on") φ.

Furthermore, Bayesians say that φ *confirms* ψ precisely if $\Pr(\psi \mid \varphi) > \Pr(\psi)$, *disconfirms* ψ precisely if $\Pr(\psi \mid \varphi) < \Pr(\psi)$, and is *neutral* with regard to ψ otherwise. There is an ongoing debate about how to define *strength* of confirmation. One of the earliest proposals (Carnap, 1950) is that φ confirms ψ (if at all) to a degree of $\Pr(\psi \mid \varphi) - \Pr(\psi)$, but many other measures have been proposed since (for discussion, see Fitelson, 1999, 2001; Fitelson & Eells, 2000; Eells & Fitelson, 2002; Crupi, Tentori, & Gonzalez, 2007; Brössel, 2013; Douven, 2021).

From axioms 3.1 and 3.2 and the definition of conditional probability, one readily derives, for instance,

Proposition 3.1 *For all* φ *and* ψ*:*

1. $\Pr(\varphi) = 1 - \Pr(\neg\varphi)$;
2. $\Pr(\varphi) \leqslant \Pr(\psi)$ *whenever* $\varphi \vdash \psi$;
3. $\Pr(\varphi) = \Pr(\psi)$ *whenever* $\varphi \equiv \psi$;
4. $\Pr(\varphi) = \Pr(\psi)\Pr(\varphi \mid \psi) + \Pr(\neg\varphi)\Pr(\varphi \mid \neg\psi)$ *whenever* $0 < \Pr(\psi) < 1$;
5. $\Pr(\psi \mid \varphi) = 1$ *whenever* $\varphi \vdash \psi$ *and* $\Pr(\varphi) > 0$;
6. $\Pr(\psi \mid \varphi) = \Pr(\psi)\big(\Pr(\varphi \mid \psi)/\Pr(\varphi)\big)$ *whenever* $\Pr(\varphi) > 0$.

While there is nothing deep about these consequences of probability theory, they are important from a psychological point of view inasmuch as it is

5. According to some Bayesians, axiom 3.1 should be strengthened so as to require that *only* contradictions receive probability 0 and *only* tautologies receive probability 1. Some also propose as a further axiom a "continuity" principle according to which, for any countably infinite inconsistent set $\{\varphi_1, \varphi_2, \varphi_3, \ldots\}$ such that $\varphi_{n+1} \vdash \varphi_n$ for all n, it holds that $\lim_{n\to\infty} \Pr(\varphi_n) = 0$. However, for the purposes of this book, we can sidestep these and other more advanced technical issues. For a detailed yet accessible discussion of such issues, see Weisberg (2011).

straightforward to test whether people's degrees of belief are in accordance with them, which gives an easy opportunity to put descriptive Bayesianism (see below) to the test. The same is true for the Bayesian notion of confirmation and in particular for measures of confirmation (Crupi, Tentori, & Gonzalez, 2007; Tentori et al., 2007).

Item 6 of proposition 3.1 is also known as "Bayes's theorem" and is of particular importance because it plays a key role in confirmation theory and therefore in what follows. To appreciate its central place in confirmation theory, let us first rewrite it naming the variables more suggestively:

$$\Pr(H \mid E) \;=\; \Pr(H)\frac{\Pr(E \mid H)}{\Pr(E)}.$$

Think of H as a hypothesis in which we are interested and of E as evidence that we gathered. Then what the theorem says is that we can determine the probability $\Pr(H \mid E)$ of that hypothesis in light of our new evidence on the basis of the likelihood H bestows upon E, $\Pr(E \mid H)$, and of how likely we deemed H and E to be before we obtained the evidence, given by $\Pr(H)$ and $\Pr(E)$, respectively. This is important because it is often not immediately clear what value $\Pr(H \mid E)$ should be assigned, whereas we often simply *know* the probabilities that we assigned to H and E prior to learning E. Determining the likelihood is often also not so hard. When H is a statistical hypothesis—which invariably is the case in the applications considered subsequently—the likelihood follows analytically.[6] For instance, if H says that a given coin is fair, then the likelihood that this hypothesis bestows on the evidence that a toss with the coin comes up heads is obviously .5; more generally, if H attributes a bias for heads of x to the coin, then $\Pr(\text{Heads} \mid H) = x$. Simpler still, if the hypothesis logically *implies* the evidence, then, by item 5 of proposition 3.1, the likelihood equals 1.

Bayes's theorem is distinct from Bayes's rule (also known as "conditionalization" or "conditioning"), which according to DC we should use to update our degrees of belief if we want to be rational. This rule says that if you have learned φ, in the sense that you have become subjectively certain of φ, and nothing stronger—nothing that logically entails φ but is not entailed by φ—then for all ψ your new degree of belief in ψ should equal the conditional degree of belief in ψ given φ that you had just before learning φ. That is, where

6. Or at least "analytically"; see again chapter 1, footnote 21.

Pr is your degrees-of-belief function *prior* to learning φ and $\Pr_\varphi$ is your new degrees-of-belief function immediately *after* learning φ, Bayes's rule requires that it hold for all ψ that $\Pr_\varphi(\psi) = \Pr(\psi \mid \varphi)$.

As said, Bayes's rule is intended to apply whenever we have learned something in the sense of having become *certain* of it. But, naturally, not all learning is like that. By briefly peeking into a badly lit room, we may become *more* certain that the color of the carpet in the room is orange even if it does not make us fully certain about that color or indeed about anything about which we were not already certain beforehand. We may receive statistical information, such as that smokers are at a higher risk of developing coronary artery disease than nonsmokers. Thereby we have clearly *learned* something, but it is not the kind of evidence to which Bayes's rule applies. Again, it has been argued that the rule does not apply when we come to learn conditional information, such as that London will be flooded if global warming continues (Douven, 2012b; see also Douven & Dietz, 2011). Bayesians have come up with different proposals for accommodating each of these kinds of evidence.

For the first kind, Richard Jeffrey (1983, ch. 11) proposed a generalization of Bayes's rule, which now commonly goes by the name "Jeffrey conditionalization," and which (roughly) makes one's new degree of belief in the target proposition (in our example, that the carpet is orange) a weighted average of one's conditional degrees of belief in the proposition given the various possible evidence propositions (perhaps the propositions that we saw an orange carpet, and its negation), where the new degrees of belief assigned to those evidence propositions after the learning experience serve as weights. James Joyce (1999, ch. 6) gives a useful overview of the variety of rules that have been proposed for accommodating learning events of the second type (see also Uffink, 1995). And Bayesian procedures for accommodating conditional information have been proposed by Mario Günther (2018) and Benjamin Eva, Stephan Hartmann, and Soroush Rafiee Rad (2020).[7]

Returning to the core principles, SC and DC: why should we accept them? The dynamic Dutch book argument and the inaccuracy-minimization argument have already been mentioned as the most commonly cited reasons for holding that Bayes's rule is the only rational update rule. Each of these arguments was in fact preceded by a version addressing SC: Ramsey and (independently) de Finetti were the first to argue that if and only if our degrees

7. See also Sprenger and Hartmann (2019, ch. 4).

of belief conform to the axioms of probability are we safe from Dutch bookies (Vineberg, 2016). And Joyce (1998) purported to show that degrees of belief that are not probabilities are not as accurate as they could otherwise be. In the following discussion, I am concerned only with the arguments in support of DC. What I have to say about these arguments does *not* imply that Bayes's rule is necessarily wrong. Rather, the claim is that Bayesians are mistaken when they hold that whenever our evidence is not of the uncertain kind for which Jeffrey conditionalization was devised, or consisting of statistical information, or conditional in form, Bayes's rule offers the only rational way to update one's degrees of belief. Sometimes the best way to update may be via Bayes's rule, but sometimes it is via some precisification of abduction and sometimes (perhaps) via some altogether different rule. So, to the extent that people are rational, we should not expect them to always follow Bayes's rule; in other words, we should expect descriptive Bayesianism, to which we turn in the next section, to be false as a general hypothesis.

As a final comment, I note that in the way presented here Bayesianism is highly subjective, stemming from the fact that axioms 3.1 and 3.2 impose only very weak constraints on the function Pr. It could be the representation of *your* degrees-of-belief function, or of mine, or of anyone else's, provided that only these functions satisfy the said axioms. But those axioms leave a tremendous amount of room for variation, so much so that some believe them to lead to a debilitating form of relativism, a relativism according to which for some piece of evidence E and some hypothesis H, E might be evidence for H from your perspective, evidence against H from mine, and neutral with regard to H from a third person's perspective. As far as Bayesianism goes, we may all be "right." But we feel queasy about the idea that whereas we just obtained experimental evidence *against* a certain hypothesis, it could have been evidence *for* that hypothesis if only we had started from a different prior probability assignment.

Not all Bayesians share this sentiment. As for instance the Bayesian statistician Dennis Lindley writes,

> I am often asked if the [Bayesian] method gives the right answer, or, more particularly, how do you know if you have got the *right* prior. My reply is that I don't know what is meant by "right" in this context. The Bayesian theory is about coherence, not about right or wrong. (Lindley, 1976, p. 359)

To the eyes of many, however, this is little more than a restatement of the problem. It is unsurprising therefore, that various authors have taken to supplement Bayesianism with a number of further (supposedly) rationality requirements. Ideally, such further requirements allow us to make sense of the idea that there is (in Lindley's terms) a right prior, so that rational persons, provided they have the same background knowledge, would have to start from the same probability assignment. Somewhat less ideally, they rule out at least whole classes of priors as *not* right.

The Bayesian literature is rife with discussion of candidates for additional rationality requirements. One well-known proposal in this regard is Lewis's (1980) Principal Principle, according to which, roughly, one's degree of belief in φ given that the objective probability or chance of φ equals x, should equal x. But, given that for many propositions the notion of objective chance makes no sense (according to some, it makes sense only for propositions that are about quantum-mechanical events), this constraint is manifestly too weak to block the subjectivism previously pointed out. A seemingly more promising proposal has been the Principle of Indifference, mentioned in section 1.2.2. In its simplest version, this says that where $\{H_1, \ldots, H_n\}$ is a set of mutually exclusive hypotheses the disjunction of which is a logical truth, then in the absence of reasons to the contrary, you ought to assign each H_i the same probability, $1/n$. For example, if all we know about a given coin is that its bias for heads is a whole multiple of .1 (so it might be 0, .1, .2, and so on, until 1), then the said principle enjoins us to assign a probability of $1/11$ to each element of $\{H_1, \ldots, H_{11}\}$, where H_i is the hypothesis that the coin has a bias of $(i - 1) \times .1$ for heads.

As also mentioned, however, the principle may appear plausible at first, but on closer inspection it is hard to maintain. The problem is that it imposes inconsistent requirements. To see this, consider the set $S = \{H_1, \ldots, H_n\}$ of mutually exclusive hypotheses whose disjunction is a logical truth, and assume $n \geqslant 3$. Then $S' = \{H_1, \bigvee_{i=2}^{n} H_i\}$ is also a set of mutually exclusive hypotheses whose disjunction is a logical truth. It is perfectly consistent to assume both that you lack reasons to favor any hypothesis in S and that you lack reasons to favor either hypothesis in S'. However, given the former assumption, the Principle of Indifference enjoins you to set $\Pr(H_1) = 1/n$, whereas, given the latter, it enjoins you to set $\Pr(H_1) = 1/2$, which you cannot consistently do.

We looked at some ideas about how an appeal to explanatory considerations might help to save the principle but found those wanting, too.[8]

3.2.2 *Descriptive Bayesianism*

Descriptive Bayesianism is the view that SC and DC are not *merely* norms—which people might fail to satisfy by a wide margin—but are also *actually* obeyed by people, or at least that people behave cognitively *as if* their aim was to obey these norms, by and large. "By and large" is an important qualification. According to axiom 3.1, we should believe every logical falsehood to a degree of 0 and every logical truth to a degree of 1, no matter how difficult it may be for ordinary mortals to detect that they are logically false or logically true.[9] No psychologist expects people to fully satisfy this requirement or regards a failure to satisfy the requirement as a token of irrationality. Similarly, whereas item 2 of proposition 3.1 appears to require that if φ entails ψ, we assign a probability to the latter at least as high as the probability we assign to the former, this can translate to a requirement of rationality only with the qualification that we be able to know, or perhaps even easily detect, the holding of the entailment relation. Again, the same point could be made with respect to item 3 of proposition 3.1: if we cannot reasonably be expected to see the equivalence between two propositions, then we cannot reasonably be criticized for assigning different probabilities to them.

That we are not logically omniscient does not exclude our being rational. But assigning a probability of, for example, .7 to the possibility of rain tomorrow and of .5 to the possibility of no rain tomorrow does betoken irrationality, according to descriptive Bayesians. That thereby we are violating the axioms of probability requires no logical omniscience or other superpowers on our

8. Decock, Douven, and Sznajder (2016) have proposed a demonstrably consistent version of the Principle of Indifference, capitalizing on Gärdenfors's conceptual spaces approach (Gärdenfors, 2000, 2014; Douven & Gärdenfors, 2020). For all we presently know, however, only certain families of concepts are representable in conceptual spaces. If so, the aforementioned version of the Principle of Indifference has limited applicability.

9. As a matter of fact, given that Bayesians usually understand the $\vdash$-relation as stronger than logical entailment and as "incorporating all of contemporary mathematics" (Howson & Urbach, 1993, p. 20) and correspondingly take $\bot$ and $\top$ to designate not only logical but also mathematical falsehoods and truths, a strict descriptive Bayesianism would require us to assign probability 0 to all mathematical falsehoods and probability 1 to all mathematical truths, on pain of irrationality, no matter if even the greatest mathematicians could not (yet) recognize them as such.

part. Similarly when it is, or should be, obvious to a person that φ entails ψ and yet she believes the former to a higher degree than the latter, or if we know that the equivalence of the two propositions is transparent to her but she believes them to different degrees. Essentially the same point could be made with respect to many other straightforward consequences of probability theory. In other words, we can distinguish between interesting and uninteresting failures to obey probability theory, and descriptive Bayesianism should be thought of as the claim that the former are rare (even if the latter are rampant).

Psychologists have mustered a welter of evidence showing that people do indeed tend to obey Bayesian norms, at least in certain tasks. Still the most impressive piece of evidence to date stems from Mike Oaksford and Nick Chater's (1994) work on the previously mentioned Wason selection task. Whereas Wason and other early psychologists of reasoning had thought of this and similar tasks as *deductive* inference tasks, Oaksford and Chater convincingly argued that they are better conceived as *probabilistic* inference tasks. Specifically with respect to the selection task, these authors showed that when thus conceived, the predominant response Wason and colleagues had found was in fact the one we should expect to find if people followed something close to an optimal Bayesian strategy for acquiring information. The core idea is that people will tend to interpret the task before them as one of discovering a statistical dependence between A cards and 2 cards, and that by turning the A card and the 2 card in front of them, they are gaining the most informative evidence regarding such a dependence; at least, that is so given assumptions about people's priors and about how best to measure information gain, which Oaksford and Chater argue to be plausible in the given task.[10] Other frequently cited evidence apparently supporting descriptive Bayesianism is found in later work by the same authors (e.g., Oaksford & Chater, 1996, 2007; Chater & Oaksford, 1999) and in the work of Joshua Tenenbaum and colleagues (e.g., Griffiths & Tenenbaum, 2006; Tenenbaum, Griffiths, & Kemp, 2006; Gopnik & Tenenbaum, 2007; Tenenbaum et al., 2011).

---

10. It is to be noted, though, that results reported in Nickerson, Butler, and Barch (in press) suggest that lack of motivation to engage with the materials and, more importantly still, a tendency to misinterpret "If there is an A on one side of the card, then there is a 7 on the other side" as "If there is an A on the visible side of the card, then there is a 7 on the hidden side" (and similarly for similar statements) may also be part of the explanation of participants' performance in typical selection-task experiments.

However, these apparent successes merit two comments. First, although there is the previously presented evidence showing that people at least in certain tasks seem to follow Bayesian prescriptions, there are also well-documented violations of those same prescriptions. Among these is the so-called conjunction fallacy (Kahneman, Slovic, & Tversky, 1982; Tversky & Kahneman, 1983), that is, the finding that people sometimes deem a conjunction more probable than its least probable conjunct—which is a violation of item 3 of proposition 3.1 above, and an interesting one at that, inasmuch as we generally have no difficulty recognizing that a conjunction entails its conjuncts. However, it has been said (e.g., by Tversky and Kahneman themselves) that this is a case in which people confuse a question about probability with one about representativeness, or in which they mistake probability for the related but different concept of confirmation (Tentori, Crupi, & Russo, 2013). That would not make the fallacy any less fallacious—given that probability is neither representativeness nor confirmation—but at least the fallacy would no longer directly contradict descriptive Bayesianism. Similarly, researchers have recorded order-effects on sequential probability judgments, effects that, normatively speaking, should not occur (see Baratgin & Politzer, 2007, and further discussion in this chapter). Here too, attempts have been made to explain these effects in ways that render them compatible with descriptive Bayesianism. I further mention the so-called central tendency effect, which seems to show a general preference for middling values, including middling degrees of belief (i.e., degrees of belief not too far from the midpoint of the scale on which they are usually measured), which again can lead people to hold degrees of belief that are not formally probabilities (Stevens, 1971). Again, there may be a Bayesian explanation for this effect (Douven, 2018). Perhaps most problematic is the base rate fallacy (Bar-Hillel, 1980), which is famously illustrated by the fact that people experiencing certain symptoms can panic if an online search informs them that some dreadful disease is typically accompanied by those symptoms, even if the prior probability for anyone contracting that disease is vanishingly small. It shows that people have a tendency to equate Pr(Disease | Symptoms) with Pr(Symptoms | Disease), in clear violation of Bayes's theorem, as stated in the previous section.

The second comment is that it is not always easy to distinguish experimentally between descriptive Bayesianism and what one might more broadly call "descriptive probabilism," according to which degrees of belief are central to our reasoning, where these degrees of belief also tend to be approximately representable as probabilities and where updates of our degrees of belief tend to approximately proceed by Bayes's rule (Over, 2009; Elqayam & Evans, 2013; Elqayam, 2018). To understand why that matters, recall from the preceding chapter that instances of EXPL could be said to be close to Bayes's rule, certainly for small values of the explanation bonus that is attributed. In any event, much of the evidence supposedly supporting descriptive Bayesianism does not and was not intended to discriminate between descriptive Bayesianism and descriptive probabilism. What is more, Oaksford and Chater (2013, p. 374) explicitly admit that "it is unclear what are the rational probabilistic constraints on dynamic inference," where by "dynamic inference" they mean updates of the kind that according to strict Bayesian norms are to be governed by Bayes's rule and related rules, such as Jeffrey conditionalization.

Evidence that people update their degrees of belief in ways inconsistent with Bayes's rule dates back at least to the 1960s and hence long predates the advent of the New Paradigm. Lawrence Phillips and Ward Edwards (1966) used a new at the time experimental paradigm to show that people's probability estimates after the receipt of new evidence differed systematically from what they should have been according to Bayes's rule. In this paradigm, participants are informed that one container (e.g., an urn or a bag) holds two different types of objects (e.g., red chips / blue chips, or black balls / white balls) in one specific ratio, and another container holds those same types of objects in a different ratio. The participants are then shown a collection of objects sampled randomly from one of the containers, without being told from which, and are asked to estimate the probability that the sample comes from the first container rather than from the second. That the estimates deviate reliably from Bayesian updates has been replicated in numerous experiments since (see, e.g., Edwards, 1968; Marks & Clarkson, 1972; Fischhoff & Lichtenstein, 1978; Schum & Martin, 1982).

These deviations from Bayes's rule could have been, and perhaps were, entirely unsystematic. In any event, none of the early experimental work addressing the descriptive adequacy of Bayesian updating probed the possibility that people's updates were influenced by explanatory considerations. Needless to say, that possibility is of immediate interest to the topic of this book, and we next look at work concerned with the possible connection between belief updates and explanation.

3.3 Explanation and Belief Change

The first psychologists to seriously look into the connection between updating and explanation were Nancy Pennington and Reid Hastie in their influential work on juror decision making (1988, 1992, 1993). This work showed the importance of the order in which evidence is presented to jurors. Pennington and Hastie's participants were reliably more inclined to judge a defendant guilty when the prosecutors presented their evidence in an order that facilitated the mental construction of an explanatory story of how the crime unfolded. These researchers also found evidence that how the different pieces of evidence impacted their participants' degrees of belief differed consistently from what descriptive Bayesianism predicted.

Also closely related to a comparison between Bayesian and explanatory reasoning, Bénédicte Bes et al. (2012) demonstrated that when participants were given information about causal relations among three random variables, alongside explicitly provided correlations among those variables, they based their probability judgments on the causal information only, ignoring the statistical information. Although they were not thereby violating Bayes's rule (which is not meant to cover the provision of statistical information, as previously mentioned), they *were* violating the Principal Principle (see previous discussion), which many Bayesians endorse.

According to Bes et al., the effect of the causal information on their participants' probability judgments is likely related to the extent to which the participants could process that information into an explanatory story. It was not part of these authors' design to ask participants to judge the explanation quality of the statements whose probabilities they were asked to estimate. Such judgments could have been illuminating and might well have revealed a strong correlation between their participants' explanation-quality judgments and their probability judgments.

That, at least, is suggested by the experimental results from work that Jonah Schupbach and I documented in Douven and Schupbach (2015a). In that paper, we compared the descriptive adequacy of Bayesianism with the descriptive adequacy of explanationism (see p. 20), that is, the claim that, in updating their degrees of belief, people take into account their explanatory judgments in a way not captured by Bayes's rule. In particular, we were interested in the following questions:

(1) How do Bayesianism and explanationism compare with regard to their descriptive adequacy? Do judgments of the explanatory goodness of hypotheses play an essential role in updating, in a way that is incompatible with the Bayesian doctrine?
(2) If explanatory judgments are found to have such a role, do conditional probabilities retain an important influence in updating alongside such judgments?
(3) What sort of explanatory judgments in particular (if any) factor into updating?

To answer these questions, Schupbach and I re-analyzed data from an experiment first reported in Schupbach (2011), which used the "bookbags-and-poker-chips" paradigm from Phillips and Edwards's earliest studies on probabilistic updating.

However, Schupbach's experiment added a twist to that paradigm. Schupbach designed a sequential probabilistic updating task that allowed him to measure the degree to which explanatory factors influenced the updating. He interviewed the participants individually, showing them at the start two urns, urn A and urn B, each of which contained forty balls. The participants were also shown that urn A contained thirty black balls and ten white ones and that urn B contained fifteen black balls and twenty-five white ones. This information stayed available to the participants, in the form of a picture of the urns' contents, during the whole interview. Then on the basis of the outcome of a coin flip, one or the other urn was chosen, outside of the participants' view, after which ten balls were drawn, one by one, from the selected urn and were lined up before the participants. The balls were drawn without replacement, which was also clearly visible to the participants. The participants were asked to answer three questions after each draw. In order, these were

(i) how well the hypothesis that urn A had been selected explained the results from the drawings so far;
(ii) how well the hypothesis that urn B had been selected explained those results; and
(iii) how probable it was, in their opinion, that urn A had been selected, in view of the results so far.

Participants had to answer the questions about explanatory goodness by marking a point on a continuous scale with anchors "extremely poor explanation" and "extremely good explanation."

To analyze the responses, Douven and Schupbach (2015a) used mixed-effects models, a type of statistical model briefly mentioned in chapter 1. Such models are becoming increasingly popular in various areas of research, including psychology (see, e.g., Baayen, 2008, ch. 7; Gelman & Hill, 2009; Zuur et al., 2009). As discussed on page 11, mixed-effects models allow you to focus on the variance in the data in which you are interested, and filter out variance in your data that results from random elements in your design. For instance, the group of participants that you have is typically a random matter. Had you run your experiment a day earlier or a day later, or had you offered to pay your participants less or more, then presumably you would have had an at least partly different group of participants, which in all likelihood would have resulted in an at least partly different data set. Specifically with respect to the research reported in Douven and Schupbach (2015a), it is reasonable to think that judgments of explanatory goodness may relate to posteriors differently for different participants. Mixed-effects models allow one to take such possible individual differences into account by treating participants as so-called random effects.

In the main part of our analysis, Schupbach and I fitted a number of mixed-effects models, each with the collected participants' responses to the third question as fixed effect and at least the objective conditional probabilities as predictor variable. The models also had random intercepts and slopes for participants. In a classical regression model there is exactly one intercept—which predicts the outcome variable when all predictor variables are 0 (or, if the variables have been scaled, at their mean)—and one slope, which predicts how the outcome variable will change depending on changes in the predictor variables. By contrast, in a mixed-effects model each random effect (participants, in Douven and Schupbach's models) can have its own intercept and slope.

Three of the mixed-effects models reported in Douven and Schupbach (2015a) are of particular interest to us. In the first of these (model MMO, in Douven & Schupbach, 2015a), objective conditional probabilities were the only predictor. A second model further included both the collected responses to the first question and the collected responses to the second question as predictors (MMOAB). The third model of interest had next to objective conditional probabilities the computed *difference* between the participants' responses to the first question and their responses to the second question as predictor (MMOD). Because including more predictors will in general lead

Table 3.1: Model comparison results concerning the main models from Douven and Schupbach (2015a).

	k	LL	AIC	ΔAIC	BIC	ΔBIC
MMO	6	282.34	−552.68	82.37	−531.32	68.12
MMOAB	15	329.13	−628.26	6.79	−574.85	24.59
MMOD	10	327.53	−635.05	0.00	−599.44	0.00

Note: *k* is the number of parameters and LL the log-likelihood. ΔAIC is the difference in AIC value with the model with lowest AIC value, and analogously for ΔBIC.

to better model fit, these models were compared using criteria that penalize for extra predictors, in particular the so-called Akaike information criterion (AIC) and the Bayesian information criterion (BIC), which weigh (in slightly different ways) model fit, as measured in terms of the log-likelihood of the model (basically, how much variance remains unaccounted for by the model), against model complexity, as measured in terms of the number of parameters estimated in the model (such as the coefficients of the fixed effects in the model and the variance of the residuals; see Akaike, 1973, and Schwarz, 1978; for useful discussion, also see Burnham & Anderson, 2002, and Sober, 2002).[11] Taking model complexity into account helps one reduce the risk of *overfitting*, that is, fitting one's model so tightly to the *available* data that it generalizes poorly to *new* data (where generalization to new data is typically what one is interested in). The results from the model comparisons are summarized in table 3.1.

AIC values and BIC values are to be interpreted as penalties, meaning that lower is better. Also, they make sense only comparatively, and then only for

11. It could be said that thereby AIC and BIC are a bit in the spirit of abduction. But note that the idea of abduction is broader, in that it is not confined to statistical model selection. More importantly, there is no reason for explanationists to commit to an unpacking of simplicity in terms of number of adjustable parameters or to the view that simplicity is the only or dominant factor determining explanation quality. Finally, whereas AIC and BIC are often presented as based on an analytical result, it is important to notice that this result depends on a number of substantive assumptions. In particular, it is assumed that the distance of a fitted model from the truth is to be measured by the Kullback–Leibler distance. As shown in chapter 7, there are alternatives to that, and a key finding of that chapter suggests that which distance to use may depend on the problem at hand. Accordingly, while to the best of my knowledge this has not been investigated, there is a real possibility that, depending on context, different information criteria, which weigh fit and complexity differently, are appropriate.

models that are fit to the same data, as is the case here. The AIC value of MMO was more than 10 higher than those of the other models, which according to Kenneth Burnham and David Anderson (2002) is to be interpreted as indicating that the former model enjoys *no* support from the data, given the availability of the other models. The pattern is the same for the BIC values.[12]

These model comparisons would seem alarming for advocates of descriptive Bayesianism, who hold that Bayesian norms also achieve greater predictive accuracy than competing norms. After all, again by the Principal Principle, Schupbach's participants after each draw should have set their degree of belief that urn A had been selected equal to whatever the objective conditional probability was that that urn had been selected, given all registered draws at that point. It is not just that this outcome failed to materialize. The bigger problem is that whereas from a Bayesian perspective it should appear puzzling that the participants' judgments of explanatory goodness reliably helped to predict their degrees of belief, by explanationists precisely this was predicted. And from the perspective of anyone committed to the broad idea of abduction, there also *should* have been an impact of explanatory considerations on people's updates.

Thus, the answer to the first research question from Douven and Schupbach (2015a) is that explanationism is predictively more accurate than Bayesianism and that judgments of explanatory goodness factor in people's updates in an essentially non-Bayesian way. At the same time, it is not as though conditional probabilities played no role. To the contrary, as Schupbach and I reported, conditional probability came out as highly significant in every model that contained it as a predictor, answering our second research question in

12. Given that the argument is between explanationists and Bayesians, it is worth rerunning the analysis using the machinery of Bayesian statistics. In Bayesian statistics, the currently recommended criterion for model comparison is the so-called leave-one-out cross-validation information criterion (LOOIC), which is, as with AIC and BIC, to be thought of as a penalty: a lower value indicates greater predictive accuracy. The Bayesian analysis showed that the equivalent of model MMO has an LOOIC value of −607.3 and the equivalent of model MMOD has an LOOIC value of −691.7. A Bayesian analysis also allows one to calculate weights for each of the models, which can be interpreted as the probability that the given model will perform best on new data, conditional on the supposition that the true model is among the ones considered (Vehtari, Gelman, & Gabry, 2017). Performing the calculations for those models showed that, in light of Douven and Schupbach's data, the model with objective conditional probabilities and difference in explanatory goodness as predictors has a weight of almost 90 percent, whereas the "Bayesian" model, with only objective conditional probabilities as predictor, has a weight of just over 10 percent.

the positive. The answer to the final question is especially noteworthy. In MMOD, the mixed-effects model with predictors (1) objective conditional probabilities, and (2) difference in explanatory goodness between the hypothesis that urn A had been selected and the hypothesis that urn B had been selected, both predictors came out highly significant. Moreover, this was the best model. So, the answer to the third question is that, to the extent that explanatory considerations influence updating, the salient considerations will be comparative, conveying the extent to which an explanatory hypothesis outperforms its competitor(s). This is noteworthy because it is exactly what one would expect to find supposing the descriptive adequacy of Bird's suggestion, mentioned in section 2.2.2, that in abductive reasoning we should heed not only the quality of the best explanation but also how much better it is, qua explanation, than the second-best explanation. We encounter more direct support for that suggestion subsequently in this chapter.

Whereas the preceding corroborates the hypothesis that explanatory considerations are at play when people update, at least in Schupbach's experimental setting, it is consistent with everything said so far that *how* explanatory considerations influence updating is unsystematic at least to the extent that it cannot easily be conceived as the result of rule-following behavior, for instance, of following something like EXPL. However, results from Douven and Schupbach (2015b) suggest that people respond to the receipt of new evidence at least somewhat as if they followed an update rule akin to EXPL.

In that paper, Schupbach and I used the objective probabilities from our earlier study as well as some probabilistic measures of explanatory goodness to compute, for each participant and each draw separately, the explanatory goodness of both hypotheses at stake in the study (that urn A had been selected, and that instead urn B had been selected). We then tried to predict the participants' updates again, now using objective conditional probabilities and computed explanatory goodness of the two hypotheses as predictors. So, we basically repeated the analysis summarized previously, but now using, instead of the *subjective* judgments of explanatory goodness that had served as predictors in the earlier analysis, the *computed* explanatory goodness values.

Adding the computed explanatory goodness values as predictors to the model with only objective conditional probabilities yielded a significantly better model. This was true in particular for two measures of explanatory goodness, to wit, Karl Popper's (1959) measure, according to which hypothesis H's power to explain evidence E is given by

$$\frac{\Pr(E \mid H) - \Pr(E)}{\Pr(E \mid H) + \Pr(E)},$$

and I. J. Good's (1960) measure, according to which H's power to explain E equals[13]

$$\ln\left(\frac{\Pr(E \mid H)}{\Pr(E)}\right).$$

In other words, if only objective probabilities are available to help us predict people's updates, we can substantially improve on the Bayesian model by making the objective probabilities do double duty and use them also as input in Popper's or Good's measure of explanatory goodness. At least this is so in the kind of updating context studied in Douven and Schupbach (2015a).

The results from Douven and Schupbach (2015b) serve further in the book as inspiration for delving deeper into the normative aspects of the debate between Bayesians and explanationists.

3.4 Just Noise?

We showed that there have been attempts to explain away previously known violations of Bayesian norms. But together these form a rather heterogeneous set of proposals. It is only very recently that authors have tried to explain those violations in a unified manner, attributing all or at least a great number of them to a cognitive mechanism we know to be active on independent grounds. Specifically, Fintan Costello and Paul Watts (2016, 2018) have sought to explain away deviations from Bayesian norms in terms of "noisy" sampling from memory. The deviations addressed by these authors were not directly related to the *updating* of degrees of belief, but there is nothing in their approach per se to suggest that it might not apply equally to registered deviations from Bayes's rule, such as those reported in the previous section. What would thereby be achieved is *not* that somehow what appear to be deviations fall into place as being in accordance with Bayesian updating after all. Instead, although there would still be evidence of a failure of descriptive Bayesianism, it would be of an utterly boring kind, not *really* brought about by some influence of

13. Douven and Schupbach (2015b) in fact used a rescaled version of Good's measure; the mathematical details need not detain us here.

explanatory considerations on reasoning but simply due to the well-recorded fact that human memory is fallible. In other words, the violations would still be real but of a kind that we should expect to observe even if people did what they could to meet Bayesian standards. If so, then this would show that descriptive Bayesianism can still be correct—to the extent that any theory of rationality can be descriptively correct, given some long-known human imperfections.

At the root of Costello and Watts's work is the eminently plausible contention that people make random errors in estimating probabilities, including estimating conditional probabilities. For example, on their account we would estimate the conditional probability that a student will pass a given exam on the supposition that he or she studies hard by sampling from memory students who worked hard for an exam and then counting among those the ones who passed the exam. But this process is error-prone: our memories are not fully dependable, and we may thus misremember a student who worked hard but failed to have passed the exam, or the other way around; indeed, we may even misremember a student to have worked hard for an exam. Costello and Watts (2018, p. 9) make the general assumption that "events have some chance $d < 0.5$ of randomly being read incorrectly." They demonstrate that for a number of probabilistic identities, such errors tend to cancel out, whereas for others they do not or even get compounded. They offer an abundance of evidence for their hypothesis and show how that helps to explain a number of well-known biases, including order effects in sequential probability judgments and the conjunction fallacy.

The aim of Costello and Watts is not to show that, contrary to what the mainstream believes, these and related biases are unproblematic. Rather, it is to identify their source, namely, the noise present in the process by which we determine probabilities. Thus, the claim is, the said biases do not refute descriptive Bayesianism, given that this position is not committed to the assumption that human memory is failsafe. Because "our reasoning processes are necessarily subject to noise" (Costello & Watts, 2016, p. 131), any alternative to descriptive Bayesianism faces the same problem that it can be accurate only up to the biasing effects of sampling errors.

As said, Costello and Watts do not consider deviations from Bayes's rule.[14] But they are right to remark that their results present a challenge that goes beyond non-Bayesian explanations of the biases that they did consider:

> [R]andom noise in reasoning can cause systematic biases in people's responses even when people are using the rational reasoning processes of standard frequentist probability estimation. To demonstrate conclusively that people are using heuristics, researchers must show that observed biases cannot be explained as the result of systematic effects caused by random noise. (Costello & Watts, 2016, p. 132)

Surely the same is true for attempts to explain biases by reference to mechanisms other than heuristics, such as attempts to attribute the findings summarized in the previous section to the influence of explanatory considerations on people's updating. Might that ostensible influence not also be in reality attributable to estimation errors?

First, there are some a priori reasons to doubt that Douven and Schupbach's results are due to the kind of sampling noise that figures in Costello and Watts' account. As said, that noise is supposed to arise from the fact that we estimate probabilities, including conditional probabilities, by sampling and counting from memory, and in that process sometimes make mistakes. Although plausible in general, it is to be recalled that in the experiment reported in Schupbach (2011), participants at all times had the drawings from the urns with their contents in front of them and were allowed to consult this memory aid as often as they wanted. At a minimum, their estimation of conditional probabilities would seem to have carried a lesser risk of being affected by noise than if these conditional probabilities had been the result of sampling from memory.

Second, adopting Costello and Watts's proposal to account for the deviations from Bayes's rule recorded by Douven and Schupbach would raise an explanatory challenge. For then where would the predictive superiority of the explanatory models come from? Given that in Costello and Watts's model the noise is supposed to be *random*, how could it moderate people's probability judgments precisely in such a way that in Douven and Schupbach's

14. That is, they do not consider deviations from what *we* are calling "Bayes's rule." They do consider what they call "Bayes rule identities," by which they mean the most direct consequences of the definition of conditional probability, including (what we call) Bayes's theorem.

analysis their explanatory judgments would come out as highly significant predictors? There does not appear to be anything in Costello and Watts's model that could account for a close link between the (putative) random noise at work in people's probability estimates and those same people's judgments of explanatory goodness.

Raising a priori doubts about the applicability of Costello and Watts's proposal to the data at issue takes us only so far. After all, while it is uncontroversial that human memory is error prone, researchers have identified various sources of noise in the nervous system that can have behavioral consequences, including causing information processing slips (Faisal, Selen, & Wolpert, 2008). So, random noise may corrupt the estimation of conditional probabilities even if that estimation does not involve any sampling from memory. And for the Bayesian purpose of explaining away the seeming influence of explanatory considerations, the exact provenance or nature of the noise is immaterial. Indeed, while Costello and Watts's assumption that the errors they measured are due to memory glitches may be plausible, there is nothing in their papers to exclude that those errors are not actually, wholly or partly, of a different origin.[15]

Thus, a better way to answer the question of whether Costello and Watts's model can account for Douven and Schupbach's findings is to try to predict the subjective updated probabilities reported by the latter authors, not by the objective conditional probabilities (henceforth called O) but instead by the "noisy" version of that predictor, transformed according to a formula given by Costello and Watts.

To do so, we first define a function that takes Costello and Watts's error parameter d as input and outputs the transform of the predictor O, to be designated as $f(\mathrm{O}, d)$, for the given value of d. The function f is based on equation 17 in Costello and Watts (2018), according to which

$$\Pr_*(\varphi \mid \psi) \quad = \quad \frac{(1 - 2d)^2 \Pr(\varphi \wedge \psi) + d(1 - 2d)\big(\Pr(\varphi) + \Pr(\psi)\big) + d^2}{(1 - 2d) \Pr(\psi) + d},$$

where $\Pr_*(\varphi \mid \psi)$ is the noisy estimate of the probability of φ conditional on ψ, and $d \in [0, .5)$ is the noise parameter. Notice that if $d = 0$, indicating that there is no noise at all, then $\Pr_*(\varphi \mid \psi) = \Pr(\varphi \wedge \psi) / \Pr(\psi) = \Pr(\varphi \mid \psi)$.

15. Nor would it matter to their overall conclusion. See the previous citation in this section, in which they refer to random noise in reasoning in general.

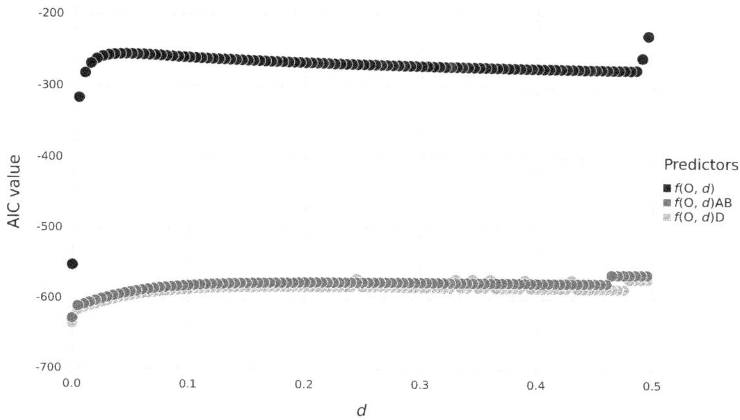

Figure 3.1: AIC values of the models with the same predictors as MMO, MMOAB, and MMOD, respectively, except that O is replaced by $f(O, d)$, for d from 0 to 0.5 in steps of 0.005.

We can then fit, for d from 0 to 0.5 in increments of 0.005, mixed-effects models like the ones specified in the previous section except that the new models have the noisy objective conditional probabilities $f(O, d)$ rather than the untransformed ones, O, as a predictor. If the deviations from Bayes's rule reported in Douven and Schupbach (2015a) are due to the interference of noise in registering frequencies, in the manner of Costello and Watts, then we should find that for some values of d the corresponding Bayesian model with $f(O, d)$ as its only predictor does best, and adding judgments of explanatory goodness as predictors should not lead to any improvement.

Figure 3.1 shows the AIC values of all the fitted models. It is immediately obvious that replacing O by $f(O, d)$, for any admissible value of d, only leads to worse fit and that in particular for no value of d does the Bayesian model—or at least the model that does not take judgments of explanatory goodness into account—come close, in terms of model quality as captured by AIC, to the models that do take such judgments into account (whether directly or indirectly via the difference in goodness between the two urn hypotheses). Comparison in terms of BIC values leads to qualitatively identical results. (See the Jupyter notebook for this chapter that is part of the Supplementary Materials; see appendix E for instructions on use.)

In short, not only is there a priori little reason to believe that Costello and Watts's proposal might also hold the key to understanding the deviations from Bayesian updating to be found in the data from Douven and Schupbach (2015a), but there is also no support from the data.

3.5 Explanatory Reasoning and Accuracy

So far, we have looked at a re-analysis of the data from Schupbach's (2011) experiment. Those data were gathered in one-on-one interviews, and so the number of participants was understandably limited (twenty-six, to be precise). That was at least mildly concerning in itself, but furthermore, because each participant had received a unique series of draws of balls from the selected urn, it gave no reliable information about individual differences, notably, about how widely the influence of explanatory factors on updating varied across participants.

To that end, Douven and Schupbach (2015a) conducted a follow-up experiment. This experiment was conducted online and had 259 participants. The participants were divided into four groups, and each group received a different one of four series of draws that had been randomly chosen from the twenty-six series that had occurred in Schupbach's experiment. Just as in that experiment, participants were informed that the balls would be drawn without replacement from an urn that had been chosen from two urns—again called "urn A" and "urn B"—on the basis of the flip of a fair coin. They were also fully informed about the contents of each urn. They were then shown, one at a time, the ten draws and asked, after each draw, to make the judgments 1, 2, and 3 as previously described (on p. 84).

Of Douven and Schupbach's 259 participants, 206 remained after exclusion of participants on the basis of a number of standard criteria (such as being nonnative speakers of English), and 167 remained after further participants were excluded for failing to answer some comprehension questions correctly. We regressed the remaining participants' degrees of belief that urn A had been selected, first onto the objective probabilities that urn A had been selected, and second onto those objective probabilities as well as the difference between (1) the judged explanatory goodness of the hypothesis that urn A had been selected and (2) the judged explanatory goodness of the hypothesis that urn B had been selected. This led to results that were qualitatively identical to those for the main experiment: the larger model vastly outperformed the "Bayesian" model with only objective probabilities as independent variable, and it did so across all conventional measures of model fit. Most notably, the Bayesian model had an AIC value of −1627.55 and the larger, "explanationist" model had an AIC value of −2422.07. (Recall that for AIC values, lower is better.)

At several junctures it was already indicated that Bayesians tend to greatly emphasize the importance of making updates minimally inaccurate. That sounds reasonable enough: we want to have the sharpest picture of reality possible, and our updating practices should help us obtain it. We go into the details of the Bayesian inaccuracy-minimization argument in chapter 4 and will find it wanting. For now, all we need to know are two things. The first is that Bayesians are really concerned with *expected* inaccuracy (i.e., the inaccuracy that, from your current perspective, you expect to achieve by updating), having left the question of *actual* inaccuracy entirely to the side.[16] The second is that the most popular measure of inaccuracy (according to some Bayesians, the only correct one) is the so-called Brier score. We formally define and discuss this scoring rule in chapter 5 and here mention only that Brier scores are penalties and that hence lower Brier scores are better.

Douven and Schupbach were interested in evaluating the descriptive adequacy of Bayesianism and explanationism; they were not concerned with the issue of accuracy. However, in Douven (2016b) I brought their data to bear on that issue, as follows. For each of the four series of draws used in Douven and Schupbach's follow-up experiment, it was known from which urn draws had come in the experiment from Schupbach (2011), in which people were interviewed individually. Specifically, three of the groups had received series of draws that had come from urn B, and the remaining group had received a series of draws from urn A. Consequently, for each of the participants in the follow-up experiment I could calculate the Brier score incurred over the ten draws. I could then run a separate regression analysis for each participant, again with degrees of belief after each update as dependent variable and objective probabilities and difference in judged explanatory goodness between the hypotheses as independent variables. The β weights (basically, standardized regression coefficients) for the independent variables thereby obtained can be interpreted as the relative weights a participant assigns to probability and explanation in determining their new degree of belief after seeing the outcome of a draw. Finally, I checked how these weights related to how accurate, in

16. Briggs and Pettigrew (2020) is an exception, but see chapter 4, footnote 17.

terms of Brier penalties, the given participant was. The question I thus sought to answer was whether taking into account explanatory considerations enables people to achieve greater accuracy, or whether it has an opposite effect, or whether there is no relation at all.

To determine the relationship between, on one hand, the weights that the participants gave to probability and to explanation and, on the other, the total Brier score that they incurred, I fitted a number of additional linear models, which had as dependent variable the total Brier scores of each participant and as independent variables one or both of these participants' β weights for probability and the β weights for explanation that resulted from the individual regression analyses mentioned previously. It turned out that the model with both the β weights for probability and those for explanation as independent variables topped the others in terms of all model comparison criteria. In that model, the slope for the probability variable was -1.91 and the slope for the explanation variable -2.38 (the regression results were all highly significant; see Douven, 2016b, for statistical details). This means that both variables had a lowering (i.e., improving) effect on Brier scores. Specifically, the best model indicates that, keeping the weight for probability fixed, every extra unit of weight given to explanation (as measured in standard deviations) lowered the total Brier score by over two points on average. The effect of attending to objective probabilities in the model was although similarly directed somewhat smaller. So, supposing that objective probabilities are available, it is certainly a good idea to attend to them, in line with what normative Bayesians would recommend. However, the more crucial finding of the analysis from Douven (2016b) was that contrary to what those Bayesians would recommend, it is an even better idea to attend at the same time to explanatory considerations, as doing so is likely to further increase the accuracy of one's degrees of belief.

3.6 Good-Enough and Second-Best Explanations

In the foregoing, we have concentrated on experimental results bearing on probabilistic versions of abduction, such as EXPL. Such versions also take the limelight in subsequent chapters. It is nevertheless worth briefly discussing experimental work in which Patricia Mirabile and I addressed a number of questions more directly having to do with categorical versions of abduction (Douven & Mirabile, 2018). We were especially (although not exclusively) interested in the descriptive adequacy of ABD2, proposed by Musgrave and Lipton, which emphasized the need for the best explanation to be good enough, and of Bird's proposed amendment of that version, which adds that the best explanation should be considerably better than its closest competitor. Specifically, we focused on the following research questions:

(1) Is the quality of an explanation a good predictor of people's willingness to accept that explanation, and is there a threshold effect in that an explanation's quality must be above a certain threshold for people to accept it?
(2) Given a potential explanation of some phenomenon, does it make any difference, in regard to the perceived quality of that explanation and in regard to people's willingness to accept it, whether or not a second explanation is introduced (and if so, why)?
(3) Given two rival explanations of some phenomenon, does the *magnitude* of their difference in quality make a difference to people's willingness to accept the better of the two?

We conducted three experiments that together were designed to answer these questions.

All three experiments drew on a fixed set of six basic scenarios, each of which presented a fact accompanied by one (in the first experiment) or two (in the second and third experiments) explanations of that fact. These explanations could vary in quality; where two explanations appeared, the explanations could vary in quality independently of each other. Participants were asked whether they accepted one of the explanations, how likely they thought the explanations were, and how good they were, qua explanations. In the third experiment, they were also asked, for each scenario separately, how confident they were in their judgment about the acceptability of the explanation.

Across three experiments, we found evidence that the quality of an explanation is a good predictor of people's willingness to accept that explanation.[17] There was also clear evidence of a threshold effect, supporting the descriptive adequacy of Lipton's proposal. Thus, the answer to the first of the preceding questions is *yes*. Interestingly, the first two experiments also showed that

17. Basically for the reasons mentioned in footnote 12 in this chapter, we ran both frequentist and Bayesian analyses for all experiments. The results of the frequentist and the Bayesian analysis were always in line with each other. See Douven and Mirabile (2018) for details.

although an explanation's probability was a good predictor of willingness to agree with the explanation, it was as such inferior to explanation quality.

It was further found that it matters for people's willingness to infer the truth of an explanation whether that explanation appears alone or side by side with a competitor, although it does *not* affect how they perceive the quality of the explanation, thus answering the first part of research question 2 in the negative and the second part in the positive.[18] The same results showed that the quality of the competitor matters to people's willingness to infer the truth of the other explanation: if an explanation was presented together with a competitor that was about as good, participants tended to be significantly less inclined to infer the truth of the former, whereas the effect of introducing a competitor tended to be smaller when that competitor was clearly inferior. That is a *yes* to the third of the research questions.

The data from Experiment 3 in Douven and Mirabile (2018), specifically participants' answers to the questions about their confidence in their own acceptability judgments, offered a good insight into exactly *why* the presence and quality of a second candidate explanation matter to people's willingness to infer to the best explanation. It could have been that those factors had somehow influenced people's *perception* of the quality of the alternative explanation. Contrast effects have long been known in perception research—a color can look brighter in the presence of a second, less bright color than the first color looks on its own, to mention a famous example—but more recently these effects have been shown to occur also in the cognitive domain and in particular to affect people's standards of judgment. For instance, it has been found that whether a consumer product is judged to be desirable may depend on which other products a potential buyer is attending to (Shoots-Reinhard et al., 2014). The dampening effect that the presentation of a second explanation had could have been due to something similar: confronted with a second possible explanation, people might no longer have found the first explanation

18. Tenney, Cleary, and Spellman's (2009) work on the notion of being beyond reasonable doubt in the context of criminal law already gave some reason to expect these results. In an experiment, these authors found that the introduction of an additional suspect by the defense in a (fictional) murder case significantly lowered their participants' preparedness to find the target suspect guilty beyond reasonable doubt. While Tenney and colleagues do not relate their results to explanatory reasoning, it is natural to think of the introduction of the second suspect as offering an alternative explanation for the events described in the case.

so compelling, which could explain why they were less inclined in that case to infer to that first explanation.

As previously mentioned, however, it was found that the perception of the quality of the explanations was largely unaffected by the presence of an alternative alongside them. Instead, the alternative lowered people's inclination to infer to the other explanation, at least when the alternative was about equally good or not much worse, because that undermined people's confidence in the inference.[19] The said dampening effect thus appeared to be akin to a seemingly unrelated result reported by Ruth Horry and Neil Brewer (2016). These authors found that when participants were briefly shown a face and then asked to identify that face in a set of faces shown simultaneously, the confidence in their choice correlated negatively with the degree of similarity between the target face and whichever other face in the set was most similar to it.

It should probably have been unsurprising that we found something similar in our experiment: given two explanations competing for bestness, the more similar in quality they are, the less confident people become in their judgment of which is best and the more hesitant they become to infer the truth of that explanation. After all, it was noted earlier that explanation quality is supposed to depend on such factors as simplicity, scope, and coherence with background knowledge. And each of those factors carries some vagueness with it. So, explanation quality being somewhat vague, it is easily understandable that people may feel that their verdicts concerning the relative quality of explanations are not entirely reliable whenever they are confronted with explanations similar in quality. This explains an empirical finding, but at the same time it provides a rationale for Bird's proposal: explanatory bestness being somewhat vague, a judgment to the effect that this or that explanation tops the others will not always be reliable, especially not if we are confronted with two or more explanations that are close in terms of quality. Then it would seem reasonable to refrain from inferring to the best explanation, or at a minimum from having much confidence in the conclusion.

19. As, following Ackerman and Thompson (2015, 2017, 2018; also Thompson, Prowse Turner, & Pennycook, 2011), we could more exactly say: it undermined their *metacognitive* confidence, that is, their confidence in (parts of) their own cognitive functioning, in this case their ability to draw the right conclusion based on explanatory considerations.

3.7 Should Philosophers Care?

This chapter has reported several empirical studies related to abduction—for the most part, probabilistic versions thereof. These studies have all been published in psychology journals and fit into the more general trend in those journals of an increasing number of papers devoted to explanation. And there is certainly more empirical work to be done. As Douven and Schupbach (2015a) remark, it is a limitation of their study that the hypotheses and contexts that they examined were quite simple and artificially stochastic in nature and that their materials did not, for instance, provide any causal-mechanical details describing how the explanandum came about. That means that the explanatory judgments that their participants gave have inevitably been of a rather shallow sort. They thus call for follow-up research using less artificial, explanatorily richer contexts. Douven and Mirabile (2018) used precisely such contexts, but they looked at categorical versions of abduction only and also did not study sequential updating.

Is there any reason for *philosophers* to be interested in the work reported in the foregoing, or in what follow-up research may bring? No doubt some will shrug their shoulders, declaring that because there is no inference from "is" to "ought," there is no reason for philosophers, who are not empirical scientists and should be concerned only with normative matters, to even look at any of the empirical data relating to abduction. Yet I believe that many philosophers *will* be interested, although their reasons may vary.

First, I mentioned the growing enthusiasm for empirically informed philosophy, which seems to be largely fueled by a concern that analytic philosophy is at the risk of devolving into the study of Alice and Bob. You know, Alice and Bob, who are constantly wondering whether they really saw Carl or rather his evil twin Dave—or was it their friend Eve in disguise? (Eve loves practical jokes; just the other day, she dressed as a painted mule.) And on it goes, until Zachary enters the picture, at which point we are all supposed to have the feeling that we are onto something deep. Thinking up such little stories can be tremendous fun and can sometimes also lead to real insights, but it can also be unsatisfactory in the long run, if only because we often just *know* that soon someone will come up with a different story, possibly starring Alfie and Beatrice, that undermines whatever lessons we thought could be learned from Alice and Bob. Thus people began to look for more solid starting points in philosophy, out of which grew empirically informed philosophy, its aim being to replace the intuitions that Alice-and-Bob stories were supposed to elicit with hard data about what laypeople, not indoctrinated by years in graduate school, thought was reasonable to say under various circumstances.

Engaging in empirically informed philosophy usually meant one of two things: sifting through the psychology journals, hunting for results that could be related to this or that philosophical dispute; or getting one's own hands dirty, by conducting experiments, typically ones that asked for ordinary people's verdicts about cases that philosophers had strong opinions about (like Gettier cases or the kind of cases figuring in the skepticism debate). This approach has led to quite a bit of bad news for philosophers lately, in that there appeared to be surprisingly little support among nonphilosophers, for claims that philosophers had variably designated as intuitive, pretheoretically obvious, natural to say, or using similar expressions meant to signal that a claim does not stand in need of any serious argumentative backing (e.g., Machery et al., 2004; Weinberg et al., 2010).

Not all philosophers agree that these findings should be taken seriously (see, e.g., Williamson, 2007, 2018). However that may be, it is certainly fortunate to have some *good* news for a change, in that with regard to abduction, the empirical results are in line with ideas that philosophers had put forward as being intuitively correct. This is true for Bird's as well as Lipton's proposals, both of which concerned a *categorical* version of abduction, but also for the idea that explanatory considerations are central to how we update our *graded* beliefs. Naturally, however, this will still give limited reason to rejoice if you deem, as some do, the said philosophical proposals completely wrongheaded.

This brings me to another reason why I expect at least some philosophers to be interested in the data discussed in this chapter. Even if there is no inference from the descriptive to the normative, it is still true that few philosophers relish attributing massive error to the folk; indeed, it has been argued that we should do so only if all else fails (Wilson, 1959; see also the quote from Lewis, 1986, on p. 58). Still, attributing massive error is what we must do, given what we examined in this chapter; at least, it is what we must do if the mainstream is right and abduction is to be repudiated as irrational. There may be considerable evidence that people sometimes behave like Bayesian agents, but there is also considerable evidence that sometimes they do not and that at least sometimes when they do not, the reason is that their belief changes are guided by perceived explanatory goodness. This puts some pressure on the mainstream to rethink its stance vis-à-vis abduction and revisit its arguments against abduction.

To be sure, there may be no way to escape those arguments. But in the following I argue that that is just not so. As a matter of fact, I want to show that the way in which those arguments fail is anything but subtle or otherwise hard to grasp—which may raise the question of why they are so widely taken for granted. On my analysis, these arguments use a kind of diversionary tactic not unlike the old magician's technique that makes the audience look *here* so that it will not pay attention to the real action (dumping a pack of cards, smothering a rabbit, stitching together a female assistent, . . .) that is going on *there*; specifically, these arguments make us focus so strongly on potential *costs* of explanatory reasoning that they make us forget to ask about potential *benefits*. This is not to suggest that the diversion was surreptitiously *planned* by the authors who came up with these arguments. It may simply be an unintended side effect of the arguments' fairly technical nature, which may have created a kind of collective blindspot.[20] Be this as it may, that we should finally start attending to abductive reasoning's advantages is a recurring message in the next chapters.

20. See the literature on attentional blindness (e.g., Chabris & Simons, 2010) for an explanation of how that may come about.

4

Facing the Challenges

While according to some, abduction is the cornerstone of scientific methodology, critics have argued that if abduction is at variance with Bayes's rule, then, like any other such rule, it is to be rejected as leading to irrational belief updates. For many years, the standard argument for this claim has been the previously mentioned dynamic Dutch book argument, which purports to show that updating by any rule other than Bayes's makes one liable to sure financial losses. However, Bayesians have increasingly come to regard their erstwhile favored argument as addressing the wrong issue, to wit, whether it is rational from a *practical* rather than an *epistemic* viewpoint to deviate from Bayes's rule. This has led some theorists to pursue a different strategy in defense of Bayes's rule, a strategy that is purported to offer a distinctively epistemic argument in support of that rule, spelled out in terms of inaccuracy minimization. Roughly, the argument is that by updating via any non-Bayesian rule, one's degrees of belief are not as accurate as they would have been had one updated via Bayes's rule.

In this chapter, I begin with showing that neither of the aforesaid arguments succeeds. The first part of the chapter argues that while current developments in mainstream epistemology may help to deflect some of the criticism that the dynamic Dutch book argument has met, the argument fails nonetheless because it rests on unfounded (and unstated) premises. The second part focuses on the inaccuracy-minimization defense of Bayes's rule, arguing that there appear to be several equally legitimate ways to interpret the notion of inaccuracy minimization and using computer simulations to show that under some of them it may be abduction rather than Bayes's rule that does best with regard to inaccuracy minimization. Chapters 6 and 7 greatly expand on the latter finding.

4.1 The Dynamic Dutch Book Argument Revisited

According to Bayesians, an agent is rational exactly if, first, at any given moment her degrees of belief obey the probability axioms—this was the content of SC discussed in section 3.2.1—and second, as she acquires new evidence, she changes her degrees of belief in accordance with Bayes's rule, which was the content of DC discussed in the same section. As before, where Pr is defined to be a probability function representing a given agent's belief system at a certain point in time, Pr_E represents the belief system that results from Pr upon learning E and nothing stronger. We can thus write Bayes's rule for belief change

$$\mathrm{Pr}_E(H) = \mathrm{Pr}(H \mid E), \tag{4.1}$$

with

$$\mathrm{Pr}(H \mid E) = \mathrm{Pr}(H)\frac{\mathrm{Pr}(E \mid H)}{\mathrm{Pr}(E)}. \tag{4.2}$$

It was pointed out that (4.2) is a form of Bayes's theorem, which follows from the Kolmogorovian ratio definition of conditional probability. By contrast, (4.1) does not follow from probability theory, which merely constrains what beliefs can be rationally held at one and the same time and is silent on *change* of belief.

For many decades, the Ramsey–de Finetti Dutch book argument has been viewed as key to the Bayesian account of rationality. We saw in chapter 3 that according to this argument we are susceptible to Dutch books—collections of bets ensuring a negative net pay-off come what may—precisely if our degrees of belief violate the axioms of probability. From this, Ramsey and de Finetti concluded that rational degrees of belief are formally probabilities.

Hacking (1967) may have been the first to observe that the Ramsey–de Finetti argument in fact does nothing to justify Bayes's rule. However, a few years after the publication of Hacking's paper, Paul Teller (1973) reported a Dutch book argument—which he attributed to Lewis—aimed at justifying Bayes's rule as the only rational update rule.[1] This dynamic Dutch book argument (as it is now called) purports to show that if a person updates by some rule other than Bayes's, she can be offered a series of bets at different

---

1. The source of the argument reported by Teller was later published as Lewis (1999).

Table 4.1: Results of updating three bias hypotheses on two consecutive heads via EXPL and, in parentheses, Bayes's rule.

	H_1	H_2	H_3	Next toss heads
Initial probabilities	.33	.33	.33	.50
First toss heads	.00 (.00)	.30 (.33)	.70 (.67)	.85 (.83)
Second toss heads	.00 (.00)	.16 (.20)	.84 (.80)	.92 (.90)

points in time such that each bet will seem fair at the time it is offered and yet jointly the bets guarantee a financial loss. What is worse, the argument continues, the person could have seen this loss coming. This vulnerability to "dynamic" Dutch books has convinced many that non-Bayesian updating is a mark of irrationality.

Van Fraassen (1989, ch. 6) gives a particularly lucid presentation of the dynamic Dutch argument that specifically targets abduction. His presentation makes use of a specific statistical model and, although not explicitly stated, models probabilistic abduction in terms of the rule we labeled "EXPL," which, as explained, first updates on incoming evidence via Bayes's rule, then adds an explanation bonus (.1, in van Fraassen's argument) to the best-explaining hypothesis, and finally renormalizes.

There is considerable leeway in defining explanatory bestness. Simplifying van Fraassen's presentation somewhat, we look at EXPL in the context of a coin-tossing model, the hypotheses of interest all pertaining to possible biases of the coin. (For brevity of notation, and unless indicated otherwise, from here on I use the label "EXPL" to designate the instance that assigns a bonus of .1 to best explanations.) We also follow van Fraassen in taking a bias hypothesis to be the best explanation of the available evidence at a given time (the outcomes of the tosses until that time) if that bias was closest to the actually observed frequency of heads in the evidence.

Consider three hypotheses concerning the bias of a given coin C:

H_1: C has perfect bias for tails.
H_2: C is fair.
H_3: C has perfect bias for heads.

The three hypotheses are assumed to be jointly exhaustive and to have equal initial probability. We now start flipping our coin and update our probabilities as the first and second toss both come up heads. Table 4.1 gives the results when we update by means of EXPL and by means of Bayes's rule (Bayes's rule updates are given in parentheses). If the dynamic Dutch book argument is correct, then it must be that a Dutch book can be made against anyone who updates her beliefs by means of EXPL. That seems to be the case indeed.

Suppose a bookie approaches you and offers to make bets on the following hypotheses concerning tosses of *C*, of which you know it is either fair or has a perfect bias either for heads or for tails:

> *A*: The first two tosses will land heads.
> *B*: The third toss will land tails.

The bookie offers two bets:

> I pays \$240 if *A* is false.
> II pays \$3,000 if $A \wedge B$ is true.

The expected values (initial probability times pay-off) of these bets are \$140 and \$125, respectively. The bookie is happy to sell the bets at these prices and you are happy to buy them, at a total cost of \$265.

Now suppose that at least one of the first two tosses does *not* come up heads. Then you have won bet I but lost bet II, which means that you receive \$240 but nevertheless suffer a total loss of \$25 (given that you payed \$ 265 for the bets). That, it would seem, is all in the game. However, suppose the first two tosses do come up heads. Then you have lost bet I but may still win bet II. However, before the coin is tossed a third time, the bookie approaches you again and now instead of proposing to sell any bets proposes to buy the following bet from you:

> III pays \$3,000 if *B* is true.

If we assume that you update by EXPL, your probability that *B* will come true is now .08, as shown in table 4.1. So you will agree to sell III for the price of \$240. However, whatever further may happen, you have now lost money. If the third toss does land tails, you receive \$3000 but have to pay the same amount; in the other case, you do not have to pay anything but will also not receive anything. Regardless, you have spent \$265 on the bets you bought and only made \$240 from selling bet III, meaning that whether *A* does or does not come true, you are bound to lose \$25. Were you to use Bayes's rule instead, nothing of this kind could happen. Using that rule you would in our example have assigned a probability of .1 to *B* had the first two tosses come up heads. Thus not $240 but $300 would have seemed a fair price to ask for III, in which case you would have had a gain of $35 (and in which case presumably no bookie would have proposed to buy III).

The foregoing argument is open to a number of immediate objections. For example, the betting concept of probability (see ch. 3, footnote 3), on which all Dutch book arguments ultimately rest, may be deemed psychologically unrealistic (see, e.g., Williamson, 1998). Perhaps even more unrealistic is the fact that the bookie is assumed to know not only your current probabilities but also your update rule. That is unrealistic because, typically, *we ourselves* do not exactly know how we update our belief states. Most importantly, it may be held against this argument that a loss in one respect can be outweighed by a benefit in another respect; the *net* effect of non-Bayesian updating could thus be positive, no matter the costs that also come with it. That is a main point to be argued for further on.

But note that all these objections grant (what is sometimes called) the dynamic Dutch book *theorem*—which asserts that updating by a probabilistic version of abduction such as EXPL or by any other non-Bayesian rule makes one susceptible to a dynamic Dutch book—and target the dynamic Dutch book *argument*—which infers from the theorem that non-Bayesian updating is irrational. Patrick Maher's (1992) is one of the few attempts to refute the former. He argues that if the bettor in the dynamic Dutch book argument looks ahead, she can see the sure loss coming and so will decline to bet. In terms of the above example, already when the bookie offers you bets I and II, you can calculate that, should the first two tosses indeed come up heads, you will prefer to sell back the bet on *B* to the bookie and thus be guaranteed to lose. This situation is easily avoided: simply do not buy bets I and II in the first place. We just noted that even if updating by some rule other than Bayes's should entail costs, it does not follow that non-Bayesian updating is irrational; it might have advantages that outweigh the costs. However, it seems that Maher's "look-before-you-leap principle" (as Earman, 1992, p. 49, calls it) opens up the possibility of non-Bayesian updating at no cost at all.

Unfortunately, as Brian Skyrms (1993b) points out, Maher's argument hinges crucially on the assumption that if (again in terms of our example) you refuse to buy the first two bets, the bookie will not propose to buy III, and this need not be so. If the assumption is dropped and the expected pay-off of selling III but not buying I and II is calculated, you will find that, although this time there is no sure loss, the *expected* loss is the same as when the first bets are bought. In fact, if the bookie should add some premium, however minute, for each transaction made, buying these bets will be preferable to only selling III.

Maher (1993) acknowledges Skyrms's point of critique and concludes from it that the dynamic Dutch book theorem is correct after all. As I argued in Douven (1999), however, Maher may be conceding too fast here, for there appears to be a strategy, building on the principle he proposed, that does offer the possibility of non-Bayesian updating free of charge. The basic idea underlying this strategy is that the advice to look before you leap can be made into a precise method for calculating the initial probability of a proposition that *corresponds to*, or is *symmetrical with*, the process of updating that we enter at the moment the evidence relevant to the truth of that proposition comes in. As subsequently shown, had you looked ahead in this sense, you would not have accepted the previously described initial bets I and II—not because you would have seen that in combination they make you vulnerable to a Dutch book even though separately they seem fair, but simply because the bets would *not* have seemed fair. The strategy comprises three parts.

We previously saw that, on a Bayesian model of learning, our present beliefs constrain the acquisition of new beliefs through equation (4.1). The first part of my proposal is to keep this on board but to take seriously a suggestion by Ramsey concerning how we arrive at (nontrivial) conditional probabilities, a suggestion that makes those probabilities depend on updating, specifically, on hypothetical updating. Ramsey (1929), in a famous footnote, suggests that to determine our probability of H conditional on E, we run through all the following in "simulation mode": (1) add E to our stock of beliefs; (2) make minimal adjustments to preserve consistency of our beliefs (if necessary); and (3) determine in that new belief state how probable we deem H. Various authors have argued that this procedure actually captures the notion of conditional probability better than the ratio definition does (see, e.g., Edgington, 1995). The simulation procedure has also been shown to be more descriptively adequate than the latter (Zhao, Shah, & Osherson, 2009). The proposal is now to bring your actual and hypothetical, or "pretend," updating practices into harmony. That will make your prior conditional probabilities reflect whichever update rule you actually happen to use. More exactly, the non-Bayesian should, on this proposal, adhere to the following:

$$\text{(4.3)} \qquad \mathrm{Pr}_E(H) = \mathrm{Pr}(H \mid E) \equiv \ldots$$

with the definition of the preferred rule (in the case of our example, EXPL) replacing the ellipsis.[2]

It may not be immediately clear how this could affect the problem the non-Bayesian faces, for conditional probabilities apparently had no part in the situation that led to the Dutch book presented previously. However, that applies only under a particular assumption. The Dutch book in this chapter was closely modeled on van Fraassen's more elaborate example. This included the way the initial probabilities of A and $A \wedge B$ were calculated. In particular, it was assumed that to calculate initial probabilities of n successive heads for a model with m bias hypotheses, one can use what is often called "the special multiplication rule":

$$\text{(SMR)} \qquad \mathrm{Pr}(E_1 \wedge \cdots \wedge E_n) = \sum_{i=1}^{m} \prod_{j=1}^{n} \mathrm{Pr}(H_i)\,\mathrm{Pr}(E_j \mid H_i).$$

Provided that the E_js are conditionally independent given each H_i, SMR is an admissible way to calculate initial probabilities. A more general method, which does not assume conditional independence, is

$$\text{(GM)} \quad \mathrm{Pr}(E_1 \wedge \cdots \wedge E_n) = \left(\sum_{i=1}^{m} \mathrm{Pr}(H_i)\,\mathrm{Pr}(E_1 \mid H_i)\right) \times \left(\prod_{j=2}^{n} \sum_{i=1}^{m} \mathrm{Pr}(H_i \mid E_1 \wedge \cdots \wedge E_{j-1})\,\mathrm{Pr}(E_j \mid H_i)\right).$$

Note that if the hypotheses and evidence statements are of the sort we considered in the dynamic Dutch book argument, the antecedents and consequents of all the conditional probabilities of any hypothesis H_i given one or more of the E_js are explanatorily dependent—the hypotheses are explanatorily relevant to the evidence statements—or so we assumed for the sake of van Fraassen's argument.

Appendix A shows the following:

2. Note that this would also allow explanationists to appropriate the Bayesian definition of confirmation (see section 3.2.1) and thus obliterates the need to come up with a definition of their own, for instance, in the manner of Conee and Feldman (2004, 2008), which has appeared to be problematic (Byerly, 2013).

Theorem 4.1 *Where $H_1, \dots, H_n$ is a set of mutually exclusive and jointly exhaustive hypotheses such that* $\Pr(H_i) > 0$, *for all i, and $E_1, \dots, E_m$ are independent conditional on H_i, for all i, then SMR and GM are equivalent methods for computing initial probabilities if, and only if, the conditional probabilities in GM are calculated according to Bayes's theorem.*

The second part of the strategy of Douven (1999) consists of the recommendation that in this kind of case, non-Bayesian updaters (e.g., EXPL updaters) use the more general method.

To illustrate the difference and also to show how the strategy helps you avoid the Dutch book, note that applying SMR in our example to compute the probability of A amounts to computing

$$\sum_{i=1}^{3} \Pr(H_i) \Pr(\text{first toss heads} \mid H_i) \Pr(\text{second toss heads} \mid H_i) =$$
$$\left(\frac{1}{3}\right)(0)(0) + \left(\frac{1}{3}\right)\left(\frac{1}{2}\right)\left(\frac{1}{2}\right) + \left(\frac{1}{3}\right)(1)(1) = \frac{5}{12} \approx .417.$$

By contrast, using GM while attending to (4.3) yields the following probability for A:

$$\left(\sum_{i=1}^{3} \Pr(H_i) \Pr(\text{first toss heads} \mid H_i)\right)$$
$$\times \left(\sum_{i=1}^{3} \Pr(H_i \mid \text{first toss heads}) \Pr(\text{second toss heads} \mid H_i)\right) =$$
$$\left(\left(\frac{1}{3}\right)(0) + \left(\frac{1}{3}\right)\left(\frac{1}{2}\right) + \left(\frac{1}{3}\right)(1)\right) \times \left((0)(0) + \left(\frac{3}{10}\right)\left(\frac{1}{2}\right) + \left(\frac{7}{10}\right)(1)\right) = .425.$$

Thus, had you followed the strategy recommended here, you would *not* have been willing to pay \$240 for a bet on A being false and indeed would have been guarded against dynamic Dutch books quite generally.

To see why this is so, consider that what the bookie in the foregoing example exploits is that your initial odds for $A \wedge B$ to $A \wedge \neg B$ predictably diverge from what your odds for $A \wedge B$ to $A \wedge \neg B$ will become if the first two tosses should land heads. This divergence can arise because if the first tosses indeed land heads, you will twice endow H_3 with a bonus, which is not accounted for when you calculate the initial probabilities of $A \wedge B$ and

$\neg A$. Because you already *know* that H_3 will get two bonuses under the given circumstances, an obvious idea is that, to ensure that the divergence cannot arise, you do account for these bonuses when you calculate initial probabilities. However, the method of calculating initial probabilities employed in the example leaves no room for that. In contrast, because using GM to calculate $\Pr(E_1 \wedge \cdots \wedge E_n)$ in a model with m hypotheses requires calculating each $\Pr(H_i \mid E_1 \wedge \cdots \wedge E_j)$, this procedure, when combined with (4.3), lets us take account of anything we take account of when we actually update on $E_1, \ldots, E_{n-1}$, should these obtain. As per Ramsey's suggestion, it is as if in calculating in this method the initial probability of a proposition, stating the occurrence of n events, we go through a process of updating hypothetically, "pushed into the fantasy" that we come to know, one after the other, the first $n - 1$ events that will occur if the proposition is true. In this process, each of the conditional probabilities, including those with consequents that possibly explain their antecedents, obtains exactly the value its consequent will unconditionally get if its antecedent should come true. That way, the bonuses get accounted for quite automatically, and the divergence the bookie exploited is avoided from the start.[3]

We have now seen how the non-Bayesian can avoid dynamic Dutch book vulnerability. The story cannot end here, however. The reason is that as de Finetti (1937, p. 68 ff) has shown, you will be vulnerable to a *static* Dutch book unless your degrees of belief satisfy, for all φ and ψ, the following double equation:

$$(4.4) \qquad \Pr(\varphi \wedge \psi) = \Pr(\varphi)\Pr(\psi \mid \varphi) = \Pr(\psi)\Pr(\varphi \mid \psi).$$

3. You may have noticed that I tacitly assume that the bettor will come to know the data in the order she expects. Of course this assumption may be false. In cases in which it is, "hypothetical" and "real" updating may still fail to match. Because non-Bayesian update rules are not generally permutation invariant, it can happen that when $E_1, \ldots, E_n$ all turn out to be true but the bettor learns about their truth in an order other than she had expected, her posterior probability for H does not equal her initial probability for H conditional on $E_1 \wedge \cdots \wedge E_n$. Such a divergence would be innocuous, however. To construct a dynamic Dutch book the divergence must be *predictable*, and in a case of the kind considered, that requirement is not fulfilled. (If the bookie were to know that the data will come in some particular order different from the one the bettor expects, he *can* predict the divergence; but then again, not even full-blooded Bayesians are safe from bookies who know more than they do.) Shafer (1985) shows in a general fashion that dynamic Dutch book constructions are themselves based on assumptions concerning the order in which the data relevant to the bets will (or is expected to) be revealed.

As is readily verified, in our account the non-Bayesian will *not* generally satisfy (4.4); for instance, for a proponent of EXPL who follows the advice given above it does not hold that

$$\Pr(H_3)\Pr(\text{first toss heads} \mid H_3) = \Pr(\text{first toss heads})\Pr(H_3 \mid \text{first toss heads}),$$

where H_3 is our earlier hypothesis that asserts coin C to have perfect bias for heads. As a result, it might seem that the recommended strategy merely trades one problem for another.

As argued in Douven (1999), however, is actually trades a problem that Maher's look-before-you-leap principle *cannot* handle for one it *can* handle. Commenting on Maher's argument against the dynamic Dutch book theorem, Earman (1992, p. 49) writes: "In essence, the decision-theoretic message is to look before you leap. Such advice is just as valid in the synchronic setting as in the diachronic or multitemporal setting." This was written before Skyrms had shown that, although possibly valid, the advice to look before you leap was unhelpful in the diachronic case. However, in the synchronic case it is clearly unproblematic: simply check, whenever you want to make a bet, that it does not lead to a sure loss in combination with bets already made. Quite evidently, if a bettor follows this principle, no static Dutch book can be made against her. Perhaps it is not so simple to follow the advice in practice, but because it may not be too difficult for the non-Bayesian to see under what circumstances problems could arise (for instance, the EXPL user must pay extra attention when joint bets are proposed on propositions that are explanatorily dependent or when a bet is proposed on a proposition that is explanatorily related to a proposition on which she already has a bet), it seems reasonable to suppose that more workable principles are available. So, the final part of the strategy proposed in Douven (1999) consists of Maher's look-before-you-leap principle or some similar principle.[4,5]

It merits stressing that in that paper I did *not* claim that coherent non-Bayesian updating cannot be achieved by other means than the strategy presented previously, nor that that strategy is necessarily the best—nor even that rationality requires coherent updating. Rather, the point of that paper was to show that the property of coherence pertains not to update rules in isolation but to packages of both epistemic and decision-theoretic principles; that an update rule that in combination with certain principles leads to Dutch book vulnerability need not do so when combined with other principles; and that this is enough to undermine van Fraassen's main argument against abduction.[6]

4. In responding to the proposal as presented in Douven (1999), Tregear (2004) objected that the look-before-you-leap principle, when applied not only to betting but to forming judgments, would lead to cognitive paralysis in that an agent would then never make *any* judgments about any of her beliefs unless she were willing to risk ending up in an inconsistent state of mind. Writes Tregear (2004, pp. 513–514): "The conditions under which judgements of fairness are not to be made are just those conditions where the set of beliefs lead into a Dutch Book. These conditions are recognised by assessing the set of beliefs and determining whether they lead into a Dutch Book. Thus, in order to decide whether a set of beliefs is inconsistent, the beliefs must be judged." However, as I said in the main text, the situations in which a Dutch book looms for the advocate of a rule like EXPL have a generic characterization (roughly, as those in which a bet is proposed on a proposition that is explanatorily connected to some other proposition or propositions on which one has bet). Correspondingly, it is possible to "assess" a set of beliefs in order to avoid a possible inconsistency in one's belief state without having to determine degrees of belief in each of the separate beliefs. As a result, the proposed method for avoiding inconsistencies does not have to lead to a total cognitive paralysis.

5. Roche and Sober (2013) have sought to buttress the conclusion of van Fraassen's dynamic Dutch book argument by arguing that explanatory considerations cannot provide the kind of probabilistic boost that they are supposed to provide according to explanationists. Specifically, they argue that the evidence E for a hypothesis H screens off from H any fact F about explanatory connections between E and H, meaning that $\Pr(H \mid E \wedge F) = \Pr(H \mid E)$. It should be noted, however, that their argument for this claim proceeds from start to end on Bayesian assumptions. If giving bonuses to best explanations makes no sense from a Bayesian perspective, why should that bother explanationists? It should also be noted that, in view of (4.3), there may be no reading of $\Pr(H \mid E \wedge F)$ and $\Pr(H \mid E)$ that is neutral in the present debate (see also McCain & Poston, 2014; Climenhaga, 2017a; Lange, 2017).

6. Some might want to note at this juncture that van Fraassen has given an argument for the claim that explanatory considerations cannot have any positive epistemic import that is more general than the argument of the bad lot and the dynamic Dutch book argument. The more general argument starts from the premise that it is part of the meaning of "explanation" that if one theory is more explanatory than another, it must be more informative than the other.

More than twenty years after that paper, it seems to me that I was too accommodating to the Bayesian back then. For instance, Maher's look-before-you-leap principle is meant to be compatible with the Bayesian principle that rational decisions maximize expected utility. I no longer believe such compatibility is required. One could, for instance, think of it as a default–interventionist principle (to borrow a term from Evans, 2007b), according to which expected utility maximization is the default but can, under circumstances (e.g., in the face of a looming Dutch book), be overridden. In fact, in view of the literature on ecological rationality that inspired some of the subsequent chapters, it is even doubtful that expected utility maximization should be assumed the default. As Herbert Simon (1982) has argued, in reality people not only *tend* to be "satisficers" rather than maximizers, they *should* be, in view of limitations on their working memory and attentional resources. And building on Simon's insights, Rakefet Ackerman and colleagues propose that a choice is rational for a given agent in a given context to the extent that it facilitates achievement of the agent's goals within the computational constraints posed by the combination of agent and context (Ackerman et al., 2020).

Be this as it may, the question of whether a response to the dynamic Dutch book argument along the foregoing lines might be able to convince Bayesians has lost some of its interest, because most of them have become

---

Further, according to van Fraassen (1989, p. 192), it is "an elementary logical point that a more informative theory cannot be more likely to be true" so that "attempts to describe inductive or evidential support through features that require information (such as 'Inference to the Best Explanation') must either contradict themselves or equivocate." The elementary logical point, he claims, is "most [obvious] . . . in the paradigm case in which one theory is an extension of another: clearly the extension has more ways of being false" (1985, p. 280). However, the problem with this argument is that in any other kind of case the "elementary" point is not obvious at all. In particular, it is not obvious in those cases in which we are confronted with empirically equivalent rivals. For instance, it is entirely unclear in what sense the Special Theory of Relativity "has more ways of being false" than Lorentz's version of the ether theory (which as previously stated is inconsistent with, yet demonstrably empirically equivalent to, the Special Theory of Relativity). Yet, it seems that the Special Theory of Relativity is superior, qua explanation, to Lorentz's theory. If van Fraassen were to object that the former is not really more informative than the latter, or at any rate not more informative in the appropriate sense, whatever that is, then we should certainly refuse to grant the premise that in order to be more explanatory a theory must be more informative. See also Leeds (1994) for an excellent discussion of van Fraassen's general argument against explanatoriness as a confirmational virtue.

wary of that argument, and indeed of Dutch book arguments generally, on unrelated grounds. In particular, the view has spread in Bayesian quarters that the argument addresses the wrong kind of rationality. When we are concerned with the rationality of degrees of belief as well as with the rationality of how we change our degrees of belief over time, we are concerned with questions of *epistemic* rather than *practical* rationality. Given that being vulnerable to cunning bookies seems primarily a practical liability, the Dutch book arguments have been said to be beside the point (Joyce, 1998). In response to this, some have claimed that Dutch book vulnerability does flag an underlying epistemic defect: it is a manifestation of the fact that a person deems a single bet or series of bets as both fair and not fair and thus is in an inconsistent state of mind.[7] Note, however, that even if this claim is true for the (static) Ramsey–de Finetti argument, the point does not carry over to the dynamic Dutch book argument. An agent may be susceptible to engage in the kind of betting over time that figures in that argument without at any one time holding inconsistent views on the fairness of any bets. Naturally, after a learning event she may regard a bet as unfair that previously she regarded as fair, but the same would have been true had she been a Bayesian learner.

But even if the dynamic Dutch book argument is all about pragmatic rationality, that in itself need not be a reason to abandon it. More specifically, it seems to me that Bayesians could try to get some mileage out of the pragmatic turn that a number of epistemologists have recently been taking (see, e.g., Fantl & McGrath, 2009; Rinard, 2017). The epistemic status of a belief has traditionally been thought to depend solely on matters that bear on the truth of the belief, such as the quality of one's evidence or whether or not one is reliably connected to what the belief is about. But over the past years, various authors have argued that the epistemic status of a belief is inextricably bound up with the believer's practical situation, in particular, with what is at stake for her in believing correctly or incorrectly. Bayesians wishing to maintain the integrity of the Dutch book defense may not want to buy into any particular one of the arguments that have been advanced in favor of this "pragmatic encroachment view," as it has been called. However, it may suffice for them to argue that the mere prominence in current epistemology of the debate on pragmatic encroachment is enough to call into question the existence of the clear-cut divide between the epistemic and the pragmatic that the critics of

7. This point is to be found in Skyrms (1987).

the Dutch book arguments are presupposing. The reason is that, it may be said, if there were such a clear-cut divide, contributions to this debate should have gone down like lead balloons. And, Bayesians may conclude, if there is no such divide, then little is left of the charge that Dutch book arguments address the wrong type of rationality.[8]

Be this as it may, there is a deeper problem with the dynamic Dutch book argument, one that remains even if the pragmatic encroachment view is endorsed and even if the strategy for avoiding Dutch bookability described previously is, for some reason I happen to overlook, to be rejected. In essence, the argument is that updating by means of a non-Bayesian rule can cost you money that would stay in your pocket if you updated by Bayes's rule. But note that many things cost money. By dining out you can, and typically will, lose money, and in a foreseeable manner: waiters tend to present you with a bill not long after you have finished your meal. Still, that the loss can be prevented by fixing your own dinner at home is not a compelling argument against dining out; if all goes well, you get something in return that is worth the expense. More concretely, Dutch book invulnerability is only one among, in principle, an indefinite number of practical interests that people may have. What if EXPL or some other non-Bayesian rule serves other practical interests better than Bayes's rule? Could the Bayesian in that case still maintain that, in view of the dynamic Dutch book argument, Bayes's rule is the only rational update rule? Surely there exist practical goals whose achievement would more than make up for running the risk of being fleeced by a Dutch bookie—especially in view of the fact that Dutch bookies exist only as fictional characters in philosophers' tales!

From here on, one of the main themes is that abduction has practical advantages over Bayes's rule—not only practical advantages, in fact, but these are the ones we start with—and that it is a major oversight of Bayesians to focus strictly on what the costs of non-Bayesian updating may be and never to ask whether those costs could be worth incurring, given what one may get in return.

---

8. They might also want to point out that, next to a possible conceptual connection between the epistemic and the practical, there is a rather mundane connection, to wit, that the kind of epistemic projects that one can pursue depends on all sorts of practicalities. Whether I am in the position to pursue any research projects at all depends on how well I can organize my life, on whether I can secure a living as a researcher, on the degree to which I succeed in protecting my research time, and so on.

4.2 Abductive Reasoning and Practical Interests

When comparing Bayes's rule with one or more probabilistic explications of abduction, I will typically do so in the kind of statistical model that van Fraassen used in his presentation of the dynamic Dutch book argument. In particular, it will be supposed that an agent or agents are receiving binomial data (i.e., independent and identically distributed binary data), the data coming in one at a time and the agent's or agents' task being to estimate the probability of "success" (e.g., the probability that the patient survives, or that the treatment is effective, or that the coin lands heads). For present purposes, let the data consist of the outcomes of tosses with the same coin, where it is antecedently known that the bias of the coin (for concreteness, the probability of heads) can take eleven possible values, ranging from 0 (certain to land tails) to 1 (certain to land heads), in increments of .1. A priori, none of these values is more likely to be the true bias than any other, so before any tosses have been observed, it is reasonable to be no more confident that the bias takes one particular value than that it takes another. In other words, where H_i is the hypothesis that the bias for heads equals $i/10$, we are looking at the set of hypotheses $\{H_i\}_{0 \leqslant i \leqslant 10}$, with each of those having initially the same probability of $1/11$.

For now, I limit myself to a comparison between Bayes's rule and EXPL (meaning, again, the $c = 0.1$ instance of the schema labeled as such in ch. 2); subsequent comparisons include other probabilistic explications of abduction. Suppose then that a Bayesian and an EXPL user are watching the same sequence of coin tosses to which the previously mentioned model pertains. In particular, both have started with a flat probability distribution over the eleven bias hypotheses. Then they start updating their probabilities for these hypotheses via their respective rules as they see the sequence unfold. It has been said that a proposition "is assertable to the extent that it has high subjective probability for its assertor" (Jackson, 1979, p. 565).[9] If so, we may ask who—the Bayesian or the EXPL user—is more likely to first be in a position

9. More recent work on assertion suggests that this can be only approximately correct (see, e.g., Williamson, 2000, ch. 11; Douven, 2006, 2009; Goldberg, 2015; Kelp, 2016), but while some of that more recent work holds that assertion requires certainty and not just high probability, empirical evidence indicates that that requirement is too strong (Kneer, 2018, 2021; Reuter & Brössel, 2018; Marsili & Wiegmann, 2021).

to assert the truth about the bias of the coin, supposing the sequence is long enough for both eventually to be in that position.

To answer this question, I ran computer simulations of sequences of coin tosses that were long enough for both the Bayesian and the EXPL user to assign a high probability to the true bias hypothesis. "High probability" is a vague term, which I somewhat arbitrarily precisified as a probability greater than .9.[10] More exactly, for each bias hypothesis I ran 1,000 sequences of 500 (simulated) tosses of a coin with the bias as specified by the hypothesis, letting the Bayesian and the EXPL user update their probabilities for the various bias hypotheses and registering who was first in assigning a probability greater than the threshold value of .9 to the true hypothesis (called "convergence" here). I was also interested in who was the first to *stably* assign to the truth a probability above that threshold, that is, who was first in assigning a probability to the truth above .9 and *staying* above .9 for the remaining tosses in the sequence of 500 tosses ("stable convergence"). The results are graphically summarized by the plots shown in figure 4.1.[11]

This figure needs a little explanation. The individual plots (i.e., the twenty colored lines) are density plots, which can be interpreted as showing the relative likelihood that the first time a probability above the threshold is assigned (in the left panel of the figure) or is stably assigned (in the right panel) to a hypothesis occurs at a given toss. For instance, in the left panel the orange line for Bayes's rule peaks at about 100 and the orange line for EXPL peaks at about 90, both peaks being quite pronounced. This indicates that in the vast majority of the simulations with a coin with bias .2, the Bayesian converged after observing about 100 tosses, whereas in the vast majority of the same simulations, the EXPL user converged after observing about 90 tosses. Looking

10. I say that this specification of the threshold for high probability is *somewhat* arbitrary because .9 is often mentioned in the literature as a threshold for acceptability; see, for instance, Kaplan (1981, p. 308), Moser and Tlumak (1985, p. 128), and Kyburg (1990, p. 64). Relatedly, it has been recommended by legal scholars (e.g., Kagehiro & Stanton, 1985) as an explication of the notion of being beyond reasonable doubt, which is supposed to be a criterion for a conviction in the practice of criminal law. Note that one could also try other arguably more sophisticated "stopping rules" here; see, for instance, Kruschke (2013a).

11. Results for biases from .6 to .9 are not displayed in figure 4.1 because they are the mirror images (barring random noise) of the results shown—as they should be, given that a bias of p for heads is equivalent to a bias of $1 - p$ for tails. For the extreme biases, convergence and stable convergence happen at a fixed toss: for a coin that always comes up heads, Bayes's rule assigns a probability to the truth that is above the threshold after 23 tosses, EXPL after 14 tosses. Obviously, convergence and stable convergence coincide in the extreme cases.

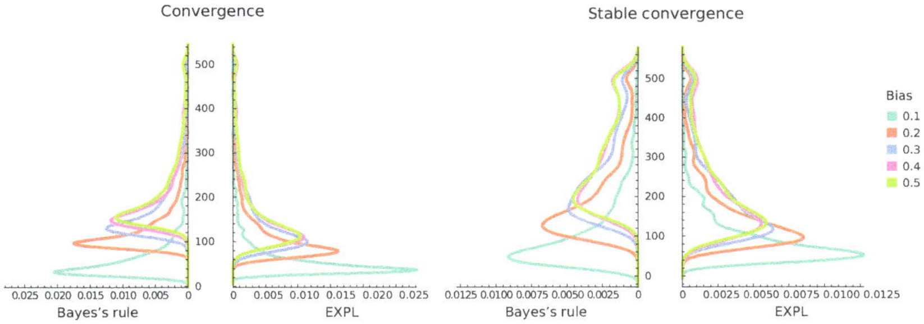

Figure 4.1: Paired density plots showing speed of convergence to the truth (*left*) and speed of stable convergence to the truth (*right*) for Bayes's rule and EXPL, for biases from .1 to .5. The x-axes show probability densities, and the y-axes show toss numbers.

in this way at the rest of the results, we see that for both convergence or at stable convergence, the peak (officially known as "the mode of the distribution") for EXPL lies typically quite a bit lower than the peak for Bayes's rule. More exactly, in a series of Mann–Whitney U tests comparing for each bias hypothesis the convergence results from Bayes's rule with those from EXPL, the latter rule each time came out as significantly faster (all $ps < .0001$) with the exception of the test for the bias = .1 hypothesis, in which there was no significant difference. Exactly the same outcome was obtained for the stable convergence results.

If this sounds unremarkable, imagine that the hypotheses concern some scientifically interesting quantity—such as the success rate of a medical treatment or the probability of depressive relapse—rather than the bias of a coin, and the tosses are observations or experiments aimed at determining that quantity. Which researcher would not want to use an update rule that increases her chances of being in a position to make public a scientific theory or a new medical treatment ahead of the (Bayesian) competition?

However, there is a possible downside to the apparent success of EXPL. The reason why the EXPL user so often beats the Bayesian in these simulations in assigning a high probability to the truth is that she is, in a clear sense, bolder in her responses to new information because she adds a bonus to the best explanation. It does not take much to see that this feature also makes the EXPL user more prone to assign a high probability to some false hypothesis:

a row of consecutive tosses producing a subsequence in which the relative frequency of heads starkly deviates from the probability for heads is more likely to push the EXPL user's probability for some false bias hypothesis over the .9 threshold than it is the more cautious Bayesian's. In other words, EXPL's greater speed can come at the expense of accuracy, so that even if EXPL may put one more rapidly in a position to assert truths, it may also put one more rapidly in a position that licenses the assertion of falsehoods.

To make this more formal, note that we can associate with each EXPL update a probability that the explanation bonus in its entirety is assigned to a false hypothesis and that this probability can be considerable. Suppose, for instance, that in our present model a coin that has, unbeknownst to the updater, an actual bias of .5 (so a fair coin) is tossed ten times. Then the probability that after the tenth toss EXPL will assign the bonus to the *true* bias hypothesis is the probability that there will be exactly five heads in the ten tosses, which can be calculated to be approximately .25. Consequently, the probability that after ten tosses a *false* hypothesis receives the bonus is about .75.

This probability goes down as the coin is tossed more often, and it is also lower for more heavily biased coins (whether toward heads or toward tails). For example, if the bias were .2 (equivalently, .8) and we were to consider the possibility that the bonus gets assigned to a false hypothesis after the 100th toss, the probability of that actually happening would be about .17; for the 1,000th toss, it is smaller than .0001. More generally, the probability of a wrong assignment goes to 0 (regardless of the bias of the coin) as the number of tosses taken into consideration goes to infinity.[12] But if we are interested in which rule should be used, not by idealized agents but by us ordinary mortals, then such limit behavior carries little significance.

12. To be entirely exact, we first note that the normal distribution gives a good approximation to the binomial distribution provided both $np \geqslant 10$ and $n(1 - p) \geqslant 10$, where p is the success probability (the bias) and n the number of trials (tosses, in our case). Excluding the extreme bias hypotheses (always heads, always tails), these conditions are satisfied for even moderately large numbers of tosses. Then the probability that the bonus is assigned to a false hypothesis after n tosses of a coin with bias p (where $0 < p < 1$) is approximately

$$1 - \operatorname{erf}\left(\frac{.05}{\sqrt{2}} \cdot \sqrt{\frac{n}{p - p^2}}\right),$$

where erf is the error function, which is defined: $\operatorname{erf}(x) = 2/\sqrt{\pi} \int_0^x e^{-t^2}\,dt$. The given expression can be shown to converge to 0 as n tends to infinity.

Consider, then, an arguably fairer comparison between EXPL and Bayes's rule. A magician has a set of eleven coins in his pocket: one coin with a perfect bias for heads, one with a bias of .1 for heads, one with a bias of .2 for heads, and so on. He entices a Bayesian and an EXPL user into playing a game against one another. In this game, the magician picks one of the coins and starts tossing it, showing each outcome to the two players. A player scores a point by raising her hand and asserting the truth about the bias of the coin. However, if the player gets the bias wrong, the point goes to the other player. If both players raise their hands after the same update and both are right about the bias, they both receive a point; if they are both wrong, neither receives a point; if one is right and the other wrong, the former receives two points. Once a player has asserted a bias hypothesis—whether correctly or incorrectly—the magician puts the coin back into his pocket and again picks a (not necessarily different) coin. Then everything starts all over. This procedure is repeated 100 times, after which the player with the highest score is declared the winner. Note that this game seems to treat the possible advantages and disadvantages of EXPL updating in an evenhanded manner: because of her boldness, the EXPL user might often be first to identify the truth, but by the same token she might earn quite a few points for her opponent. Who is more likely to win the game, the EXPL user or the Bayesian? Or are their chances of winning equal?

To answer these questions, I ran 1,000 simulations of the game. In these simulations, the EXPL user *always* won and typically did so by a wide margin. To be more exact, the mean of the differences between the EXPL user's score and the Bayesian's score was 27 (± 9); the smallest difference between the two players' score at the end of a game was 5, and the greatest was 52.

This is not meant to demonstrate the superiority of EXPL. Sets of trick coins of the variety required for playing this game are hardly more common than Dutch bookies. Moreover, nothing said here excludes the possibility that there are circumstances under which Bayesian updating offers practical advantages more consequential than Dutch book invulnerability. Instead, the point is to draw attention to a hidden assumption in the Dutch book approach, to wit, that Dutch book invulnerability trumps all other practical goals we may have. Absent an argument for this assumption, the Dutch book approach does not pose a threat to EXPL or any other probabilistic explication of abduction, even granting that the pragmatic pervades the epistemic.

4.3 Abduction and Our Epistemic Goal

The relation between the pragmatic and the epistemic is immaterial to an approach to justifying Bayes's rule and more generally the tenets of Bayesianism that has emerged over the past twenty years or so. Various theorists, disconcerted by the apparently pragmatic focus of the Dutch book arguments, have sought to justify the Bayesian tenets in (supposedly) strictly epistemic terms. Specifically, they have sought to show that those tenets are most conducive to the achievement of our epistemic goal, as spelled out for graded beliefs.

This development owes much to Joyce (1998), in which a first attempt is made to formulate an epistemic goal in terms of graded beliefs and which argues that for every degrees-of-belief function that violates the probability axioms, there is a degrees-of-belief function that obeys those axioms and is closer to the epistemic goal. Joyce thought that this vindicated the synchronic part of Bayesianism, according to which degrees of belief ought to be probabilities, though in Joyce (2009) he admits that the earlier argument was not airtight. For present purposes, however, what matters is mostly Joyce's conception of our epistemic goal as pertaining to graded beliefs.

Among mainstream epistemologists, who tend to be concerned first and foremost with categorical beliefs, it is almost universally held that our epistemic goal is to believe all that is true and only what is true (see, e.g., Lehrer, 1974, p. 202; BonJour, 1985, p. 8; Foley, 1993, p. 19). Until recently, there was no equally clear conception of an epistemic goal in terms of graded beliefs. This changed when Joyce proposed an epistemic goal for graded beliefs that was explicitly meant to be analogous to the aforementioned epistemic goal for categorical beliefs. The proposal is that our system of degrees of belief ought to be *gradationally accurate*, and in fact as gradationally accurate as any other system of degrees of belief that we might adopt. The notion of gradational accuracy is defined in technical terms, in particular in terms of so-called scoring rules, briefly mentioned in section 3.5 and discussed in detail in the next chapter. But the basic intuition underlying it is clear enough, to wit, that the higher one's degree of belief in a true proposition is, the more accurate one is, *ceteris paribus*, and also the lower one's degree of belief in a false proposition is, the more accurate one is, *ceteris paribus*.

In the same vein, Hannes Leitgeb and Richard Pettigrew (2010a, 2010b) have tried to give a nonpragmatic justification not only of the synchronic part of Bayesianism but also of its diachronic part, that is, of Bayes's rule.[13] They share Joyce's view of our epistemic goal, which they put as follows (2010a, p. 202):

Accuracy: An epistemic agent ought to approximate the truth. In other words, she ought to minimize her inaccuracy.

Also like Joyce, Leitgeb and Pettigrew make the notion of accuracy precise by reference to scoring rules. They then argue that Bayesian updating minimizes expected inaccuracy. More exactly, they argue that if, and only if, an agent updates via Bayes's rule, she minimizes the expected inaccuracy of her post-update probability function, where the expectation is minimal according to her pre-update probability function (Leitgeb & Pettigrew, 2010b, pp. 249–250).[14]

What Leitgeb and Pettigrew aim to show for Bayes's rule is in an important respect *dis*analogous with what Joyce aims to show for the synchronic part of Bayesianism. While Joyce argues that an agent whose degrees of belief are not probabilities fails to minimize *actual* inaccuracy, Leitgeb and Pettigrew argue that an agent who updates by a rule other than Bayes's fails to minimize *expected* inaccuracy. Nothing that they say precludes the possibility that some non-Bayesian rule outperforms Bayes's rule with respect to actual inaccuracy minimization. More importantly, Leitgeb and Pettigrew are concerned only with the inaccuracy of the immediate post-update belief state. But an equally—if not more—legitimate question one may ask about update rules is which of them will lead to the most accurate belief state *in the long run*. Indeed, given that the epistemic goal for graded beliefs is supposed to be analogous to the epistemic goal as discussed by mainstream epistemologists, the long-run question is the more natural one to ask. After all, in mainstream epistemology the epistemic goal is mostly conceived as an *ultimate* goal to which all our epistemic endeavors are geared and in light of which they are to be assessed.[15]

13. For similar attempts at vindicating Bayes's rule, see Rosenkrantz (1992) and Greaves and Wallace (2006).

14. To be still more specific about the expectation: it is the weighted average, taken over all worlds w that the agent deems epistemically possible before the update, of the inaccuracies of the post-update probability function if w were the actual world, with the pre-update probabilities for the worlds w serving as weights.

15. See, for instance, Latus (2000, p. 29), who writes that in asking what our epistemic goal is "we are interested in what we want the overall result of our various ways of 'finding

To see just how unconnected the two are, let us for a moment set aside the idealizations commonly made in discussion about updating and note that, in science, we rarely just *happen* across useful data. Typically, we must actively search for data, and in the many areas of science that rely on experimentation even *produce* our data. Because our time is limited, as is our funding, we constantly have to make decisions as to which instruments (telescopes, microscopes, etc.) to construct, which expeditions to undertake, which experiments to run, and so on. Such decisions will be informed by which hypotheses we deem most promising. Had we deemed hypothesis H promising, and had we wanted to compare that with the hypothesis currently dominant in our field, we might have run a different set of experiments from those that we actually did, given that in fact we deemed H' more promising than H and were mainly interested in comparing H' with the received doctrine. Which hypothesis or hypotheses that we deem most promising and most worthy of spending our limited resources on, will at least in part depend on how probable they appear to us, compared with their most direct rivals. If, for example, a Bayesian update makes H more probable than H', whereas the opposite will be the case if we update via some non-Bayesian update rule, then our decision to use one of these rules may put us on a very different research path with very different downstream consequences from the path encountered if we had decided to use the other rule. Which of these paths will eventually lead us to the more accurate representation of the world will have nothing to do with which of the rules minimizes expected inaccuracy of the piece of evidence now lying before us.

The general problem for the inaccuracy-minimization approach to which this points is that Accuracy permits a number of different interpretations. For instance, it can be interpreted as demanding that every single update minimize expected inaccuracy, as Leitgeb and Pettigrew do, or that every update minimize actual inaccuracy, or that every update be aimed at realizing the long-term project of coming to a minimally inaccurate representation of the world, even if individual updates do not always minimize inaccuracy

out' . . . to be." Goldman (2010, section 2) also argues that (what he calls) epistemic systems are to be evaluated in terms of their accuracy, which he phrases in terms of degrees of truth-possession, which in turn is defined by reference to a scoring rule. However, he is very clear that what he has in mind is long-term accuracy: "Epistemic system E is better than epistemic system E^* iff conformity to E would produce (in the long run) a higher total amount of degrees of truth-possession than conformity to E^* would produce" (p. 194).

or expected inaccuracy. This is problematic because none of these can offhand be rejected as not a legitimate epistemic goal, and that it is by no means obvious that if rule R is most conducive to the realization of one goal and rule R' is most conducive to the realization of a second, then $R = R'$, or that at least the rules are equivalent in that they always yield the same output, given the same input.

In fact, there is not just the question of whether the goal of inaccuracy minimization is meant to pertain (only) to the "next update" or is rather meant to be a long-term project. Even if we assume that it is conceived as a long-term project, there are still further distinctions that can be made. One notable distinction concerns whether we should aim to have a minimally inaccurate probability function in the long run, *however* far in the future that may be, or whether it is better to have a moderately accurate probability function in the shorter run. It is certainly imaginable that, for purely epistemic reasons, one might opt for the latter, supposing that there is a choice to be made. We are, after all, *curious* about the truth (Hempel, 1965, p. 333; Foley, 1987, p. 11), and curiosity typically comes with a sense of urgency. As a result, we might prefer an update rule that is more likely to take us fairly close to the truth in a reasonably short time span over one that is more likely to take us *extremely* close to the truth in the long run but less likely to take us even fairly close to the truth in, for example, the medium-long run.[16,17]

16. Douven (2010, section 4) makes a parallel point in terms of the qualitative notion of truth approximation for update procedures that take into account peer opinions.

17. Briggs and Pettigrew (2020) present an inaccuracy-minimization argument assuming that we aim to be maximally accurate over a lifetime. They do not argue for this assumption nor say how it relates to the discussion of our epistemic goal in traditional epistemology (which Joyce took as a starting point). On its face, the assumption is implausible, if only because, realistically speaking, it would make sense to sacrifice some accuracy along the way for the benefit of having a *highly* accurate belief state in the somewhat longer run. See, for instance, Galison (1997) on how scientists with partly different backgrounds (e.g., theoretical physicists and engineers) must sometimes be willing to settle on a kind of common language (a *pidgin*, as Galison's refers to it) in order to be able to communicate at all, even though the language allows none of the collaborators to be, by their own lights, entirely accurate. All the while, the ultimate goal of the collaborative effort is that the accuracy of our representation of this or that part of the world be enhanced. Those of us who are educators will be familiar with a similar phenomenon. In explaining to freshmen concepts or theories or standpoints that are new to them, there is sometimes no other route to a first understanding than to cut some corners, skip some details, and omit some qualifications, all for the good of *ultimately* achieving an exact understanding of those concepts, theories, or standpoints some time later in the curriculum. More generally, note that at a minimum it is not a priori that aiming to be as accurate as

We could go on for some time disambiguating Accuracy in this way, if only because speed and accuracy of convergence to the truth can be traded off in an indefinite number of ways. However, our aim here is not to catalog all reasonable precisifications of Accuracy but rather to show that the inaccuracy-minimization argument poses no real threat to abduction. Leitgeb and Pettigrew argue that Bayesian updating minimizes the expected inaccuracy of our next belief state relative to our current belief state and thus may be said to be most conducive to *one* particular epistemic goal. Even granting this, it does little to undermine abduction as long as it is understood that abduction (or a particular explication of it) is most conducive to some other epistemic goal or goals, where possibly the different epistemic goals can both (or all) be regarded as explications of the same broad idea that we should strive for a maximally accurate belief system.

At this juncture, Bayesians might try to argue that expected inaccuracy minimization of our next belief state as judged from our present one trumps any other epistemic goal that we may have. The prospects of this tactic are about as bleak as the prospects of showing that Dutch book invulnerability trumps any other practical goal that we may have; bleaker still, for it would seem absurd to claim that it is epistemically more important to have an update rule that minimizes *expected* inaccuracy than to have one that minimizes *actual* inaccuracy. This seems especially true in view of Robbie Williams's (2012, p. 835) observation that expected inaccuracy minimization defenses of Bayes's rule raise the question of "why [you should] trust an *outdated* belief state to tell you how to fix your beliefs now you have new information."

---

possible overall offers the best guarantee of ending in a highly accurate belief state. And Briggs and Pettigrew make no attempt to show that there is such a guarantee. There is an even more fundamental problem with their argument. The argument is supposed to show that a non-Bayesian updater could come to see that there is an alternative prior from which she might have started from that in combination with the strict use of Bayes's rule would have made her more accurate over the course of her epistemic life than she now is, because she did *not* start from that prior and is committed to a non-Bayesian update rule. Aside from concerns about the idealizing assumptions that go into their argument, it is to be noted that for all that Briggs and Pettigrew show, adopting the alternative prior might from the person's current perspective appear to amount to "believing crazy stuff." More generally, why should it bother the non-Bayesian that she could have been more accurate than she actually is if only she had started from degrees of belief which she simply never had and which perhaps would never have seemed reasonable to her? The point of Briggs and Pettigrew's argument is that a Bayesian would never find herself in the kind of situation just sketched, but to the extent that it is unclear what is so dreadful about that kind of situation, the argument fails.

Alternatively, Bayesians might try to show that Bayes's rule does best with respect to *all* epistemic goals. This may be hard to establish in an a priori manner, supposing that among the relevant goals are ones whose realization requires actual inaccuracy minimization. How well a rule does in terms of truth approximation may plausibly depend not only on the rule but also on which world is the actual one. For example, even if the advocates of abduction have so far failed to produce a reason for believing that best explanations tend to be true, we might happen to inhabit a world in which—perhaps as a brute fact—best explanations do tend to be true. Worse yet (for the Bayesians), simulations that I performed, which compare Bayes's rule and EXPL in terms of gradational inaccuracy in the setting of our earlier statistical model, give reason to doubt that Bayes's rule is the most conducive to the realization of every epistemic goal.

In these simulations, gradational inaccuracy is measured by means of the Brier scoring rule. Using the same statistical model that was used in the previous simulations, I again ran 1,000 simulations of a sequence of 1,000 coin tosses for each possible bias. Also as previously, the Bayesian and the EXPL user updated their degrees of belief after each toss. I registered in how many of those 1,000 simulations the EXPL user incurred a lower Brier penalty than the Bayesian at various reference points. The outcomes of these simulations are displayed in table 4.2. They show that, at each reference point, EXPL is usually the winner (it incurs the lowest penalty). Hence, at least in this kind of context, EXPL appears a better choice than Bayes's rule.

Or maybe not—for I also calculated averages, taken over the 1,000 simulations for each bias value, of the penalties incurred by the EXPL user and the Bayesian at the designated reference points. Table 4.2 gives for each of these points the mean of the EXPL penalties minus the mean of the Bayes's rule penalties (in parentheses). As shown by those numbers, if there is a difference between the two means, it is always in favor of Bayes's rule. The reason for this is manifest from the spread of the simulation outcomes (not represented here), which shows that, although EXPL wins in most instances, it is typically by a relatively small margin, whereas in some of the runs in which EXPL loses, it incurs considerably greater penalties than does Bayes's rule. This in turn is due to the previously noted fact that the EXPL user reacts in a bolder fashion to the evidence than does the Bayesian and is therefore more easily led astray by rows of tosses that produce subsequences with deviating relative frequencies. Indeed, inspection of individual runs shows that such subsequences,

Table 4.2: Results of 1,000 simulations of sequences of 1,000 coin tosses for each bias value: the columns give the number of simulations in which EXPL incurred a *lower* Brier score than Bayes's rule after 100, 250, . . . , tosses; in parentheses, the mean of the EXPL penalties minus the mean of the Bayes's rule penalties, taken over the 1000 simulations (rounded to two decimal places).

Bias	100	250	500	750	1,000
.0	1,000 (.00)	1,000 (.00)	1,000 (.00)	0 (.00)	0 (.00)
.1	923 (.02)	993 (.00)	1,000 (.00)	1,000 (.00)	1,000 (.00)
.2	748 (.10)	937 (.02)	991 (.00)	1,000 (.00)	1,000 (.00)
.3	717 (.11)	918 (.03)	984 (.00)	998 (.00)	999 (.00)
.4	690 (.11)	865 (.06)	968 (.01)	990 (.00)	997 (.00)
.5	679 (.12)	879 (.05)	972 (.00)	992 (.00)	998 (.00)
.6	660 (.14)	904 (.03)	977 (.00)	992 (.00)	995 (.00)
.7	698 (.12)	907 (.03)	978 (.01)	995 (.00)	1,000 (.00)
.8	754 (.09)	947 (.02)	996 (.00)	999 (.00)	999 (.00)
.9	910 (.04)	990 (.00)	1,000 (.00)	1,000 (.00)	1,000 (.00)
1.0	1,000 (.00)	1,000 (.00)	1,000 (.00)	0 (.00)	0 (.00)

while they push both agents off the right track, impact the EXPL user much more severely than the Bayesian, and that in turn has a noticeable effect on the differences between the Brier penalties that they incur.

From these simulations, each of the two update rules under consideration comes out doing better in different respects; this further buttresses the point of this section. There seems to be no clear answer to the question of whether it is better, epistemically speaking, to use an update rule that in general achieves greater accuracy than other update rules, even if typically not *much* greater accuracy; or to use an update rule that is less likely than another update rule to ever make one *vastly* inaccurate, even though it *typically* makes one *somewhat* more inaccurate. Naturally, we would prefer a rule that offered the best of both worlds, or better yet, of *all* worlds—in other words, a rule that was most conducive to any epistemic goal that we might have. But there may be no rule that fits this bill, and in any event, the previously described simulations give reason to believe that Bayes's rule is not that rule. To be sure, these simulations do not warrant a more positive verdict about EXPL either. But the aim of this section is not to make a case for EXPL or even for the broader idea of abduction. Rather, it is to highlight and question a hidden assumption in the inaccuracy-minimization defense of Bayes's rule, to wit, that there is but one way in which an update rule can be said to minimize inaccuracy, or at least only one that matters epistemically.

Bayesians might seem to have an easy response to the foregoing. Previously in this chapter, it was argued that what makes the dynamic Dutch book argument objectionable may be not so much the fact that the argument addresses the wrong kind of rationality—practical rationality—but rather that it focuses on just one possible practical concern to the neglect of all others, apparently without good reason. If pragmatic considerations can be legitimately invoked in the present discussion, then one might try to argue that even if the goal or goals that Bayes's rule serves best are not privileged from an epistemic viewpoint, they *are* privileged from a practical one. Specifically, one might try to argue that if the results of the above simulations have some general validity and do not hold only for the particular type of statistical model considered here, then the fact that updating via Bayes's rule leads on average to a lower penalty is enough to justify this rule. And indeed, it does follow from the said fact that were one to keep betting on consecutive tosses with a given coin, each time posting betting odds in accordance with one's degrees of belief at that time, then one would maximize one's expected payoff by updating via Bayes's rule.

But it would be false to think that payoffs and scoring rule penalties must always be so strictly related as in this example. To see how they can come apart, suppose that you are the owner of one of two engineering firms in a city. The two firms are in ongoing competition for contracts from the local authorities. As a rule, a contract goes to the firm that proposes the better solution to whatever the engineering problem is that the authorities want to be solved. Given that the engineers employed by your firm are about as competent as the engineers employed by the other firm, which firm comes up with the better plan typically depends on the accuracy of the information on which they base their proposals. Note that under these circumstances, being able to base one's proposal on information that is just slightly more accurate than the information available to the competition is enough to get the contract. By contrast, if the available information is less accurate, it is immaterial whether it is only slightly less accurate or much less accurate: one will not get the contract either way; in this case, a miss is as good as a mile.[18] Again supposing the results of our simulations to have some general validity, they suggest that if you and your team update via EXPL and the competition updates via Bayes's rule, then you are likely to get the vast majority of the contracts. Once in a while—very rarely—a contract will go to the other firm. When it does, the information on which your proposal was based was probably much more inaccurate than the information on which the rival plan was based. However, this has no financial consequences beyond the fact that you miss out on the contract, which would also have happened if the information available to you had only been slightly less accurate than the information available to the other firm.

At this point, it has come to appear that there are circumstances under which it is better to update via EXPL, practically speaking, given that it greatly increases the chance of ending up with a (perhaps only slightly) more accurate belief system than one would have ended up with had one updated via Bayes's rule. Naturally, this is only one sense in which one can aim to minimize one's inaccuracy, and the foregoing considerations in fact suggest that from a pragmatic viewpoint, there may be no single best update rule. Depending on the circumstances and on what exactly one's interests are, Bayes's rule may serve one's interests best, or EXPL may do so, or perhaps yet another rule may do so. In epistemology, contextual approaches have been much in the limelight lately. I am not aware of anyone suggesting that which update rule use by may depend on context, and yet that is one of the main claims of this book.

Consider the previously described situation in which you own one of two engineering firms. There, EXPL clearly seemed the better option when the choice was between that rule and Bayes's rule. Or consider an example from Douven (2017b), in which you are wondering about the source of the persistent rumor that Mozart did not die from natural causes but was poisoned, a rumor that is almost certainly false (Zegers, Weigl, & Steptoe, 2009). Actually it was Mozart himself who started the rumor that he had been poisoned, shortly before his death (Robbins Landon, 1999, p. 148 ff). If you did not know this already and if I had not told you, you might out of sheer curiosity have wanted to find out about the source of the rumor. And if you were really curious, you might have wanted to find out quickly. But if in the course of

18. We shall see more of this in Chapter 6.

your investigations of the matter, before you eventually found out the truth, you had come to mistakenly hold that it was Mozart's wife who started the rumor or that is was one of the newspapers that first reported his death, that might have done little to no harm: accuracy does not matter much in this kind of case, in which being mistaken is unlikely to have any dire consequences, if at all. Here, too, EXPL might be the rule of choice, given how much faster than Bayes's rule it is, even if it is on average not quite as accurate as that rule.

By contrast, to use another example from Douven (2017b), suppose that you are the producer of roulette tables and want your tables to be extensively tested before they leave the factory. Suppose further that the testing proceeds by rolling a ball 1,000 times and registering whether it lands on red or on black. You do not want to sell a table with a bias, as that might damage the reputation of your brand. Here, you might want to use Bayes's rule for updating your beliefs about the fairness of a table on the basis of the outcomes of the rolls of the ball. Accuracy matters greatly here; speed is not much a concern.

This suggestion of context dependence is explored further in subsequent chapters. In fact, we go beyond the simulations presented in this chapter along two different lines. In chapter 6, we consider more realistic situations and compare in them a greater number of update rules than we have done here. In that chapter we also embed the question of which update rule to use in a broader discussion about the notion of rationality, paying special attention to recent developments in cognitive psychology and cognitive science that indeed make rationality an inherently context-bound notion.

In chapter 7, we address another limitation of the simulations described in this chapter, to wit, that they ignore that we typically learn as members of a collective. As social epistemologists have been emphasizing for some time, we will obtain an incomplete and possibly distorted picture of epistemology if we fail to recognize the extent to which our beliefs are shaped by, and even rooted in, our interactions with others. Taking the lessons from social epistemology seriously, chapter 7 looks at communities of agents who update their opinions by taking into account both the evidence that they receive directly from the world and the opinions of their epistemic peers.

4.4 Summary

We have considered the currently prevailing arguments for the claim that when it comes to rational update rules, Bayes's rule is the only game in town. We have argued that neither the dynamic Dutch book approach nor the inaccuracy-minimization approach succeeds in challenging the rationality of updating via probabilistic versions of abduction. We proceeded by laying bare some unsupported hidden assumptions of these arguments. At the most general level, the problematic (tacit) assumption is that Dutch book invulnerability and / or expected next-step inaccuracy minimization are *all* that we should care about when it comes to judging update rules. As briefly mentioned in chapter 1, if you are an epistemic consequentialist and hold that update rules should be judged by their fruits, then do not cherry-pick: judge the rules by *all* their fruits, the sweet and the sour.

With respect to the dynamic Dutch book argument: if updating via Bayes's rule has certain practical advantages in comparison with other rules, a version of EXPL may have different, possibly more important, practical advantages than does Bayes's rule. And that is to grant that non-Bayesian updating does imply Dutch book vulnerability—something that may have seemed plausible only because of another, more specific hidden assumption taken aboard by Bayesians, to wit, that update rules are to be judged on their own and not in combination with other epistemic or decision-theoretic principles with which they must interact if they are to have any consequences at all.

Equally, if Bayes's rule minimizes inaccuracy in some sense, a version of EXPL may minimize inaccuracy in another sense, in which for all anyone has shown, no single one of these senses can be said to be privileged in that it captures *the* epistemic goal. That expected next-step inaccuracy minimization trumps any other kind of inaccuracy minimization, or indeed any other possible benefit of non-Bayesian updating, is again an assumption for which no Bayesian has cared to argue.

Note that empirically the Bayesian claim of inaccuracy minimization was implausible already in light of the re-analysis of the data from Douven and Schupbach (2015a), discussed in section 3.5, which showed that Douven and Schupbach's participants tended to have more accurate degrees of belief the more weight they had given, in their updating, to explanatory considerations. We can further mention Nathan Berg, Guido Biele, and Gerd Gigerenzer's (2016), for which 125 academics were asked for their degrees of belief concerning the usefulness of prostate cancer screening. These authors found literally zero correlation between the degree to which their participants conformed to Bayesian principles and the actual accuracy of their degrees of belief. Even more ironic, they found that the participant who conformed most closely to those principles was also the one whose degrees of belief were most inaccurate.

It is one thing to argue that the criticisms leveled against non-Bayesian updating in general, and probabilistic versions of abduction in particular, are baseless, but it is another to defend specific forms of non-Bayesian updating, such as, most relevantly, updating via EXPL and kindred rules. Much of the remainder of this book is devoted to defending forms of probabilistic abduction. First, however, I would like to point out yet another hidden assumption of the inaccuracy-minimization argument. We saw that this argument was explicitly meant to be *epistemic* in nature and thereby to improve on the older defense of Bayes's rule in terms of dynamic Dutch book vulnerability, which many had come to regard as posing only a *pragmatic* problem. We also saw that the inaccuracy-minimization argument relies crucially on a scoring rule as a formal measure of inaccuracy. For the argument to be purely epistemic in nature—as it is supposed to be—scoring must be a purely epistemic matter. The next chapter challenges this assumption, arguing that there are inextricably pragmatic aspects to scoring: how best to measure inaccuracy may vary from one context to another, and the decision to use this or that scoring rule may partly depend on what our practical concerns are. This also leads me to challenge the claim, taken for granted so far, that Bayes's rule minimizes at least expected next-step inaccuracy.

5
A Closer Look at Scoring

We have encountered the concept of scoring in relation to measuring inaccuracy, a couple of times, but have skipped formal details so far. After making the concept precise, this chapter starts with an observation that may at first seem only tangentially related to our current discussion but that actually raises serious doubts about two claims central to the inaccuracy-minimization argument, to wit, that this argument is of a strictly epistemic nature, and that Bayes's rule minimizes expected next-step inaccuracy.

5.1 Standard Scoring Rules and Truthlikeness

Scoring can easily appear a technical and purely formal concept, but the underlying idea has been familiar since elementary school. Indeed, those of us who are teachers engage in scoring quite frequently when we grade our students' exams or essays. In general, grading is not particularly hard. Suppose that a teacher gives an exam containing ten questions, each of which has a straightforward yes / no answer and each of which contributes equally to the final score. Then if a student gets nine questions right, the reasonable grade for this student appears to be 90 percent or whatever that translates to in the teacher's grading system (e.g., an A in the US public school system, or a 3.6 in a 4.0-based system).

Naturally, when questions do not have straightforward yes / no answers, grading can be more complicated. A student may get none of the questions completely right and yet show a good understanding of the relevant subject matter. For example, a student may turn in an exam that gives clear evidence of her mathematical acumen, on one hand, and of a certain inattentiveness in carrying out simple algebraic operations, on the other. Most mathematics teachers would not want to fail such a student but would rather encourage her to address her sloppiness. What mark to give in such a case may not be entirely straightforward and may be to some degree subjective.

Grading can be more complicated still. Think of grading essays, for which there are usually no more-or-less right answers. Grading also becomes more complicated when we are dealing with questions that ask for a probability estimate of some future event. Such questions may not occur so frequently on a school exam (though see Bickel, 2010), but they are the bread and butter for a variety of professionals, ranging from financial analyst to physician, from football coach to engineer, and from foreign policy advisor to weather forecaster. Suppose that one weather forecaster predicts rain for tomorrow with a probability of .25, whereas a second predicts rain for tomorrow with a probability of .75. If tomorrow stays dry, neither forecaster can be said to have been wrong, for neither predicted rain with certainty. Yet from a pretheoretic perspective, it appears that the former forecaster did a better job than the latter; that forecaster was, we would like to say, the more accurate of the two.

How to turn this intuition into a difference in grades (or scores) for the two forecasters—supposing that our only concern is their predictions of rain tomorrow—might also be deemed a partly subjective matter. But researchers from various quarters think otherwise and have sought to answer the foregoing question by providing fully objective formal rules. More specifically, proposals for how to assess the forecasters, and for how to assess the quality of probability estimates generally, have come in the form of scoring rules. So far, I have mentioned only the Brier score, which was the historically first, dating to the 1950s, but by now a bewildering variety of scoring rules exists (Brier, 1950; Rosenkrantz, 1981, ch. 2; Cooke, 1991, chs. 8 and 9). Given that these rules may lead to different qualitative verdicts (e.g., on which weather forecaster did best), and given that, as mentioned, this kind of scoring is supposed to be objective, the question has arisen of which of the many available scoring rules today we should rely on in practice.

Theorists have almost invariably tried to answer this question by arguing that this or that scoring rule uniquely satisfies certain standards of "goodness" (Winkler & Murphy, 1968). There is debate about this issue because there is no generally shared conception of goodness in the relevant sense and hence no agreement on which criteria to impose on scoring rules. Nonetheless, it is fair to say that most theorists are, or have been, in favor of one of two scoring rules, either the Brier rule[1] (e.g., Brier, 1950; Rosenkrantz, 1981; Joyce, 1998; Selten, 1998; Greaves & Wallace, 2006; Leitgeb & Pettigrew, 2010a) or the logarithmic (or "log") rule (e.g., Good, 1952; Bernardo, 1979; Bernardo & Smith, 2000; Bickel, 2007, 2010; Levinstein, 2012; McCutcheon, 2019).

To state these rules formally, let $\{H_i\}_{i=1}^{n}$ be a *hypothesis partition*, that is, a set of mutually exclusive and collectively exhaustive hypotheses, and let δ_{ij} (for $i,j \in \{1,\ldots,n\}$) be the Kronecker delta, which equals 1 if $i = j$ and equals 0 otherwise. Suppose that a person assigns probabilities $\mathbf{p} = (p_1,\ldots,p_n)$ to the elements of $\{H_i\}_{i=1}^{n}$, with p_i her probability that H_i is true. Then, assuming that the true hypothesis is H_j, the Brier score for this person equals $\mathcal{B}_j(\mathbf{p}) := 1/n \sum_{i=1}^{n}(\delta_{ij} - p_i)^2$, whereas her log score equals $\mathcal{L}_j(\mathbf{p}) := -\ln(p_j)$.[2] As previously mentioned, we tend to conceive of scoring rules as assigning *penalties*, as measuring how *bad* a probability distribution is. But this is just a convention; if we wish, we can take the negative of any given rule's scores and think of them as *rewards*, as measuring how *good* a probability distribution is. Note that on the conventional conception of scores, lower scores are better; on the alternative, lower scores would be worse.

Given a scoring rule $\mathcal{S}$ and a probability distribution $\mathbf{p}$ on a hypothesis partition, the **p**-*expectation* of the $\mathcal{S}$-score (or **p**-expected $\mathcal{S}$-score, for short) for a second probability distribution $\mathbf{p}^*$ on the same partition equals $\mathbb{E}_{\mathbf{p}}[\mathcal{S}(\mathbf{p}^*)] := \sum_{i=1}^{n} p_i \mathcal{S}_i(\mathbf{p}^*)$. A scoring rule $\mathcal{S}$ is said to be *proper* if and only if for all $\mathbf{p}$, $\arg\min_{\mathbf{p}^*} \mathbb{E}_{\mathbf{p}}[\mathcal{S}(\mathbf{p}^*)] = \mathbf{p}$, and *strictly proper* if and only if each of these minima is unique. Scoring rules were originally proposed for eliciting probabilities (Cooke, 1991, p. 121), and when used for that purpose, they should not give a person an incentive to announce probabilities that she does not actually hold. This is why many theorists regard strict propriety as an important requirement for scoring rules.

Both the Brier and the log score are known to be strictly proper. They also both achieve their minimum of 0 when probability 1 is assigned to the truth, which is generally regarded as another important desideratum. Nevertheless,

1. Or the quadratic scoring rule, which is a generalization of the Brier score; see the subsequent discussion.

2. With regard to the Brier score, note that as long as one makes comparisons given a fixed number of hypotheses, division by that number is not strictly necessary and should be considered optional. It thus appears unproblematic to equate, for most practical purposes, the Brier score with what Rosenkrantz (1992, ch. 2) calls the "least squares" score (which is, for instance, also what Leitgeb & Pettigrew, 2010a, and Pettigrew, 2016, do).

Table 5.1: Probability assignments to hypotheses A, B, and C.

	David	Emma	Frank
Alice wins (A)	.1	.5	.3
Bob wins (B)	.5	.1	.3
Charlotte wins (C)	.4	.4	.4

the following example brings out that the rules can lead to very different verdicts.

Suppose that there will soon be an election for a new president of your university. There are three candidates in the running: Alice, Bob, and Charlotte. Your colleagues David, Emma, and Frank hold different views on which of the candidates is most likely to become the new president. Specifically, their relevant probabilities are as given in table 5.1. Suppose that Charlotte wins the election. Then David and Emma have the same Brier score: $(.1^2+.5^2+.6^2)/3 \approx .21$; Frank's Brier score is lower and thus better: $(.3^2 + .3^2 + .6^2)/3 = .18$. Nonetheless, all of them have the same log score, to wit, $-\ln(.4) \approx 0.92$.

That David and Emma do equally well on either scoring rule seems as it should be: David and Emma both assign a probability of .1 to one of the false hypotheses and a probability of .5 to the other, and it is difficult to see (at least given the information provided here) how it might matter that they do not assign these probabilities to the same hypotheses.

That Frank's Brier score is lower than David's and Emma's illustrates the general fact that, for any hypothesis partition, given a particular probability assigned to the true hypothesis, one minimizes one's Brier score by assigning equal probabilities to the remaining hypotheses. How reasonable is this?

According to advocates of the log rule, this is not reasonable at all. Why should we care—they ask—what probabilities a person assigns to any hypothesis other than the truth? In their view, given that David's, Emma's, and Frank's probabilities for C are the same, they should be assigned the same score (Winkler, 1969; Bernardo & Smith, 2000, p. 72; Bickel, 2010, pp. 347–348). The log rule is in fact known to be the only strictly proper scoring rule that guarantees this outcome; it is, in more official terminology, the only strictly proper scoring rule that is *local* (McCarthy, 1956).

In the debate about scoring, various other intuitions in favor of or against either of these rules have been called upon. For instance, Eric Bickel (2010,

p. 348) notes that the log rule, but not the Brier rule, ensures that *higher* probability assignments to the truth will result in *lower* penalties.[3] He regards this as a compelling reason to prefer the log rule over the Brier rule. By contrast, Reinhard Selten (1998, pp. 49–50) prefers the Brier rule because he thinks that in some situations the log rule is more sensitive to small differences between probability assignments than is warranted by intuition and in other situations, not sensitive enough.[4]

That different authors have given different weights to intuitions regarding scoring has partly to do with the fact that they have focused on different examples, and the problem is that the features of the examples that have been used to make one or the other rule look appealing are not always generalizable.

To see how different examples may steer our intuitions in different directions, note that in our own example there is no sense in which either of the false hypotheses (whether A or B) is closer to the truth than the other, again given the information provided. That does not make this hypothesis partition special. The designated feature is not altogether general, however, and by assuming otherwise one might be implicitly favoring some scoring rules over others. More generally put, our hypotheses pertain to a so-called discrete random variable whose possible outcomes lack an intrinsic ordering, making it a *nominal* variable. But discrete random variables also come in another variety: next to nominal variables there are also *ordinal* variables, whose possible outcomes can be naturally ordered. Furthermore, sometimes our hypotheses concern *continuous* variables, which as the name suggests, can be measured on a continuum (for example, temperature or weight). These distinctions are directly relevant to scoring.

3. That the Brier rule cannot guarantee this is a direct consequence of the general fact mentioned two paragraphs back.

4. Selten instead prefers the Brier rule, mainly because, as he proves, it is the only scoring rule (up to positive linear transformations) that satisfies each of what he considers to be four important desiderata for such rules, which Selten presents as axioms. According to the first axiom, the ordering of the hypotheses should not influence the score. According to the second, the score should not be affected by the introduction of an additional hypothesis that receives zero probability. The third axiom is the requirement of strict propriety. Finally, the fourth axiom concerns a type of situation that we do not consider, namely, when a probability assignment is scored in light of another probability assignment rather than in light of the truth of one hypothesis; the axiom requires that, in this situation, the score should be the same regardless of which probability assignment is considered to be the "true" one.

Suppose that David, Emma, and Frank were wondering how well a student of theirs is going to do on an exam, and the probabilities given in table 5.1 are the probabilities that the student will receive an A, a B, or a C, possibilities that we may take to be described by hypotheses A, B, and C, respectively. These hypotheses now refer to possible outcomes of an ordinal variable: it is better to get an A than a B, which in turn is better than a C. From this, there plausibly emanate relations of truthlikeness among our hypotheses. If B turns out true, then the other hypotheses would seem equally far from the truth. But if either A or C turns out true, then, although false, B would be closer to the truth than whichever is the other false hypothesis.

To illustrate the connection with scoring, suppose that C is true. Would we still want to agree with the Brier score that David and Emma do equally well, given their probabilities for the hypotheses at issue, or even agree with the log score that all three colleagues do equally well? Although the three colleagues assign the same probability to the truth, we are tempted to say that David still is closer to the mark, given that he assigns a higher probability than the others to the false hypothesis that is closest to the truth (viz., B) and a lower probability than the others to the false hypothesis that is most distant from the truth (A).

An entire program in the philosophy of science is devoted to making the notion of truthlikeness (or "verisimilitude") formally precise, and numerous measures of truthlikeness exist.[5] Here, we will not commit ourselves to any particular such measure and just note that all measures of truthlikeness currently advocated in the literature will do for the purposes of this chapter.

The idea that truthlikeness may matter to scoring has been mentioned in the literature, but usually only to be set aside as a problem that can easily be dealt with by using the so-called quadratic scoring rule (also known as "weighted least squares metric"). If H_j is the true element of hypothesis partition $H = \{H_i\}_{i=1}^{n}$, this rule assigns a penalty of $\mathcal{Q}_j(\mathbf{p}) := \sum_{i=1}^{n} w_i(\delta_{ij} - p_i)^2$ to someone whose probabilities for the elements of H are given by $\mathbf{p}$. The only general constraints on the weights w_i are that $w_i > 0$ for all i and that $\sum_{i=1}^{n} w_i = 1$, so that we actually have a *schema* here, with the Brier score as the special case in which all hypotheses are weighted equally.

5. See, for instance, Niiniluoto (1987, 1998, 1999b, 2020), Schurz (1987, 1991, 2011, 2014), Kuipers (2000, 2001, 2014, 2019, 2020), Zwart (2001), and Cevolani, Festa, and Kuipers (2013).

For instance, Roger Rosenkrantz (1981, ch. 2) calls the previously mentioned property of the Brier score to penalize more heavily when the probabilities assigned to the false hypotheses are unequal (ceteris paribus) "attractive when all false alternatives are regarded as 'equidistant' from the truth." He further states, however, that "[w]here false answers are not equally far from the truth and we wish to weight them differently, we can use the weighted least squares metric." In the same vein, Hilary Greaves and David Wallace (2006, p. 628) claim that the quadratic scoring rule "can take account of the value of verisimilitude . . . by a judicious choice of the [weights]"; specifically, their proposal is to assign weight to a proposition depending on the extent to which it represents "a set of 'close' states" (Greaves & Wallace, 2006, p. 628).

As it stands, however, this proposal will not work. Which set or sets of states are close depends on which state is *actual*; equivalently, how far from the truth a false alternative is depends on which hypothesis is *true*. And the weights of the quadratic scoring rule lack this dependence. To be sure, if we know which hypothesis is true, the dependence can be built in by hand. For instance, given that (we said) the student will receive a C, so that *A* is more distant from the truth than *B*, we can weight *A* more heavily than *B*. That might give the desired result, and probably this is the kind of use of the quadratic scoring rule that the aforementioned authors had in mind.

But which weights are we to assign when we want to use the rule *not* knowing the truth, as when we would like to calculate our *expected* score? To calculate expected scores, we consider all possibilities of truth, calculate for each individual possibility the penalty that we would incur were that possibility to be actual, and then take a weighted average of those penalties, the weights being our probabilities for the possibilities. In using the quadratic scoring rule, however, there is a second set of weights involved. And the problem is that while truthlikeness relations shift from one possibility to the next—for instance, in the possibility in which the student receives a C, hypothesis *C* is closer to the truth than hypothesis *A*, but in the possibility in which the student receives a B, hypotheses *A* and *C* are equidistant from the truth—the weights attributed by the quadratic scoring rule stay the same whichever possibility is considered. Consequently, a set of weights that adequately reflects truthlikeness relations under one supposition of where the truth lies may well fail to do so under another such supposition.

To avoid this problem, we may adapt the quadratic scoring rule by *doubly*, instead of *singly*, indexing each weight, where the additional index then refers to the true hypothesis, thereby obtaining what we shall call a *verisimilitude-sensitive scoring rule*, or VS rule for short. Where H_j is the true element of hypothesis partition H and w_{ij} is the distance from H_i to the truth, this rule imposes a penalty of $\mathcal{V}_j(\mathbf{p}) := \sum_{i=1}^{n} w_{ij}(\delta_{ij} - p_i)^2$ on someone assigning probabilities $\mathbf{p}$ to the elements of H. (Again, this is really a schema, yielding different rules for different weighting functions.)

To make the difference between the rules vivid, suppose that David, not knowing which grade his student will get, wishes to calculate his expected score, given his probabilities $\mathbf{d} = (.1, .5, .4)$ for hypotheses A, B, and C. First assume that he uses an instance of the quadratic scoring rule. Without loss of generality, let $w_A = .1$, $w_B = .3$, and $w_C = .6$. David will then find that[6]

$$\begin{aligned}\mathbb{E}_{\mathbf{d}}[\mathcal{Q}(\mathbf{d})] &= .1\big((.1)(.9^2) + (.3)(.5^2) + (.6)(.4^2)\big)\\ &+ .5\big((.1)(.1^2) + (.3)(.5^2) + (.6)(.4^2)\big)\\ &+ .4\big((.1)(.1^2) + (.3)(.5^2) + (.6)(.6^2)\big) = 0.228.\end{aligned}$$

Now suppose that David instead assumes a VS rule, for instance, that assigns a weight of .1 to the truth and that weights the other hypotheses proportionally to their distance from the truth. Supposing also that he uses a weighting function that reflects the ordering of the hypotheses—so that if A is true, then B is closer to the truth than C, and so on—but that is otherwise arbitrary, David's expected score is

$$\begin{aligned}\mathbb{E}_{\mathbf{d}}[\mathcal{V}(\mathbf{d})] &= .1\big((.1)(.9^2) + (.3)(.5^2) + (.6)(.4^2)\big)\\ &+ .5\big((.45)(.1^2) + (.1)(.5^2) + (.45)(.4^2)\big)\\ &+ .4\big((.6)(.1^2) + (.3)(.5^2) + (.1)(.6^2)\big) \approx 0.123.\end{aligned}$$

We see how the doubly indexed truthlikeness weights of the VS rule vary per possibility—as they must do, for the reason previously mentioned—while the singly indexed weights of the quadratic scoring rule stick to their propositions across all three possibilities.

6. To be entirely precise, instances of the quadratic scoring rule and the VS rule would have to be embellished with a superscript or subscript to indicate the weighting function that is being assumed. We will not be so fussy, however.

5.2 Truthlikeness and Inaccuracy Minimization

In the previous chapter, we criticized the Bayesian inaccuracy-minimization argument for focusing on a quite limited notion of expected inaccuracy, to wit, expected next-step inaccuracy. Suppose that this *is* the only notion of inaccuracy that we should care about, and also that, in adjudicating among update rules, we should care *only* about inaccuracy in this sense. Then a crucial question to be asked about our new rule is whether it still supports the Bayesian argument. To show that it does not, we suppose that David is about to receive some new piece of information E about the student from the previous example. How E is probabilistically related to hypotheses A, B, and C is specified by the following probability distribution over the set of possible worlds relevant to the present problem:

$$\begin{array}{llllll} \Pr(\{w_{AE}\}) & = .01 & \Pr(\{w_{BE}\}) & = .25 & \Pr(\{w_{CE}\}) & = .1 \\ \Pr(\{w_{A\overline{E}}\}) & = .09 & \Pr(\{w_{B\overline{E}}\}) & = .25 & \Pr(\{w_{C\overline{E}}\}) & = .3 \end{array}$$

Here I am using the notation $\overline{E}$ to designate the negation of E, and w_{AE} is the possible world in which both A and E are true, and so on. Probabilities of hypotheses can be derived from this specification by summing the probabilities of the worlds in which they hold true; for instance, according to this specification, David's probability for A equals $\Pr(\{w_{AE}\}) + \Pr(\{w_{A\overline{E}}\}) = .01 + .09 = .1$, which is as it should be.

Once David has actually received the new information E, he should, according to Bayesians, conditionalize on it, which would result in the following new probability distribution over the same set of possible worlds:

$$\begin{array}{llllll} \Pr_E(\{w_{AE}\}) & \approx .028 & \Pr_E(\{w_{BE}\}) & \approx .694 & \Pr_E(\{w_{CE}\}) & \approx .278 \\ \Pr_E(\{w_{A\overline{E}}\}) & = .0 & \Pr_E(\{w_{B\overline{E}}\}) & = .0 & \Pr_E(\{w_{C\overline{E}}\}) & = .0 \end{array}$$

To calculate David's expected VS score, assuming the weights we earlier assumed and otherwise following Greaves and Wallace (2006, pp. 459–460), or Leitgeb and Pettigrew (2010b, pp. 249–250), we set $x = .028 \approx \Pr_E(\{w_{AE}\})$, $y = .694 \approx \Pr_E(\{w_{BE}\})$, and $z = .278 \approx \Pr_E(\{w_{CE}\})$ in the following:

$$(5.1)\quad .01\big(.1(1-x)^2 + .3y^2 + .6z^2\big) + .25\big(.45x^2 + .1(1-y)^2 + .45z^2\big) + .1\big(.6x^2 + .3y^2 + .1(1-z)^2\big),$$

which yields an expected score of 0.034. The all-important question now is whether this is the minimum VS penalty David could achieve by updating (somehow) on E. And the answer is that it is not: minimizing the preceding function subject to $x + y + z = 1$ yields 0.032, where this minimum is reached at $(.097, .703, .201)$.

To be sure, there may be no "natural" update rule via which David could have arrived at these probabilities, given his initial probabilities and the evidence that he received. But that is beside the point. Greaves and Wallace (2006) compare a Bayesian update with an update via some unspecified other rule, and they explicitly refrain from offering an "intuitive rationale" (p. 614) for the latter. Their aim is to show that Bayes's rule does better, in terms of expected (next-step) inaccuracy minimization, than that other rule, so that "considerations of intuitive plausibility need not be invoked in order to outlaw [that rule]" (Greaves & Wallace, 2006, p. 614).

It is also worth looking at natural alternatives to Bayes's rule and see how, on the present score, Bayes's rule compares to those. Of course, there are natural alternatives to Bayes's rule in which we are *particularly* interested, to wit, probabilistic versions of abduction, one of which—EXPL—we already encountered in previous chapters. Does Bayes's rule do at least better than EXPL in terms of expected VS penalty minimization? We saw that EXPL is much like Bayes's rule except that it assigns a bonus to the best-explaining hypothesis, after which it renormalizes the probabilities (so that they are probabilities properly so called). Given that the bonus gets assigned after the Bayesian update, which is the first step in EXPL updating, it makes sense to divide it over the possible worlds consistent with the evidence, which in our case is only one world.[7] So, suppose that David adheres to EXPL, at least in the present context, and the explanation bonus that he assumes is .1. Suppose further that, for whatever exact reason, he deems hypothesis A worthy of this bonus once he receives the new information E. Then, after assigning the bonus and renormalizing, his doxastic state is specified by

$$\begin{array}{llllll} \Pr_E(\{w_{AE}\}) & \approx .239 & \Pr_E(\{w_{BE}\}) & \approx .543 & \Pr_E(\{w_{CE}\}) & \approx .217 \\ \Pr_E(\{w_{A\overline{E}}\}) & = .0 & \Pr_E(\{w_{B\overline{E}}\}) & = .0 & \Pr_E(\{w_{C\overline{E}}\}) & = .0 \end{array}$$

7. If there were more than one, it would seem reasonable to divide the bonus in proportion to the worlds' probabilities.

To calculate David's current expectation (i.e., before his receipt of E) of the VS penalty incurred by an EXPL update on E, we substitute again the preceding values into (5.1) in the way that we did previously (so setting $x = .239$, etc.). This yields an expected VS penalty of 0.037, which is actually higher than the VS penalty that David expects to incur if he updates via Bayes's rule. But was this bound to happen? Or did David just make an unfortunate choice of truthlikeness weights? What if he had chosen a different value for c, the bonus to be assigned to best explanations? What if he had deemed hypothesis B the best explanation instead of hypothesis A?

The intuition behind VS rules is that although it is bad in general to assign a positive probability to a false hypothesis, it is worse the further the hypothesis is from the truth. So, suppose that a VS rule's weights *reflect truthlikeness in a minimally adequate sense* precisely if hypotheses are weighted as a function of their distance from the truth, with hypotheses further from the truth weighted more heavily than hypotheses closer to the truth. We can then ask generally whether by updating via EXPL one is guaranteed to incur a higher VS penalty than by updating via Bayes's rule, provided the weights associated with the rule reflect truthlikeness in a minimally adequate sense.

Keeping all probabilities as they are above, I had the computer search for an answer by letting it go through combinations of weights that reflect the truthlikeness relations among our hypotheses A, B, and C in a minimally adequate sense. I found that looping over 1,000 such combinations is typically enough to output a set of weights that fulfil our requirement while yielding a VS penalty for EXPL that is lower than that for Bayes's rule. For instance, if we change the weights in (5.1) as follows:

$$\begin{aligned} .01\big(.0198(1-x)^2 + .4835y^2 + .4967z^2\big) \\ + .25\big(.9798x^2 + .0202(1-y)^2 + .9798z^2\big) \\ + .1\big(.4967x^2 + .4835y^2 + .0198(1-z)^2\big), \end{aligned}$$

then we obtain an expected score of 0.0469 for the Bayesian update versus an expected score of 0.0467 for the EXPL update. We do not know, of course, whether these weights reflect David's truthlikeness judgments for the hypotheses at issue, but they *might*.[8]

8. Appendix E shows how one can obtain such weights and also how one can, more systematically, find weights that will maximize the extent to which the EXPL update does better than the Bayesian update.

Alternatively, suppose that we keep David's weights as they are in the example, but instead of following van Fraassen in setting $c = 0.1$ we set $c = 0.05$. In that case, David's post-EXPL-update probabilities for $\Pr_E(\{w_{AE}\})$, $\Pr_E(\{w_{BE}\})$, and $\Pr_E(\{w_{CE}\})$ would have been .146, .610, and .244, respectively. It is easily verified that with these probabilities, David's VS penalty equals 0.033, which is better than the penalty he would incur if he updated by Bayes's rule (which equaled 0.034, as previously shown).

Finally, keeping the weights as they are and also keeping $c = 0.1$, one easily calculates that if David had assigned the bonus to hypothesis B, his relevant probabilities would have been $\Pr_E(\{w_{AE}\}) \approx .022$, $\Pr_E(\{w_{BE}\}) \approx .761$, and $\Pr_E(\{w_{CE}\}) \approx .033$, and those would have yielded a score of 0.033, which is again lower than his Bayesian-update score.

It has thus come to appear that unless Bayesians have some good reason for dismissing the new rule, their inaccuracy-minimization argument, which was limited in value to begin with (as argued in the previous chapter), unravels entirely. It is simply not true that Bayesian updating is guaranteed to minimize your expected inaccuracy. Nor does EXPL come with such a guarantee, but nobody has ever claimed as much. At least if we want to be serious about truthlikeness relations among the hypotheses of interest and hold that VS rules can legitimately account for such relations, then there may be *no* updating procedure that is guaranteed to minimize inaccuracy, always and everywhere. That would in effect dovetail nicely with much that follows in later chapters.

5.3 Truthlikeness and Impropriety

But perhaps there *is* a good reason for rejecting VS rules. Bayesians might say that while VS rules offer a seemingly straightforward way to take into account truthlikeness relations, they face a devastating objection. We can start to see what Bayesians might find so objectionable by considering that relative to his current probabilities, David minimizes his expected quadratic score by having those very probabilities; formally, $\arg\min_{\mathbf{p}} \mathbb{E}_{\mathbf{d}}[\mathcal{Q}(\mathbf{p})] = \mathbf{d}$. This is no coincidence. It is known that not only the Brier score but *all* instances of the quadratic scoring rule schema are proper, and even strictly proper (Rosenkrantz, 1981, ch. 2). On the other hand, relative to David's current probabilities, 0.123 is *not* the minimum VS score he can incur, for minimizing

$$.1\big(.1(1-x)^2 + .3y^2 + .6z^2\big) \\ + .5\big(.45x^2 + .1(1-y)^2 + .45z^2\big) \\ + .4\big(.6x^2 + .3y^2 + .1(1-z)^2\big),$$

subject to $x + y + z = 1$, yields 0.118. Hence, the VS rule that David assumes is improper. This might not be a matter of great concern. Perhaps it is again due to David's just having made an unfortunate choice of truthlikeness weights. Perhaps we should again let the computer search for better weights. But this time the problem is more fundamental, because we have

Theorem 5.1 *Every VS rule whose weights reflect truthlikeness in a minimally adequate sense is improper.*

For a proof, see appendix B. In other words, it is not just that David was unlucky in the weighting function that he picked; he could have picked none that would not have led to the same problem of impropriety, or at least none that is minimally adequate.

How damning is this result for the class of VS rules? Whereas the mainstream holds that impropriety is *totally* damning, I would like to argue that the answer should be the following: it all depends on the purpose of our scoring.

As mentioned, scoring rules were initially meant for *eliciting* probabilities. To serve that purpose, they better not encourage the subject whose probabilities one would like to elicit to lie about those probabilities. But if we are assessed by means of an improper scoring rule, then by revealing our actual probabilities we may expect to incur a larger penalty than if we pretend to have different probabilities, as was shown in the case of David. That can make it disadvantageous to be truthful.

It is generally recognized that scoring rules may also be used for the purpose of *self-assessment*. If David adheres to the VS rule considered previously, and he wonders about the accuracy of his current probabilities, he may conclude that he should replace these probabilities with those found to minimize his expected penalty. That appears problematic; as Sarah Moss (2011, p. 1057) notes, it implies that self-assessment by means of an improper scoring rule "could motivate you to raise or lower your credences *ex nihilo*, in the absence of any new evidence whatsoever."

In response, it could be argued that finding out that our probabilities do not minimize our expected penalty is new information, so that when we thereupon adapt those probabilities we are *not* acting in the absence of new evidence. This would in effect be an instance of what Lombrozo (2020) calls "learning by thinking," occasioned by what she calls an "observation generated inside the head." But some might want to reply that the evidence should, in an intuitively clear sense, bear on the hypotheses to which our probabilities are assigned. The intuitively clear sense may be difficult to pin down formally, but let that pass.[9] Because using a VS rule, or any other improper scoring rule because self-assessment seems to face a threat more severe than the one toward which Moss has pointed.

Previously we found that given David's probabilities in table 5.1, he would actually minimize his expected penalty by raising his probability for A to .146, raising his probability for B to .548, and lowering his probability for C to .306. Ironically, if he shifts his probabilities accordingly, he will find that he actually minimizes his expected penalty by setting his probabilities for A, B, and C to .164, .593, and .243, respectively. It does not end there, for relative to *these* probabilities, David minimizes his expected penalty by setting his probabilities for the hypotheses to .166, .630, and .204. And it goes on. And on? That would seem to prevent David from ever having stable probabilities for A, B, and C unless he decides to arbitrarily stop self-assessing at some point.

However, the iterative minimization process, should David engage in it, turns out to reach a fixed point. To be entirely precise, David would eventually arrive at probability assignment $\mathbf{d}^+ = (.083, .833, .083)$,[10] a probability

9. In this connection, it is also worth mentioning that, at least according to some influential Bayesian statisticians (Gelman & Hill, 2009; Gelman & Shalizi, 2012, 2013; Kruschke, 2013b), raising or lowering probabilities in the absence of the kind of evidence with "direct bearing" is accepted as a legitimate practice, most notably, as resulting from a so-called posterior predictive check in which a statistical model may be rejected because it is found unsatisfactory (according to informal criteria) in light of simulated data. (More exactly, in a posterior predictive check we use a fitted model to generate synthetic data and then check whether those data are "similar enough" to the data that were used for fitting the model; see, e.g., Gelman & Hill, 2007, p. 158.) If rejected, the model is to be replaced by a new one, which requires, among other things, a specification of new prior probabilities. The simulated data that can motivate this kind of model revision—including probability revision—is presumably not the kind of new evidence that Moss has in mind.

10. Depending on the software that one uses to calculate the fixed point and the different floating-point formats the software uses, the number of steps that David needs to reach this fixed point can differ. Using the FixedPointList function from *Mathematica*, the number

assignment that is "strongly self-recommending" (Greaves & Wallace, 2006, p. 619) in the sense that $\arg\min_{\mathbf{p}} \mathbb{E}_{\mathbf{d}^+}[\mathcal{Q}(\mathbf{p})] = \mathbf{d}^+$. (For completeness, we note that David's expected penalty at the fixed point equals 0.057, whereas for his initial probabilities it was 0.118, as shown.)

We get a first sense of what may be the real problem here if we repeat our calculations for Emma and Frank and find that they arrive at exactly the same fixed point as David! Not only that; varying the weighting function a little, we find that the three colleagues now arrive at a different fixed point but that this fixed point is again the same for all three of them. What is more, this is what we find no matter the colleagues' initial probabilities: if they iteratively adapt their probabilities on the basis of VS score minimization, they all arrive at a fixed point, and if they use the same weighting function, they arrive at the *same* fixed point!

We can illustrate this graphically by noting that vectors in the standard unit simplex (or probability simplex) of dimensionality $n - 1$ can be interpreted as probability distributions on an n-element hypothesis partition, with the ith vector component representing the probability of the ith hypothesis (de Finetti, 1962).[11] Figure 5.1 shows three two-dimensional simplexes, in each of which the initial probability assignments of David, Emma, and Frank given in table 5.1 are represented by medium-sized dots. The smaller dots represent their consecutive probability assignments while they go through the VS-rule-governed minimization process, ending with the final assignments, which are represented by big dots. The difference among the minimization processes represented in the three simplexes is only that the VS rules that are assumed in them assign different truthlikeness weights: the left simplex represents the process that relies on the VS rule that was assumed in the previous example;

found was 442; using a custom-built function in Julia, it was 428. However, that the process would reach a fixed point (if perhaps not after 428 or even 442 steps) is guaranteed by theorem 5.2, to be stated.

11. Interpreting such simplexes requires a bit of practice. Some points have a straightforward interpretation. For instance, in our example, the vertex labeled (1, 0, 0) corresponds to the point where all probability is assigned to hypothesis A, and similarly for the other vertices. Also, the geometric center of the simplex corresponds to the flat distribution. Furthermore, any point on an edge connecting two vertices represents a distribution in which the hypothesis associated with the unconnected vertex receives zero probability. More generally, given a point in the simplex, the probability assigned to the hypothesis associated with a vertex is the shortest distance from the point to the edge opposite the vertex, divided by the sum of the lengths of the shortest distances from the point to the various edges. (This generalizes to simplexes of any dimensionality.)

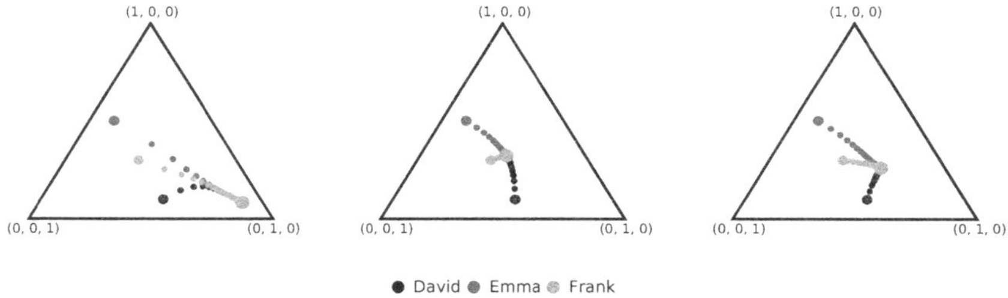

Figure 5.1: Three probability simplexes show that David, Emma, and Frank all reach the same fixed points (large dots) by iteratively adapting their probabilities on the basis of VS score minimization and illustrate that these points depend on the weights assumed by the given VS rule. Initial assignments are marked by medium-sized dots, intermediate assignments by small dots, and final assignments by big dots.

the processes shown in the other two simplexes used VS rules with slightly different weighting functions (the details are unimportant here).

To see that this finding is again no coincidence, consider the following:

Theorem 5.2 *Let S be the standard unit $(n-1)$-simplex, let $\mathbf{p}$ and $\mathbf{p}^*$ range over vectors in S, and let $m\colon S \longrightarrow S$ be defined as follows:*

$$m(\mathbf{p}) \quad := \quad \arg\min_{\mathbf{p}^*} \sum_{i=1}^{n} \sum_{j=1}^{n} p_i w_{ij} (\delta_{ij} - p_j^*)^2$$

with δ_{ij} the Kronecker delta, and with $w_{ij} > 0$ for all i, j, and $\sum_{i=1}^{n} \sum_{j=1}^{n} w_{ij} = 1$. Then (1) there is a $\mathbf{p}^+ \in S$ such that $m(\mathbf{p}^+) = \mathbf{p}^+$; and furthermore (2) there is only one such $\mathbf{p}^+$, and (3) it depends only on the w_{ij}.

See appendix C for a proof. Further observe that for any VS rule $\mathcal{V}$

$$\arg\min_{\mathbf{p}^*} \mathbb{E}_{\mathbf{p}}[\mathcal{V}(\mathbf{p}^*)] \quad = \quad \arg\min_{\mathbf{p}^*} \sum_{i=1}^{n} \sum_{j=1}^{n} p_i w_{ij} (\delta_{ij} - p_j^*)^2$$

with the w_{ij} being provided by the weighting function of the given rule. In other words, the function by which David, Emma, and Frank iteratively update their probabilities in the represented processes is an instance of the function m defined in theorem 5.2. Thus, in light of this theorem it is unsurprising that their probability assignments reach a fixed point and that despite their different initial probability assignments, this fixed point is the same for the three of them, given that, as we assume, they use the same VS rule.

Surely we have hit upon an absurd result here, which offers a compelling reason to refrain from self-assessment by means of a VS rule. As Robert Winkler (1996) notes, however, scoring rules not only have (what he calls) an ex ante use; they can also be used ex post for evaluating performance. Potential problems with elicitation or self-assessment are beside the point if one is going to use an improper scoring rule for assessing people's performance without disclosing that rule to them. And often there will be no requirement for disclosure. An investment company might hope to identify job applicants with special capacities to predict the stock market by scoring their answers to test questions via some custom-built scoring rule without informing the applicants about the rule being used. Similarly, if a television network wants to hire a new weather forecaster and is now retroactively analyzing, by means of an improper scoring rule, the performance of various candidates being considered for the job, it will make no difference whether or not the network makes publicly known how it is conducting the analysis.[12]

As an aside, I mention an interesting real-life example of this kind of usage, which is found in recent work on forecasting carried out by a group of psychologists from various American universities (Mellers et al., 2015; Tetlock & Gardner, 2015). These researchers have organized, over a period of several years, a number of prediction tournaments, mostly concerning geopolitical questions. They found that some otherwise ordinary people were much more accurate forecasters than even professional intelligence analysts. A key objective of the research was to determine what distinguishes the most accurate forecasters from the rest of the population. The researchers used a number of different scoring rules for evaluating their participants' performance, including the Brier score but also the so-called AUROC, which is known to be an improper scoring rule (for details see, e.g., Agresti, 2007, ch. 5, or Hastie,

12. It is entirely consistent with everything said here that there are still other purposes that scoring rules can serve that may again require propriety. For instance, Roche and Shogenji's (2018) claim that we should measure informativeness in terms of inaccuracy reduction but that inaccuracy should then be measured by a proper scoring rule is not in conflict with the claim that scoring rules can serve purposes that do *not* require propriety.

Tibshirani, & Friedman, 2009, ch. 9). Given that the participants were never told what the evaluation process consisted of, the use of an improper scoring rule in that process will not have affected their responses.[13]

In short, depending on the purpose of our scoring, improper scoring rules appear to be perfectly admissible. In particular, when we are evaluating a probability distribution that is *given*—we are not eliciting the probabilities—from an *objective* standpoint—we are not assessing our own probabilities—I can see no impediment to using an improper scoring rule. Arguably, it is this objective evaluation of given probabilities that is at play in the Bayesian inaccuracy-minimization argument. The argument is not meant to apply only to *elicited* probabilities nor does it assume that we frequently, or even ever, assess our own probabilities in terms of inaccuracy minimization—indeed, we do not. It thus appears that the conclusion from the previous section—that the claim that Bayes's rule minimizes expected next-step inaccuracy is not generally tenable—still stands.

To end this chapter, it is worth pointing out that verisimilitude sensitivity *need* not come at the expense of propriety, for there is a scoring rule that is both verisimilitude sensitive *and* proper. This is the so-called ranked probability score (RPS), first proposed by Edward Epstein (1969) and shown to be strictly proper by Allan Murphy (1969). This rule has received hardly any attention from philosophers nor is it widely discussed outside philosophy (O'Hagan et al., 2006, p. 169).[14] Rather than comparing a probability distribution **p** on a partition of hypotheses with the vector **v** of truth-values of those hypotheses, it compares the cumulative distribution function of **p** with the cumulative distribution function of **v**. Given a partition $\{H_i\}_{i=1}^{n}$ and a probability distribution **p** on this partition, and supposing H_j to be true, the

13. Although in this research both proper and improper scoring rules were used for the purposes of selection, one could also use an improper scoring rule to select participants while at the same time scoring them via a proper scoring rule to determine their compensation in the experiment. Letting participants know how they will be compensated will then encourage them to post their true probabilities, while the improper scoring rule—the use of which is *not* disclosed to the participants—may still yield more useful information.

14. To the best of my knowledge, Konek (2016) and Vassend (in press) contain the only references to the rule (actually, the continuous version of the RPS rule) in the entire philosophical literature. I am not aware of any mention of the rule in the psychological literature.

ranked probability score associated with **p** is defined

$$\mathcal{R}_j(\mathbf{p}) \quad := \quad \frac{\sum_{k=1}^{n}\left(\sum_{i=1}^{k} p_i - \gamma_{kj}\right)^2}{n-1},$$

where $\gamma_{kj} = 1$ if $k \geqslant j$, and $\gamma_{kj} = 0$ otherwise.

To illustrate, given David's probabilities **d** and still supposing C to be the true hypothesis (whence $j = 3$), David's ranked probability score is

$$\mathcal{R}_3(\mathbf{d}) \quad = \quad \frac{(.1-0)^2 + (.1+.5-0)^2 + (.1+.5+.4-1)^2}{2} \quad = \quad 0.185.$$

In the same way, we find that Emma's ranked probability score equals 0.305, and Frank's, 0.225.[15] In the example in which the three colleagues' probabilities are about what grade their student will receive, these outcomes make perfect sense: as previously stated, David appears to do best in this case because he assigns a higher probability to the false hypothesis that is closer to the truth than to the one further from the truth. By the same token, Emma would seem to do worst, given that she does exactly the opposite. Frank steers a sort of middle course between his colleagues in assigning the two false hypotheses equal probability.

Bayesians might particularly like the RPS rule, for it is easy to verify that, at least in the example from the previous section (on p. 144), their favored rule does minimize expected RPS penalty (see appendix E). It may well do so generally, but I do not have a proof of that. Naturally, though, supposing

15. Because, as noted, the RPS rule is strictly proper, it satisfies Selten's third axiom (see footnote 4 in this chapter). To see that it also satisfies his fourth axiom, note that for comparing a probability assignment $(p_1, \ldots, p_n)$ with a "true" probability distribution $(p_1^*, \ldots, p_n^*)$, the RPS rule takes this form:

$$\frac{(p_1 - p_1^*)^2 + ((p_1 + p_2) - (p_1^* + p_2^*))^2 + \cdots + ((p_1 + \cdots + p_n) - (p_1^* + \cdots + p_n^*))^2}{n-1}.$$

The symmetry required by the fourth axiom then follows from the fact that the addends in the numerator are all squared. Furthermore, that David's and Emma's ranked probability scores are different, as seen in the main text, is enough to show that the rule does *not* satisfy Selten's first axiom. Finally, to show that neither does it satisfy the second axiom, we can add to the partition consisting of hypotheses A, B, and C the hypothesis that the student will receive a C−, where this has zero probability for David. Keeping his probabilities for A, B, and C as they were, David's ranked probability score then becomes (approximately) 0.243, and hence the addition of the zero-probability alternative did affect the score.

Bayes's rule does minimize expected inaccuracy vis-à-vis the RPS rule, that does not militate against the family of VS rules. Indeed, it is not even true that in the presence of the RPS rule, we have no use for VS rules. VS rules are still valuable because they are flexible in a way that the RPS rule is not. In particular, given the latter rule, relations of truthlikeness are completely fixed by the *ordering* of the hypotheses in the partition, and such orderings will not always reflect intuitive judgments of truthcloseness.[16]

Consider this example: Two football teams that are about equally strong and whose past five encounters have all been draws just played another match, in which neither team managed to score. Then the prediction that the match would end in a 0 : 4 away win appears to be quite a bit further from the truth than the prediction that the match would end in a 0 : 1 win. After all, the former much more than the latter would suggest a strongly dominant away team, which would have been a surprise in view of the relative strengths of the teams and the outcomes of their past encounters. On the other hand, a 0 : 24 away win prediction would hardly be further from the truth than a 0 : 21 away win prediction: we would find these end results about equally stunning and might say that both are "about as far from the truth as can be."[17]

Or consider any value representable in a conceptual space (in the manner of Gärdenfors, 2000), for example, a color shade. We might want to order different hypotheses about this value on the basis of distances in color space (CIELAB space or CIELUV space; see Fairchild, 2013). It is known, however, that although such distances correlate well with human similarity judgments

16. Thanks to Ilkka Niiniluoto for bringing this to my attention; also to William Roche for particularly helpful comments here.

17. Some might want to hold that we are to measure distance from the truth here in terms of the difference in goals scored. In my opinion, it is more reasonable to look at how different the various mentioned non-actual worlds (the world in which the match ends in a 0 : 1 win, the world in which the match ends in a 0 : 4 win, and so on) are from the actual world. And given what we know about the teams, our world would, as mentioned, have to be rather different from the actual world for the match to end 0 : 4 while it would not have to be very different for the match to end 0 : 1. By contrast, for the match to have ended 0 : 21, something entirely out of the ordinary would have had to occur, and whatever that would have been, it would have been about equally compatible with a 0 : 24 end result. For instance, if all players who normally play for the home team had been suspended, and the coach of that team had to line up their most inexperienced players, then a devastating loss would be explainable—but the explanation would be about as good in the case of a 0 : 21 end result as it would be in the case of a 0 : 24 end result. (Thanks to Theo Kuipers and Ilkka Niiniluoto for helpful discussion here.)

at a short range, they stop doing so at a longer range (Shepard, 1987). Thus, if the actual value of the color that we are looking for is some shade of red, then the hypothesis that it is a particular shade of orange may be intuitively closer to the truth than the hypothesis that it is a particular shade of yellow, and that will be reflected by the ordering of hypotheses. By contrast, the hypothesis that the actual color is a particular shade of blue would appear as far from the truth as the hypothesis that it is a particular shade of green, and that judgment is quite compatible with the particular shade of blue being closer in color space to the actual shade than the particular shade of green.

It is hard to see how one could accommodate the intuitions at play in these examples only in terms of an ordering of hypotheses. However, one can find refined measures of truthlikeness in the literature that do allow one to go beyond merely ordering hypotheses in terms of truthlikeness and to express truthlikeness relations among the aforementioned hypotheses in a way that does justice to the aforementioned intuitions (see, e.g., Niiniluoto, 1984, ch. 7, 1987, ch. 12; Kuipers, 1992, 2000, ch. 12). And while the RPS rule cannot take such refined measurements into account, VS rules *can*.

Roger Cooke (1991, p. 121) remarks that one of the main questions concerning scoring rules is whether "the score reward[s] those features that we would like subjective probability assessments to display." Much of the debate on scoring has been premised on the assumption that the desirable features are fixed across contexts and are independent of which goal or goals a scoring rule is to help achieve. I have argued that this assumption is false, and that different scoring rules may be called for in different contexts and also for different goals.[18] This undercuts the Bayesian claim that, in contrast to the dynamic Dutch book argument, the inaccuracy-minimization argument only involves a notion of epistemic rationality.

I have in particular emphasized the importance of taking truthlikeness relations into account in our scoring, if such relations are present. Following suggestions from Greaves and Wallace led us to the family of VS rules, which *are* sensitive to truthlikeness, but which also spelled trouble for the Bayesian inaccuracy-minimization argument, in that it is not generally the case that Bayesian updates minimize expected inaccuracy when inaccuracy is measured by a VS rule. So, lest Bayesians have a decisive reason for dismissing VS rules,

18. To my knowledge, the only other authors explicitly open to this possibility are Levinstein (2017) and Schurz (2019).

nothing is left of what most of them now consider to be the main argument in favor of Bayes's rule.

6
The Ecological Rationality of Abduction

Google wasn't that much faster than Yahoo, but it didn't need to be. All it needed to be was a little bit faster and the rest is history.

—Stanley Druckenmiller

6.1 The Justification of Abduction

To this point the book has been concerned with staving off the main criticisms leveled against abduction. Nothing presented so far amounts to a *justification* of abduction. Although, as we have discussed, there is strong evidence that in some contexts people do update their beliefs by abduction, most of what was presented in the foregoing chapters is compatible with the thought that the effect of explanation on updating is no more than the manifestation of a bias. From here onward, I aim to convince you that updating one's beliefs, or degrees of belief, on the basis of explanatory considerations can be the right thing to do. Specifically, the claim argued is that explanation-based updating allows people to strike the best balance between being fast and being accurate learners.

By doing so, I expand on work begun in chapter 4, which compared Bayes's rule and EXPL along the dimensions of speed and accuracy. There, the comparison served only to rebut the claim that any form of non-Bayesian updating is to be rejected for failing to minimize inaccuracy. No conclusion was drawn regarding whether Bayes's rule or EXPL would be the more rational update rule to use. And what conclusion *could* have been drawn? The results from the earlier comparison appeared to be mixed. EXPL was found to be better at picking up the signal in the data, but when noise occasionally took a pattern that made it look like a signal, it was the first to be misled, sometimes quite dramatically.

If the question whether Bayes's rule or EXPL is the rational pick appears hard to answer, then, I submit, that is only because we are committed to an outdated "universalist" conception of rationality, which implies among other things that there must be exactly one right way to update one's beliefs or degrees of belief. Instead, I aim to follow recent work in psychology and cognitive science that argues for a conception of *ecological* rationality, an arguably more realistic and useful conception of rationality that leaves open the possibility that, at different times and for different people, different practices may count as rational. Authors advocating this conception have laid out clearly how one should go about investigating the rationality of cognitive practices. Following their guidance, I use computer simulations implementing an agent-based optimization technique to compare Bayes's rule with EXPL and similar probabilistic forms of abduction. The results will show that in some stylized but still realistic contexts, abduction, and not Bayes's rule, is the rational choice of update rule. But before embarking on a new defense of abduction, I should briefly explain why I believe that previous defenses fall short.

6.2 Previous Defenses of Abduction

For reasons already noted (see the last paragraph of chapter 2), not many today would want to subscribe to a conception of truth that posits a necessary connection between explanatory force and truth—for instance, because it stipulates explanatory superiority to be necessary for truth, as do some pragmatist conceptions of truth. At a minimum, that makes the prospects of an a priori defense of abduction dim. And indeed, all better-known defenses of abduction are of an empirical nature in that they appeal to data supposedly supporting the claim that (in some form) abduction is a reliable rule of inference.[1,2]

A particularly well-crafted defense of this sort was developed by Boyd in the 1980s (see Boyd, 1984, 1985, 1990). It starts by underlining the theory-dependence of scientific methodology, which comprises methods for designing experiments, for assessing data, for choosing between rival hypotheses, and so on. For instance, in considering possible confounding factors from which an experimental setup has to be shielded, scientists draw heavily on already accepted theories. These inform them about which confounds to reckon with and which steps can be taken to prevent those confounds or to control for them. The argument next calls attention to the apparent reliability of this methodology, which after all has yielded and continues to yield impressively accurate theories. In particular, by relying on this methodology, scientists have for some time been able to find ever more instrumentally adequate theories. Boyd then argues that the reliability of scientific methodology is best explained by assuming that the theories on which it relies are at least approximately true. From this and from the fact that these theories were arrived at mainly by abductive reasoning, he concludes that abduction must be a reliable rule of inference.

Boyd's is a *brave* attempt at defending abduction. It does not engage in the kind of appeasement politics of those who seek to show that abduction can be regarded as serving a convenient helper function for Bayesians, as discussed in section 1.2.2, but rather aims to show that abduction is a mode of inference in its own right and as such forms the core of scientific methodology. But critics have accused Boyd's argument of being circular (e.g., Laudan, 1981; Fine, 1984). Specifically, it has been said that the argument rests on a premise—that scientific methodology is informed by approximately true background theories—that in turn rests on an abductive step for its plausibility. And the reliability of this type of inference is precisely what is at stake. At a more

1. Biggs and Wilson (2017) do think abduction can be justified a priori. But their argument involves the claim that conditionals of the sort "If water and H_2O are spatiotemporally coincident, then water is identical with H_2O" can be justified a priori by means of what they call "a hypothetical form of abduction" (p. 750). I fail to see, however, what support there is for the reliability of that form of abduction.

2. Even though Devitt (1991, p. 111) does not think abduction is a priori, he still finds it so obviously correct that he thinks no defense is called for. Naturally, this was written before Bayesian epistemology became dominant and even came to appear "obviously correct" to many.

general level, Jonathan Vogel (2008, p. 531) has argued that the belief that an epistemic rule *R* is reliable cannot be justified by the application of *R*.

To this, Psillos (1999, ch. 4) has responded by invoking a distinction credited to Richard Braithwaite, to wit, between *premise circularity* and *rule circularity*. An argument is premise circular if its conclusion is among its premises. A rule-circular argument, by contrast, has a conclusion that asserts something about an inferential rule that is used in the same argument. In Psillos's analysis, Boyd's argument is rule circular but not premise circular, and rule-circular arguments, Psillos contends, need not be viciously circular (even though a premise-circular argument is always viciously circular). To be more precise, in his view, an argument for the reliability of a given rule *R* that essentially relies on *R* as an inferential principle is not vicious, provided that the use of *R* does not guarantee a positive conclusion about *R*'s reliability. Psillos claims that in Boyd's argument, this proviso is met. Although Boyd concludes that the background theories on which scientific methodology relies are approximately true on the basis of an abductive step, the use of abduction itself does not trivialize his conclusion. After all, granting the use of abduction does nothing to ensure that the best explanation of the success of scientific methodology is the approximate truth of the relevant background theories. Thus Boyd's argument still stands, or so Psillos maintains.

Even if the use of abduction in Boyd's argument might have led to the conclusion that abduction is *not* reliable, the argument's rule circularity might still be worrisome. Suppose that some scientific community relied not on abduction but on a rule that we may dub "Inference to the Worst Explanation" (IWE), a rule that sanctions inferring to the *worst* explanation of the available data. We may safely assume that the use of this rule would lead mostly to the adoption of very unsuccessful theories. Nevertheless, the said community might justify its use of IWE by dint of the following reasoning: "Scientific theories tend to be hugely unsuccessful. These theories were arrived at by application of IWE. That IWE is a reliable rule of inference—that is, it usually leads from true premises to true conclusions—is surely the worst explanation of the fact that our theories are so unsuccessful. Hence, by application of IWE, we may conclude that IWE is a reliable rule of inference." Whereas this would be an utterly absurd conclusion, the argument leading up to it cannot be convicted of being viciously circular any more than Boyd's argument for the reliability of abduction can (if Psillos is right). It would appear then that there must be something else amiss with rule circularity.

It is fair to note that, for Psillos, the fact that a rule-circular argument does not guarantee a positive conclusion about the rule at issue is not sufficient for such an argument to be valid. A further necessary condition is "that one should not have reason to doubt the reliability of the rule—that there is nothing currently available which can make one distrust the rule" (Psillos, 1999, p. 85). And there is plenty of reason to doubt the reliability of IWE; in fact, the previous argument *supposes* that it is unreliable. Two questions arise, however. First, why should we accept the additional condition? Second, do we really have *no* reason to doubt the reliability of abduction? Certainly *some* of the abductive inferences we make lead us to accept *falsehoods*. How many falsehoods may we accept on the basis of abduction before we can legitimately begin to distrust this mode of reasoning, at least enough to fail Psillos's extra condition? No clear answers have been given to these questions.

Be this as it may, even if rule circularity is neither vicious nor otherwise problematic, one may still wonder how Boyd's argument is to convert a critic of abduction, given that it relies on abduction. But Psillos makes it clear that the point of philosophical argumentation is not always to convince an opponent of one's position. Sometimes the point is, more modestly, to assure or reassure oneself that the position that one endorses, or is inclined to endorse, is correct. In the case at hand, we need not think of Boyd's argument as an attempt to convince the opponent of abduction of its reliability. Rather, it may be thought of as justifying the rule from within the perspective of someone who is already sympathetic toward abduction. This is an important insight to which I come back in chapter 8.

There have also been attempts to argue for abduction in a more straightforward fashion, to wit, via enumerative induction (e.g., Harré, 1986, 1988; Bird, 1998; Kitcher, 2001; Douven, 2002a). The common idea of these attempts is that every newly recorded successful application of abduction—like the discovery of Neptune, whose existence had been postulated on explanatory grounds, as discussed in section 2.1—adds further support to the hypothesis that abduction is a reliable rule of inference, in the way in which every newly observed black raven adds some support to the hypothesis that all ravens are black. Because it does not involve abductive reasoning, this type of argument is more likely to also appeal to disbelievers in abduction.

One problem with the argument is precisely that it relies on induction, whose status has been questioned as much as that of abduction has. Did David Hume (1748/2006) not show that a justification of induction is unattainable?

According to him, a *deductive* justification is not forthcoming because induction is not *necessarily* reliable, whereas an *inductive* justification is ruled out on grounds of circularity.

However, abduction was unknown to Hume, at least as an explicitly recognized mode of inference, so he could hardly have considered the idea of induction and abduction taking in each other's wash. Prima facie, it certainly makes sense to think that we might be able to base a defense of induction on abduction *and* vice versa. It would appear, after all, that we have reason to be confident in the reliability of induction because the registered successes of induction are *best explained* by its reliability, while conversely we have reason to be confident in the reliability of abduction because the registered successes of abduction provide strong *inductive evidence* that abduction is reliable. There is no apparent rule circularity here, nor any other kind of circularity.

Alas, this defense would fare no better than Psillos's defense of abduction. Consider the following duo of inferential principles: IWE as previously introduced and Wesley Salmon's (1963) rule of counterinduction, according to which from the observation that of all Φs observed so far x percent were Ψs it follows that of all Φs yet to be observed $100 - x$ percent will be Ψs. Although both principles are patently nonsensical, we could argue for them in tandem, as follows. We can ask why counterinduction fails, or would fail, so dismally, and then surely the worst answer we could give is that it is a reliable inferential principle. But then IWE licenses the inference that the rule of counterinduction is reliable indeed. That is helpful only if we can justify IWE. And we can, namely, by invoking counterinduction. After all, we can be sure that IWE would, if actually in use, have let us down time and again. By counterinduction, that means that it would do just great in the future!

The kind of construction used here, in which one hypothesis (the reliability of IWE) serves as an auxiliary in helping to confirm a second hypothesis (the reliability of counterinduction) while the second hypothesis serves as an auxiliary in helping to confirm the first, has received some attention from confirmation theorists under the heading "bootstrap confirmation" (e.g., Glymour, 1980; Christensen, 1983, 1997; van Fraassen, 1983; Douven & Meijs, 2006; Douven & Kelp, 2013). The proposal of bootstrap confirmation came with an important additional condition, however: given a theory, thought of as a set of hypotheses, it should not only be the case that each hypothesis is confirmed by the evidence supposing one or more of the other hypotheses as auxiliaries—or even more debatably (Douven & Meijs, 2006), supposing itself as an auxiliary—but it should *also* hold that the fact that parts of the theory are allowed to help other parts in obtaining support from the evidence should not lead to *trivialization*, meaning that it should not shield the theory from disconfirmation. We can gain support for a theory only if we put that theory at risk of being disconfirmed.

In the case at hand, however, the nontrivialization condition is not going to help, for the joint assumption of IWE and counterinduction does *not* prevent the occurrence of disconfirming evidence. Our reliance on counterinduction in this argument does not by itself prevent disconfirmation of IWE, given that our judgments of explanation quality might be so badly off that by going for the worst explanation we often end up endorsing what is by more reasonable standards the best explanation, in which case IWE might have yielded good results, and an application of counterinduction would have led us to dismiss IWE. Similarly, we may assume IWE and yet live in a hopelessly counterinductive universe, in which case counterinduction might have worked and the worst explanation of that would have been that counterinduction is unreliable, so that IWE would have led to a disconfirmation of counterinduction.

But even if it is difficult to make the deficiency of the bootstrap defense of the combination of abduction and induction formally precise, that such a defense *is* deficient is still clear from the fact that, by parallel reasoning, we could argue for a pair of what we said are patently nonsensical inferential principles.

Recent progress on the justification of induction raises hope that a defense of abduction may proceed via induction, but that a defense of induction then does not need to appeal to abduction in turn but can be accomplished by recruiting *meta-induction*. In a series of papers (Schurz, 2008b, 2009, 2012, 2017), and most comprehensively in his 2019 book, Schurz argues that Hume, and all authors who have since studied the question of the justification of induction, have erred in assuming that such a justification must involve a demonstration of the *reliability* of induction.[3] Instead of focusing on its reliability, we should focus on the question of its *optimality*. Suppose in all possible circumstances we could only be worse off by *not* relying on induction.

3. See also Schurz and Thorn (2016) and Thorn and Schurz (2019); see Sterkenburg (2019, 2020), Douven (in press), and Schurz (in press) for discussion.

Then obviously we would be justified in relying on induction. In fact, doing so would appear rationally mandated in that case.

Taking his cue from work on prediction with expert advice (in particular, Cesa-Bianchi & Lugosi, 2006), Schurz first shows that meta-induction, understood as induction over object-inductive methods, does have an analytic justification, consisting of a proof that, in all possible circumstances, the meta-inductive strategy that aggregates the predictions of the available object-inductive methods according to their registered successes is at least as accurate as and under weak additional assumptions even more accurate than any of those object-inductive methods. Whereas this gives us a justification only of meta-induction, the justification of object-induction now follows meta-inductively from the fact that object-inductive methods are known to have served us better than any non-inductive methods have. It is important to be clear about how this proposal differs from the naïve inductive justification of induction, which was already considered by Hume and rejected by him and most authors since. Although the naïve justification also starts from the registered successes of our inductive methods, it then applies *object*-induction to *object*-inductive methods. That presupposes an antecedent warrant for applying object-induction, but whether we have that warrant is precisely the question at issue. By contrast, Schurz applies *meta*-induction, which is *provably* optimal, to *object*-inductive methods. No circularity is involved in this argument.

Schurz's work on induction marks a major step forward, and I do think that it can be instrumental in defending abduction as a rational mode of inference. Granting this, there is still a concern about the *scope* of an inductive justification of abduction. Such a justification presupposes that we can agree on what counts as a successful application of the rule. In the case discussed in section 2.1 of the head and headless body that were found close to each other in Jackson, Mississippi, the medical examiner's report confirming what many had already inferred could be regarded as an undisputed piece of evidence for the reliability of abduction. But abduction is also used to infer the existence of unobservable entities—as in Johnson's "discovery" of the electron—and unobservable processes and mechanisms, as in Darwin's "discovery" of the mechanism of variation and selective retention. I put scare quotes around "discovery" in the previous sentence precisely because scientific antirealists will deny that we are in a position to claim that the electron has been discovered, or that the mechanism of variation and selective retention has been discovered.

More generally, scientific antirealists will want to remain agnostic regarding existence claims for any unobservable entities or processes. It would seem that, as a result, we can perhaps inductively justify applications of abduction as long as those applications concern strictly observable matters, but that we could not go beyond that without facing the charge of question-begging. For surely antirealists will argue that we are already relying on abduction if we cite any evidence presupposing the existence of unobservables as supporting the reliability of abduction also when it pertains to hypotheses postulating the existence of unobservables. Worse yet, by much the same reasoning, the Cartesian skeptic could challenge even the alleged evidence in the head-in-Jackson case, given that it already presupposes the existence of an external world, which the skeptic will refuse to grant.

This is a *serious* concern, insofar as the prime philosophical use of abduction has been in the context of underdetermination claims, discussed in section 2.3. And both the skepticism debate and the scientific realism debate were said to turn on such claims. The scientific antirealist is willing to grant a greater epistemic access than the skeptic, but both hold that it is more limited than people ordinarily think: the data we do have access to (sensory impressions in the case of the skeptic and data about observables in the case of the scientific antirealist) do not allow us to draw any conclusions about an external world (according to the skeptic) or about the unobservable part of the world (according to the scientific antirealist), precisely because, to their eyes, any such conclusion would be underdetermined by the evidence—that is, by what, from their perspective, can be legitimately designated as such. That is where opponents have always wanted to wheel in abduction: perhaps there are rival hypotheses that are equally compatible with the evidence, but the external world hypothesis, and scientific realism, offer better *explanations* of the evidence. Even if that is so, however, the skeptic and the antirealist will remain unfazed as long as they see no grounds for believing abduction to be a reliable rule of inference, and the question is how we are to show, without begging any questions against either the skeptic or the antirealist, that it is.

We will return to this question in chapter 8. Right now we are still facing a more fundamental challenge. For the broadly Bayesian arguments considered in chapter 4 charge abduction not just with having limited applicability but also with being inapplicable *tout court*, on pain of irrationality. We saw that the arguments failed: they proceeded by claiming that abductive reasoning could be costly but then forgot to ask whether it might be able to deliver goods that are worth the costs (granting the costs are real to begin with). At the same time, we have only scratched the surface when it comes to whether we can indeed get benefits from abduction that we cannot get from Bayes's rule nor perhaps from any other rule, and more broadly, whether it could ever be rational to reason abductively. I now want to turn to these questions.

It is important, however, to be clear about the notion of rationality that I assume in the following. Thinking about rationality was long dominated, and in philosophy still *is* dominated, by the assumption that norms of rationality should be *universal*: they should apply to everyone alike in every imaginable situation. The debate then was about what those universal norms were. As explained in section 3.2, the once popular view that the norms of rationality are the rules of logic was replaced by the psychologically more plausible Bayesian view that sees rationality as essentially about managing uncertainties. Although this was a step forward, various authors have argued that the Bayesian ideal of rationality, as formalized by the principles stated in section 3.2.1, is still woefully unrealistic. In particular, it has been argued that this ideal is unachievable for limited beings like us (see, e.g., Giere, 1988, ch. 6; Arkes, Gigerenzer, & Hertwig, 2016) and that it is not even clear what steps could be taken to *approach* it (Earman, 1992, p. 56). Such concerns have led various researchers to doubt the validity of Bayesianism as a basis for human rationality and to pursue alternatives more closely in touch with evolution-oriented theories of cognition.

This alternative approach came to fruition most prominently in the development of Simon's (1982) celebrated theory of bounded rationality and Gerd Gigerenzer and collaborators' related theory of ecological rationality (e.g., Gigerenzer & Goldstein, 1996; Gigerenzer et al., 1999; Gigerenzer, 2000; Goldstein & Gigerenzer, 2002; Gigerenzer, Hertwig, & Pachur, 2011; Todd & Gigerenzer, 2012; Todd & Brighton, 2016; Schurz & Hertwig, 2019). A more recent proposal in this vein is Elqayam's (2011, 2012) account of grounded rationality.[4]

Although these accounts differ in their details, they have the important commonality of taking into consideration the various biological and cognitive limitations to which humans are subject, as well as the environment or environments in which we, both individually and socially, operate and pursue

4. For a related account, called "resource rationality," see Griffiths, Lieder, and Goodman (2015).

our personal interests. The said limitations impose constraints on rationality that are to some extent universal—we are *all* finite beings, with finite computational powers, finite memory capacity, and so on—but the environment in which we are to act is not the same for everyone and may be different for each of us at different points in time; the same holds for the cognitive tools available to us and for the goals that we pursue. Gigerenzer and colleagues and also Elqayam argue that, on their accounts, there can be no universally applicable rationality standards that we might be able to pin down a priori. Whether a person's behavior qualifies as rational depends on whether the behavior facilitates achievement of the person's particular goal or goals in the context at issue and given the resources available to that person in that context.

As Elqayam has pointed out, the term "behavior" in the present conception of rationality is to be understood broadly, so as to include *cognitive* behavior. Concomitantly, the person's goals may be partly or wholly cognitive ones. She thus sees her notion of grounded rationality as in accordance with Jonathan Evans and David Over's (1996) proposal to conceive of instrumental (pragmatic, goal-oriented) rationality as primary and as subsuming epistemic rationality, given that among a person's goals may be obedience to certain systems of epistemic norms (like classical logic or the laws of probability).[5] But grounded rationality goes beyond a mere allegiance to instrumental rationality in that its emphasis on the role of context and on individual differences strongly implies that "instrumental rationality cannot be reduced to any one-size-fits-all normative framework" (Elqayam, 2012, p. 46; also Elqayam & Evans, 2011).

Adopting this more psychologically oriented approach to rationality makes it even clearer that as argued in chapter 4, the dynamic Dutch book argument against non-Bayesian update rules cannot simply be dismissed on the grounds that it deals with the wrong notion of rationality: epistemic rationality is a *species* of pragmatic rationality. It is then equally clear, however, that a verdict about the rationality of a given update rule cannot be reached without considering possible users of such rules *as located in a given context, with their specific goals in that context and their specific cognitive capacities.* Only then will it make sense to judge the use of a particular update rule as

5. That epistemic grounds are a special type of instrumental grounds has also recently been argued in Steglich-Petersen and Skipper (2019) and, from the standpoint of social epistemology, in Dyke (in press).

rational, or as more rational than the use of some alternative rule, and the judgment will have to be based on a cost–benefit analysis of the rule or rules within that context. To put this another way: on the present conception of rationality, the question of whether a given update rule is rational or not cannot be answered in the abstract. Given a person and an environment, we can ask which rule is most likely to help the person succeed; and given an update rule, we can ask when and for whom it is most likely to be helpful.

Hal Arkes, Gerd Gigerenzer, and Ralph Hertwig (2016, p. 33) are very specific about how one should investigate the rationality of a cognitive strategy. Their proposal is to begin by identifying (1) the goal of an individual or a group; (2) the strategies available to the individual or group for achieving that goal; and (3) the structural properties of the individual's or group's environment. In a second step, one then determines which strategy or strategies are most likely to help the individual or group achieve her / its goal in the given environment. Note how different this procedure is from trying to lay down rationality criteria a priori.

6.3 Simulating Explanatory Reasoning

To show that updating via some probabilistic version of explanatory reasoning can be rational, possibly more so than updating via Bayes's rule, I follow the steps from Arkes and colleagues' proposal. More concretely, I define an environment, agents with a goal in that environment, and a variety of update rules available to the agents and then use computational simulations to determine which rule or rules is / are most effective in realizing the agents' goal in that environment.

6.3.1 Setup

In chapter 2, I stated EXPL as a schematic probabilistic explication of abduction, which gives some positive bonus to best explanations, though I have used the label mostly to designate what is actually only a specific instance of the schema with that label, the instance in which the explanation bonus c was 0.1. From now on, I use the label again to refer generically to instances of the schema with $c \in (0, 1)$ and so to instances that can be said to embody some probabilistic form of explanatory reasoning. To these we add the instances of two further schemata, inspired by the finding from Douven and Schupbach

(2015b) that Good's and Popper's measures of explanatory goodness (stated in section 3.3) were useful in predicting participants' responses from one of the studies reported in Douven and Schupbach (2015a).

Any instance of EXPL gives all credit for explanatory goodness to the hypothesis that explains the evidence best and gives nothing to any other hypothesis. The aforementioned objective measures of explanatory goodness can help to give content to the claim that probabilistic versions of abduction may instead credit each hypothesis separately, in proportion to that hypothesis's explanatory power. EXPL was seen to work very much like Bayes's rule except that it adds a bonus point to the best explanation and then normalizes to obtain probabilities again. We cannot quite follow this procedure if we want to credit all hypotheses separately, at least not if the crediting is to be done on the basis of Good's or Popper's measure. In particular, given evidence E and a set $\{H_i\}_{i \leqslant n}$ of mutually exclusive and jointly exhaustive hypotheses, we cannot first update the hypotheses on E via Bayes's rule, then add $\mathcal{M}(H_i, E)$ to H_i (with $\mathcal{M}$ a measure of explanatory goodness), and finally normalize. The measures of objective goodness considered have a range of $[-1, 1]$ (see ch. 3, footnote 13) with 0 being the neutral point. This means that actually they can assign bonus points as well as malus points; where a hypothesis is an extremely poor explanation of the evidence, they can even assign a malus point of -1, which when added to the hypothesis's probability could result in a negative value unsuitable for "normalizing" to a probability.

Therefore, the rules to be considered will first update the hypothesis's probability via Bayes's rule, then calculate the hypothesis's explanatory goodness according to a given objective measure, next add or subtract a percentage of the hypothesis's probability in proportion to its explanatory goodness, and finally renormalize. More formally, the rules that we consider are instances of the following schema:

$$(6.1)\quad \Pr'(H_i) = \frac{\Pr(H_i)\Pr(E\,|\,H_i) + c\Pr(H_i)\Pr(E\,|\,H_i)\mathcal{M}(H_i, E)}{\sum_{j=1}^{n}\left(\Pr(H_j)\Pr(E\,|\,H_j) + c\Pr(H_j)\Pr(E\,|\,H_j)\mathcal{M}(H_j, E)\right)},$$

where Pr and Pr′ are as defined previously. What I call "Good's rule" is obtained from this by substituting Good's measure for $\mathcal{M}$, and what I call "Popper's rule," by substituting Popper's measure for the same expression. Note that, as stated, these rules are also mere schemata from which specific instances are derived by fixing a particular value for $c \in (0, 1)$, which determines what percentage of H_i's probability after a Bayesian update on E is added in proportion to this hypothesis's power to explain E. Here, too, we would obtain Bayes's rule by setting $c = 0$.[6]

The rules compared in the following—the available strategies, in the terminology of Arkes, Gigerenzer, and Hertwig (2016)—are Bayes's rule and instances of EXPL, Good's rule, and Popper's rule, where for all these instances $c \in (0, 0.25)$. The setting in which these rules are compared is, although stylized, not at all unrealistic. It concerns medical doctors at an intensive care unit (ICU) who try to determine what exactly is wrong with the patients who are rushed in, and who, on the basis of the information that they obtain from the tests administered, must choose an intervention. They are under some pressure to act fast: as time passes, the probability that the patient will die increases, and making the right intervention *decreases* that probability. On the other hand, making the wrong intervention *increases* probability of death.

Within this broad setting, we look at two more specific environments. In the first—the "Weibull environment"—we assume that for all patients brought in, probability of death can be roughly modeled by the cumulative density function (CDF) of some Weibull distribution. Such distributions are characterized by two parameters, a shape parameter k and a scale parameter l, and the associated CDF is given by

$$F(x; k, l) = \begin{cases} 1 - \exp(-(x/l)^k) & \text{for } x \geqslant 0, \\ 0 & \text{otherwise.} \end{cases}$$

More exactly, we assume that the probability of death for a given patient can be modeled in terms of an instance of this schema, where for each patient, k and l are chosen randomly, with $k \sim \mathcal{U}(0.5, 5)$ and $l \sim \mathcal{U}(50, 250)$. The resulting distribution is commonly designated as "Weibull(k, l)."

In the second environment—the "Gamma environment"—we assume instead that probability of death can be modeled by the CDF of some Gamma distribution. These distributions are also characterized by a shape parameter k

6. In terms of Vassend's (in press) distinction between inferential and predictive updating, Good's and Popper's rules are inferential update rules whereas EXPL is a predictive update rule. In the same paper, Vassend distinguishes between legitimate and nonlegitimate update rules, with Good's and Popper's rules falling into the former category and EXPL into the latter. It is to be noted, however, that he makes commutativity—in the sense that the order in which we evaluate different pieces of evidence should make no difference to the overall result—a necessary condition for legitimacy, and that is a debatable assumption (Lange, 2000).

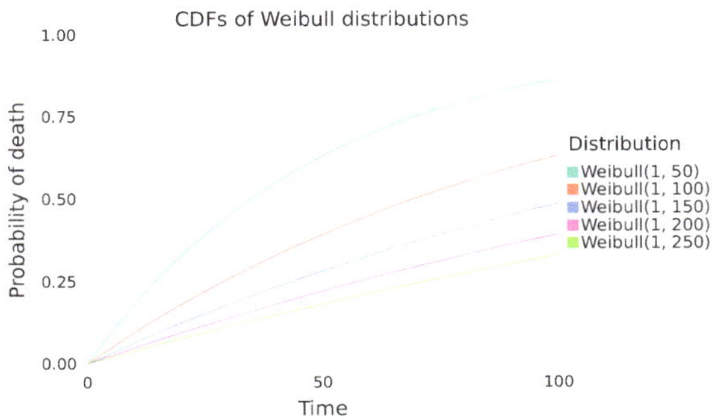

Figure 6.1: Examples of Weibull CDFs that give the probability of death of a patient as a function of time after admission into the intensive care unit.

and a scale parameter l, and the associated CDF is

$$F(x; k, l) \quad = \quad \frac{1}{\Gamma(k)} \gamma \left(k, \frac{x}{l}\right),$$

where the function Γ is a generalization of the factorial function and γ is the so-called lower incomplete gamma function.[7] The resulting distribution is commonly designated as "$\Gamma(k, l)$." Figures 6.1 and 6.2 show some examples of CDFs of Weibull and Gamma distributions, respectively.

Also associated with each patient are two parameters $a \geqslant 1$ and $b \geqslant 1$ that indicate the effect on probability of death of the right intervention and the effect on that probability of a wrong intervention, respectively. Where p_t is the probability of death of a given patient at time t, performing the right intervention at t lowers the probability of death to p_t/a whereas performing a wrong intervention at t raises that probability to $(b + p_t - 1)/b$. Figure 6.3 illustrates these effects for the case where $a = b = 2$, both for a specific Weibull distribution (left) and for a specific Gamma distribution (right).

We further assume that what is wrong with the patient can be expressed in terms of one parameter, α; we imagine that at the time the patient is rushed into the ICU her relevant medical status is known up to the value of this parameter. At that time, the only thing known about α is that it can take a value in $\{0, .1, .2, \ldots, 1\}$, with none of these values initially more likely

7. Specifically, $\gamma(a, x)$ is defined to be $\int_0^x t^{a-1} e^{-t} \mathrm{d}t$.

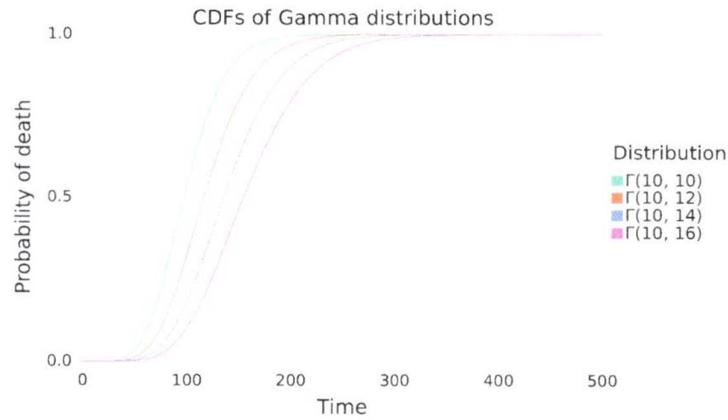

Figure 6.2: Examples of Gamma CDFs that give the probability of death of a patient as a function of time after admission into the intensive care unit.

than any other. To estimate the value of α, the doctor has to rely on the test results that she receives, with one new result coming in per unit of time. The results are either "positive" or "negative," and the tests are probabilistically independent of each other and all have the same (unknown) probability of

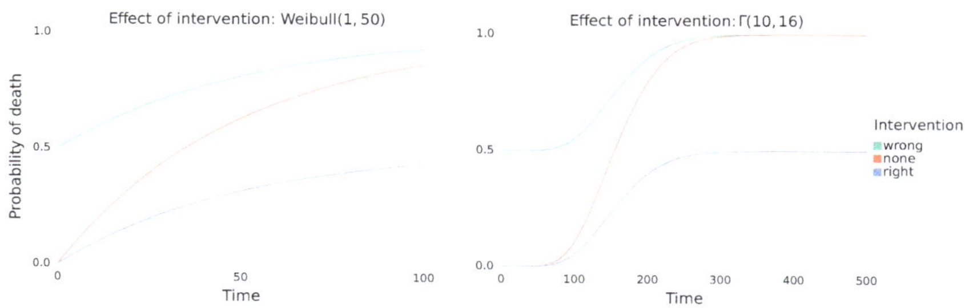

Figure 6.3: Examples of the effect of right and wrong interventions for a Weibull distribution (*left*) and for a Gamma distribution (*right*). In both panels, the orange graph is the probability of death of the patient over time if no intervention is performed; the green graph gives, for every point in time, the probability of death of the patient if at that point in time a wrong intervention is performed; the blue graph does the same for the correct intervention.

being positive. The hypothesis that $\alpha = x$ is to be interpreted as implying that the probability for any given test turning up positive equals x.

Stephen Pauker and Jerome Kassirer (1980) and Benjamin Djulbegovic and colleagues (2014) found that the so-called threshold model, according to which physicians should decide to administer treatment precisely if the probability of disease is above a specified threshold, accurately predicts decision making in clinical practice (see also Djulbegovic & Elqayam, 2017). In line with this finding, we assume that the doctor must be at least 90 percent certain about a hypothesis before she is prepared to intervene. It is further stipulated that only if the doctor comes to believe the true hypothesis to a degree above that threshold will she perform the correct intervention; coming to believe a false hypothesis to a degree above the threshold will lead to an incorrect intervention, with any incorrect intervention having the same detrimental effect (as specified by the parameter b) on the probability that the patient will survive.

By now, we have described all structural properties of the types of environment in which the doctor is to operate as well as the various strategies (update rules) available to her in this context. Assuming that the doctor's goal is to save her patients' lives, which rule should she use to update her degrees of belief for the various hypotheses concerning α (that $\alpha = 0$, that $\alpha = .1$, and so on)? And might the answer depend on whether she is in a Weibull or in a Gamma environment?

We saw earlier that in a similar probabilistic model, updating via one instance of EXPL tended to yield high-probability assignments to the truth faster than did Bayes's rule. On the other hand, we also noted that we should expect that instance of EXPL to have a higher error rate, in that it has a greater tendency to assign high probability early to a false hypothesis than Bayes's rule does; the argument given for that did not depend on the exact value of c (the bonus value assigned to best explanations) and so generalizes to all instances of EXPL. For reasons previously mentioned, speed of convergence and accuracy are both important in the doctor's setting: with every unit of time that passes, the probability that the patient will die goes up; but if the doctor acts upon a false hypothesis, her intervention will make the patient's prospects even worse. So the question just raised can be translated: Which update rule offers the best trade-off between speed and accuracy in the context at hand?

6.3.2 Method

We can think of this question as a constrained optimization problem.[8] Even though all relevant features of the setting have been characterized mathematically, this problem appears too complex for analytical methods to give much guidance. However, we can resort to an optimization technique known as "agent-based optimization," which is a form of genetic programming (Holland, 1975; Koza, 1992; Bäck, 1996; Yu, Yao, & Zhou, 2012; Kochenderfer & Wheeler, 2019, ch. 9). This technique takes its inspiration from the principles of natural selection and genetics, letting agents represent different solutions to a given problem, determining their "fitness" level (according to some criterion of fitness deriving from whatever problem needs to be solved), and then selecting, either deterministically or stochastically, the fittest agents, which are retained and / or allowed to reproduce in some predetermined way and thereby provide the input population for the next round, in which the competition for survival or reproduction starts again. This can be repeated over and over, possibly as often as is needed to obtain a fixed point at which all agents represent the same solution (see, e.g., Barbati, Bruno, & Genovese, 2012; also various papers in Czarnowski, Jędrzejowicz, & Kacprzyk, 2013). Agent-based optimization has been applied with notable success to optimization problems in a broad variety of fields, including chemistry, economics, medicine, operations research, psychology, and robotics. (See, for instance, Sarkar & Modak, 2005; Dhanalakshmi et al., 2011; Heris & Khaloozadeh, 2011; Douven, 2019b; and, for a general overview, Coello Coello, Lamont, & Van Veldhuizen, 2007, ch. 7.) As subsequently shown, using this technique has the additional advantage of allowing us to shed some light on how evolution may have contributed to shaping our inferential practices and in particular may have led to the adoption of certain forms of explanatory reasoning.

I have conducted the same agent-based optimization procedure for each of the two environments introduced previously. I first give an informal description of this procedure. The procedure started with a population consisting of 200 "medical doctors," with fifty of them using Bayes's rule, fifty using

8. *Constrained* optimization because we are not looking for the optimal update rule per se (optimal in the context) but the rule among those that we are *considering* that best fits the context. This makes the current approach different from Anderson's (1990, 1991) rational analysis, which starts with basically the three steps from Arkes and colleagues summarized at the end of the previous section and then adds as an additional step a requirement to look for the optimal procedure *tout court* for achieving the individual's or group's goal.

an instance of EXPL, fifty using an instance of Good's rule, and fifty using an instance of Popper's rule; for the non-Bayesian updaters, the value of the explanation bonus c was chosen randomly per doctor, with $c \sim \mathcal{U}(0, 0.25)$. Then each doctor was assessed on the basis of treating one hundred patients, where the relevant characteristics of the patients (survival probability, effects of right and wrong interventions, and value of α) were chosen randomly and separately per patient in the way specified above (so, for instance, in the Weibull environment, the survival probability was based on a randomly chosen Weibull distribution, with the parameters falling within the indicated bounds).

For each patient, the doctor had 100 units of time available. Per unit of time, the doctor received the outcome of a test, which was positive with a probability determined by the value of α as randomly chosen for the agent. At time 0, the doctor deemed all eleven value hypotheses for α equally likely, and probabilities were updated at each following time step on the basis of the test result received at that step and using the update rule associated with the doctor. The doctor intervened as soon as the probability for one hypothesis exceeded the threshold of .9. If that probability was assigned to the *true* hypothesis, the doctor received a score determined by the probability of death associated with the *right* intervention at the time the probability crossed the threshold; if the doctor assigned a probability above the threshold to a *false* hypothesis, she received a score determined by the probability of death associated with the *wrong* intervention at the time the probability crossed the threshold; and if *no* hypothesis was assigned a probability above the threshold during the 100 time steps, the doctor received the score of 1 minus the probability of death at the 100th time step.

After each doctor had treated her one hundred patients, her total score was calculated, and the 50 percent fittest doctors (the doctors with the highest average patient survival rate) were determined. These one hundred doctors were retained for the next generation, and they were also allowed to replicate by having a copy of themselves in that generation. In every simulation, the previous steps were repeated for 250 generations, and in total fifty simulations were run.

To make this formally precise, we think of doctors as ordered triples $\langle r, b, \Pr^{r,b} \rangle$, where r is one of the rules defined above; b is a bonus value, which is either 0, for Bayes's rule, or some random value in $(0, 0.25)$ for the other rules; and $\Pr^{r,b} = \langle \Pr_0^{r,b}, \ldots, \Pr_{100}^{r,b} \rangle$ is the sequence of the doctor's degrees-of-belief functions, starting with the degrees-of-belief function $\Pr_0^{r,b}$ that the doctor has when the patient is brought in, before receiving any test results, and with $\Pr_t^{r,b}$ being the doctor's degrees-of-belief function at time $t \in \{1, 2, \ldots, 100\}$, when she has received, and sequentially updated on, $\mathcal{E}_t = \langle E_1, \ldots, E_t \rangle$, with E_t the test result received at time t, and where $\Pr_t^{r,b}$ comes from $\Pr_{t-1}^{r,b}$ by updating on E_t via rule r with bonus value b (conventionally set to 0 in case the rule is Bayes's rule).

Patients can be thought of as ordered quintuples $\langle \alpha, k, l, a, b \rangle$ with all parameter values chosen randomly (within certain bounds) per patient. The first parameter, α, concerns what is wrong with the patient (think, for instance, of the value of some blood parameter), and it is unknown at the time when the patient is brought in; the only thing known at that time is that $\alpha = h$ for some $h \in \{0, 0.1, \ldots, 1\}$, with each of those possibilities equally likely at that time; thus, $\Pr_0^{r,b}(\alpha = h) = 1/11$, for all h, r, b. The true hypothesis concerning α gives rise to the evidence (the test results) that the doctor receives during the time of the treatment, in the precise sense that $\alpha = h$ if and only if the long-run frequency of positive outcomes equals $h \times 100$ percent. The k and l parameters are the shape and scale parameters of the Weibull distribution in the Weibull environment and of the Gamma distribution in the Gamma environment; as explained, these parameters determine the patient's probability of death as a function of time, modeled through the cumulative density function of the relevant distribution.

Finally, as previously explained, a and b determine the effect of the intervention (if any) by the doctor on the patient's probability of death. Where $p_t^{k,l}$ is the probability of death of a given patient at time t, as determined by k and l, performing the right intervention at t lowers the probability of death to $p_t^{k,l}/a$, and performing a wrong intervention at t raises that probability to $(b + p_t^{k,l} - 1)/b$.

A doctor intervenes at t if and only if at t her degree of belief in one of the hypotheses concerning α comes to exceed the threshold value of 0.9. If she intervenes, she makes the *right* intervention if what she believes in to a degree above the threshold is *true*, and a *wrong* intervention if it is *false*.

If the right intervention was made at time t, the doctor received the score of $1 - p_t^{k,l}/a$; if a wrong intervention was made, the doctor received the score of $1 - (b + p_t^{k,l} - 1)/b = (1 - p_t^{k,l})/b$; if no hypothesis was assigned a probability above the threshold during the 100 time steps, and so no intervention was made, the doctor received the score of $1 - p_{100}^{k,l}$. More exactly, given a patient characterized by $\langle \alpha, k, l, a, b \rangle$ and a doctor characterized by $\langle r, b, \mathrm{Pr}^{r,b} \rangle$, the doctor's score, s, for this patient equaled

$$s\big(\langle \alpha, k, l, a, b \rangle\big) \quad =$$

$$\begin{cases} 1 - p_t^{k,l}/a & \text{if } \exists h\, \exists t\, \forall t' < t\colon \ \big(\mathrm{Pr}_t^{r,b}(\alpha = h) > .9 \wedge \mathrm{Pr}_{t'}^{r,b}(\alpha = h) \leqslant .9 \wedge \alpha = h\big), \\ \big(1 - p_t^{k,l}\big)/b & \text{if } \exists h\, \exists t\, \forall t' < t\colon \ \big(\mathrm{Pr}_t^{r,b}(\alpha = h) > .9 \wedge \mathrm{Pr}_{t'}^{r,b}(\alpha = h) \leqslant .9 \wedge \alpha \neq h\big), \\ 1 - p_{100}^{k,l} & \text{otherwise.} \end{cases}$$

To repeat, after the doctor had treated her one hundred patients, her average score was calculated, and after this had been done for all 200 doctors, the one hundred doctors with highest scores were selected and copied and thereby came to constitute the population of the next generation, which went through the preceding selection process again, and so on, 250 times.

Computational details are not described here and are instead given in the Jupyter notebook that is part of the Supplementary Materials belonging to this chapter. The program for the simulations was written in Julia. In appendix E, I explain how the code for the simulations can be downloaded and used. Because Julia code reads almost like pseudocode, readers should be able to benefit from going through that code.

6.3.3 Results

I begin by describing the outcomes for the Weibull environment. Figure 6.4 gives a first impression of the kind of results obtained in this environment. It shows, for four randomly chosen simulations, how the population evolved through the 250 generations. One can think of each of the four plots as a series of 250 stacked histograms, where these histograms show how many tokens of each agent type were present in the corresponding generation.

We can immediately make a number of observations. Most notably, explanatory reasoning prevails in three of the four examples, in some form, and in the one example in which Bayesians (in the sense of doctors using Bayes's rule) are still in the running in the last generation, it appears that they are nevertheless losing the battle and might well have disappeared completely if we had not terminated the evolutionary process after 250 generations. And while there is no single explanation-based update rule that always trumps the others, EXPL users and Popperians do clearly better than users of Good's rule, who disappear quickly in all examples. Bayesians also disappear relatively

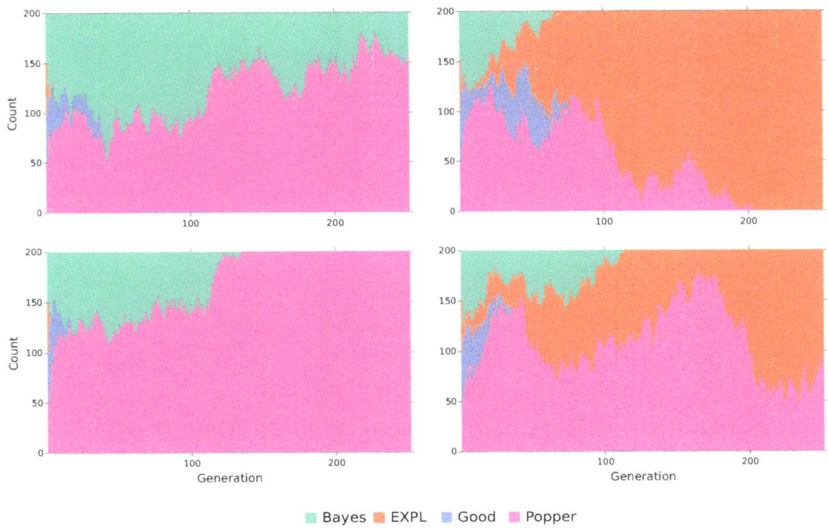

Figure 6.4: For the Weibull environment, counts of agent types per generation for four randomly chosen simulations.

quickly in three of the four examples. Also note that EXPL users appear to be critically endangered during the initial stages of all shown simulations. In the examples in which they perish, they do so *very* quickly, but where they make it through the initial stage, they become quite competitive, in one simulation even pushing out of competition all other types of agents.

As for the overall results, we first note that of all 50 × 200 = 10,000 agents in the last generations, 5,952 were Popperians, 2,246 were EXPL users, 1,554 were Bayesians, and 248 were users of Good's rule. Thus, users of Good's rule are indeed largely outcompeted into extinction. Furthermore, although among the explanation-based rules there is none that always comes to dominate the field, there is a clear sense in which Popper's rule is the winner, with EXPL being a distant second. The left panel of figure 6.5 plots the percentage of agents of each type of agent for every generation, averaged over all simulations. We note a strong tendency for the EXPL users to drop dramatically in number shortly after the start. We also see that their fate remains in serious jeopardy for some time after the start—if they do not perish completely, as the examples showed may happen. But when they manage to survive that

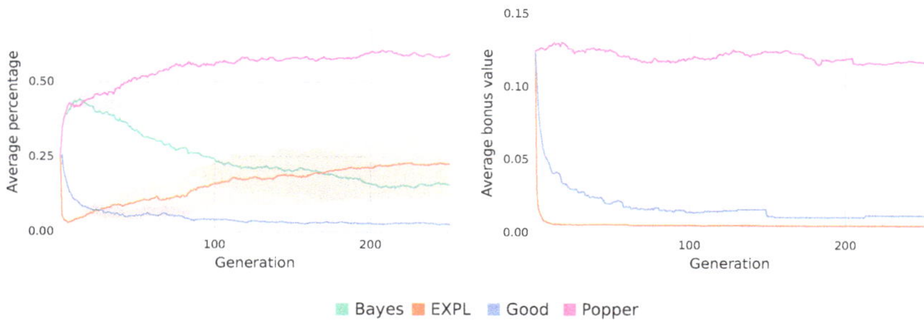

Figure 6.5: Percentage of agents per type of agents and per generation, averaged over the simulations for the Weibull environment (*left*), and mean bonus values per type of agents and per generation, averaged over the same simulations (*right*), both shown with 95 percent confidence bands.

critical phase, they are at least somewhat competitive against the Popperians. Bayesians appear competitive only for a very short while, after which they tend to fall back.

The right panel of figure 6.5 may suggest an explanation of why Popperians do better than EXPL users. This panel plots the mean bonus values in each generation for the three groups assigning such values. Because for the agents in all three groups the initial bonus values c were randomly drawn, with $c \sim \mathcal{U}(0, 0.25)$, all simulations start with an average bonus value of around 0.125 for each group. We see that the evolutionary process almost immediately drives down the average bonus value associated with the EXPL group to a level at which, as the plot also shows, it then stays for all further generations. Popperians, on the other hand, appear to get by with an average value close to 0.125: last-generation Popperians assign bonuses that are on average still close to the initial average value. So, Popperians may have an early advantage over EXPL users for which the latter never can really make up.

Turning to the Gamma environment, figure 6.6 shows how the evolutionary process unfolded in four randomly chosen simulations conducted in this environment. In these four examples convergence occurs very fast, in that one type of agents becomes dominant after fifty or even fewer generations. We further see that in these examples explanatory reasoning *always* prevails, with EXPL outcompeting all others in three of the four simulations. On the other hand, in one simulation EXPL users disappear again shortly after the

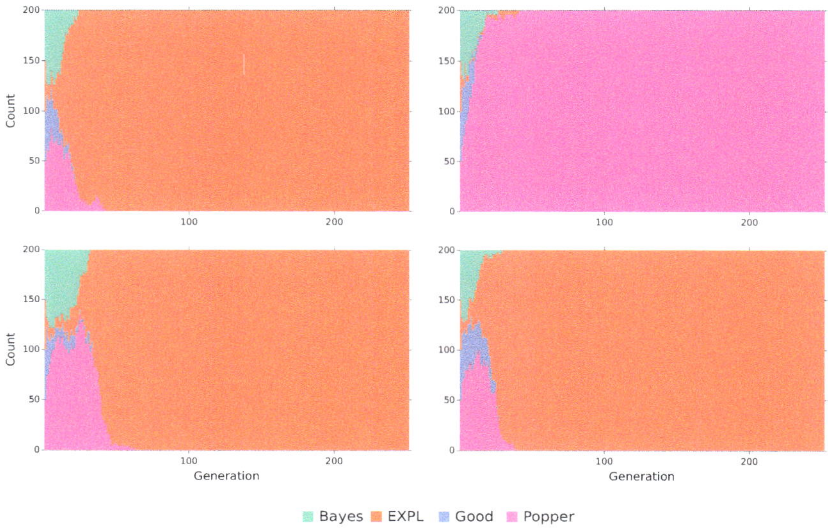

Figure 6.6: For the Gamma environment, counts of agent types per generation for four randomly chosen simulations.

start, leaving the field open to the Popperians. Even in the examples in which they end up as winners, EXPL users struggle at the beginning, similarly as in the previous simulations.

An analysis of the overall results shows that of the 10,000 agents in the last generations literally none are Bayesians or users of Good's rule. EXPL users form a clear majority, with a total presence of 7,554 agents in the last generations, the remaining 2,446 agents being Popperians. Thus, here EXPL is the clear winner, with Popper's rule being a distant second. The left panel of figure 6.7 gives a more complete picture, showing that the observations just made about the examples in figure 6.6 generalize: EXPL users appear to hang on by their fingertips for a short while, but if they are able to hold on, they tend to beat the entire competition. By contrast, although Bayesians and users of Good's rule do a bit better on average than EXPL users during the initial phase of the evolutionary process, they all go down the drain not long after.

In the right panel of figure 6.7 we see that the explanation of why EXPL users are endangered at first may again be that they have to drive their average

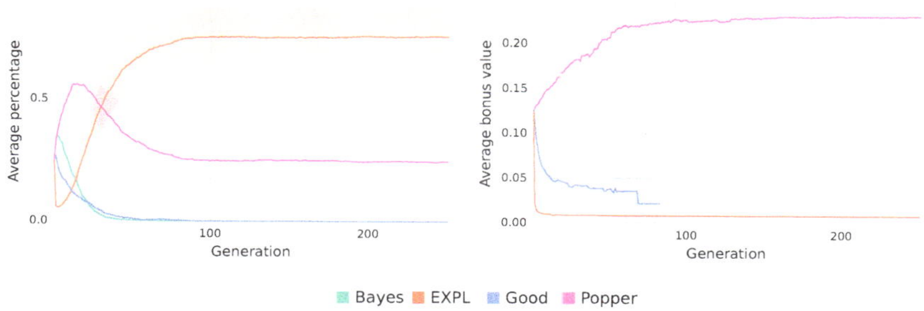

Figure 6.7: Percentage of agents per type of agents and per generation, averaged over the simulations for the Gamma environment (*left*), and mean bonus values per type of agents and per generation, averaged over the same simulations (*right*), both shown with 95 percent confidence bands.

bonus value down in order to become competitive. The same plot suggests an explanation of why in this environment Popperians do *not* outperform EXPL users: whereas in the Weibull environment the former's initial bonus values were fine from the start, at least on average, in the current environment their struggle for existence leads them to drive their bonus values up by, again on average, almost as much as the EXPL users drive theirs down. That the Popperians go through this process of adjusting their bonus values more slowly than the EXPL users may be why the former do worse, from an evolutionary viewpoint.

Next, let us compare the fitness levels—the scores on the fitness function, which measures the likelihood of patient survival—of the optimal solutions in the two environments with the corresponding fitness level of a Bayesian doctor. In the Weibull environment, the winner in the previous simulations was Popper's rule. Assuming that same environment, we run 10,000 simulations in each of which a Popperian treats one hundred patients, where the doctor always assumes the mean bonus value for last-generation Popperians in the previous simulations (which was .117, ± 0.021). We repeat this procedure with a Bayesian in place of a Popperian. And then we repeat all the foregoing but now assuming a Gamma environment and replacing the Popperian with an EXPL user, given that our simulations showed that EXPL users do best in this environment, and letting the EXPL user assign as a bonus to best explanations the mean bonus value for last-generation EXPL users in the said simulations (.008, ± 0.001). In the new simulations in the Weibull environment, the Popperian doctor turns out to save an average of 84.1 (± 2.3) percent of the patients she treats, and a Bayesian doctor saves an average of 84.0 (± 2.3) percent of the patients she treats. Conducting a one-tailed t-test shows the difference to be significant ($t = 3.13, p < .001$), although the effect size is small (Cohen's $d = 0.04$). In the new simulations in the Gamma environment, we find that the EXPL user saves an average of 88.9 (±2.3) percent of her patients, and her Bayesian colleague saves an average of 88.1 (± 2.2) percent of patients. A one-tailed t-test shows the difference again to be significant ($t = 24.15, p < .0001$), although here, too, the effect size is small (Cohen's $d = 0.34$).

Although the small effect sizes could make these results look rather unexciting, consider that, on average in a Weibull environment, of every 10,000 patients admitted to the ICU, a doctor of the type that we found to do best in this environment using a bonus value that is optimal for the environment would save ten extra lives as compared to a Bayesian doctor. In a Gamma environment, the corresponding number of extra lives saved by the type of doctor best suited to operate in this environment is eighty. Numerically, these may not be large differences. Yet it is hard to imagine a doctor who would *not* in the environments at issue prefer Popper's rule and, respectively, EXPL (with suitable bonus values) over Bayes's rule—or to imagine a patient who would *not* prefer to be treated by a doctor using an optimal rule for the given environment.

Also note the following: we may conceive of the evolutionary algorithm that we used as only a means for answering the question of which rule would be ecologically most rational in the specific environments that we considered—the answer being that a doctor would be best off on average if she updated by Popper's rule in a Weibull environment and by EXPL in a Gamma environment, using the appropriate bonus value for the given rule in the given environment, and that therefore it is rational to use Popper's rule in a Weibull environment and EXPL in a Gamma environment. However, we can also think of the evolutionary algorithm as modeling somewhat realistically how, under evolutionary pressure, agents have come to prefer the use of this or that update rule in specific types of environment because of the relative competitive advantage that the rule offers. To reiterate a frequently made point, from an evolutionary perspective small differences may be as advantageous as larger ones: often enough, a prey animal can escape a predator by being just a *little* faster than some other individuals in its group. That, I submit, best explains why Popperians do so much better than Bayesians in a Weibull setting, even though a Popperian doctor is only *slightly* "fitter" than her Bayesian colleague.

The broader evolutionary perspective further suggests that perhaps the most important advantage of the explanation-based update rules is that they offer an opportunity for adaptive learning—by increasing or decreasing the bonus for explanatory goodness—that Bayes's rule does not provide. How much a benefit this can be is especially clear if we focus on Bayesians versus EXPL users as represented in figure 6.5. Whereas the latter do quite a bit worse at the beginning, they can adapt to the needs of the context (as a group, not at the individual level, at which bonus values are fixed). As the figure shows, they were usually able to do this fast enough to avoid extinction and even to outnumber their Bayesian competitors. This is important, inasmuch as long-term survival is typically not just a matter of being well adapted to the environment in which one operates but also of being able to adapt quickly to a new environment.

The point of the foregoing is not that one should always prefer as an update rule some instance of Popper's rule in a Weibull environment and some instance of EXPL in a Gamma environment. Instead, it is to buttress Elqayam's previously quoted remark that it would be a mistake to expect a one-size-fits-all norm of rationality. According to the advocates of Bayes's rule, we are to follow it in each and every context. Not doing so would make us irrational, as the dynamic Dutch book and inaccuracy-minimization arguments are supposed to show. To refute this claim, it is enough to specify some context in which we are better off by following a non-Bayesian update rule. That leaves open the possibility that there are contexts in which Bayes's rule helps us achieve whatever it is we want to achieve more quickly or reliably or efficiently than would any other rule—which is fine, given that my aim was to show that in *some* contexts using a form of explanatory reasoning is ecologically more rational than using Bayes's rule and hence that Bayes's rule is not defensible as a *universally* valid principle of rationality; my aim was not to show that Bayes's rule is defensible under *no* circumstances, and that we should *always* adhere to some explanation-based update rule. Note that it is therefore not a flaw of the previous simulations that they relied on assumptions (about the characteristics of the patients, the time available for intervention, the threshold of certainty for action, and so on) that were to some extent arbitrary.[9] That would be a problem only if my aim had been to argue for a replacement of Bayes's rule by a different supposedly universally valid update rule.

6.4 Concluding Remarks

Chapter 3 canvassed evidence indicating that the way people change their degrees of belief is influenced by explanatory factors and that this may cause them to violate Bayesian norms of reasoning. Why would they do that, if Bayes's rule has all the virtues its proponents allege it to have? The standard arguments in support of Bayesianism suggest an answer in terms of a bias, of something regrettable, even if perhaps not completely avoidable.

But upon close scrutiny in chapters 4 and 5, those arguments turned out to be fallacious. The key observation was that even granting that failure to minimize next-step inaccuracy and / or Dutch bookability are bad, nothing follows about abduction unless it has been ruled out that some compensation can be had in return. In this chapter, I have argued that people's tendency to rely on explanatory considerations in their belief updating can have clear advantages indeed and that, more generally, abductive reasoning is rational in some environments—rational in a sense that lets us focus on how well a cognitive strategy (e.g., an update rule) is adapted to the local environment in which it is deployed rather than on how well the strategy complies to certain internal standards such as consistency or probabilistic coherence.

The problem for the traditional Bayesian approach to updating, and to rationality in general, that we have aimed to highlight is not that it posits an ideal that we would not even know how to approximate. Instead the problem is that, at least in some situations, the use of Bayes's rule is simply *not* ideal. The computer simulations that were reported gave a straightforward example of this: in some contexts, use of Bayes's rule led to the survival of fewer patients, on average, than the use of a version of abductive reasoning. It was also seen that what helped some of those versions come out on top in the given contexts is that they are highly adaptable by having an adjustable parameter, which allows for contextual fine tuning. Bayes's rule lacks this functionality. More generally, because we inhabit an ever-changing world and can within a relatively short time span find ourselves in contexts that

9. However, see footnote 10 in chapter 4 on why .9 is a reasonable threshold value.

pose very different challenges, reasoning should not just be well adapted to the context that we are in at a given point in time; it should also be easily *adaptable* to whichever context we may, just moments later, find ourselves in (Schurz & Thorn, 2016). Insisting on the unique rationality of Bayes's rule would make us needlessly inflexible. If we are instead permitted to take on board instances of EXPL, Good's rule, Popper's rule, and other rules perhaps (including Bayes's rule), that would greatly facilitate quick adaptation, by switching from EXPL to Popper's rule, for instance, and / or by adjusting the explanation bonus.

It is fair to say that the approach to rationality that is prominent in much of the recent psychological literature, as we have been assuming in the preceding, has been concerned mainly with arguing for the ecological validity of so-called fast and frugal heuristics rather than that of the kind of higher-level cognitive principles that were compared in this chapter. A key insight of the work of Gigerenzer and his various collaborators is that the use of simple heuristics often leads to better outcomes than the use of more complicated principles of reasoning. However, nothing in Gigerenzer et al.'s work suggests that the notion of ecological rationality applies *only* to heuristics. Indeed, we showed in the previous section how a clear sense can be given to the notion of one update rule matching the environment better than another update rule.[10]

The main method used in this chapter was agent-based modeling. The framework used for the simulations is highly flexible and allows for almost endless variations. One could explore still further distributions for modeling probability of death, different ways of modeling the effects of an intervention, richer and more varied sources of information for agents to update on, and also different modes of agent reproduction (Bäck, 1996, ch. 2). Studying such variations provides a path to probing the *robustness* of update rules, which indicates how broadly applicable they are (Gigerenzer, 2001, p. 47). Part of the

10. Admittedly, the concerns of computational intractability that Gigerenzer and others have voiced over Bayesian principles would seem to apply equally to the explanation-based rules studied in this chapter; in fact, given that they appear slightly more complex than Bayes's rule (in view of the additional computational work needed to assign bonuses), the latter principles may give even more cause for concern in this respect. But for all that has been said, the formal rules that we have studied may in actuality be implemented by informal heuristics or may at least be approximated by them. And to shed light on the informal heuristics by means of computer simulations, there may be no alternative to relying on such formal approximations. (See Schurz & Thorn, 2016, and Schurz, 2019, for various other examples of the same approach within the paradigm of ecological rationality.)

conception of rationality that we have been assuming is that rational agents are able to pick the right tools for the right situation; just considering updating on incoming information, this may include knowing when best to use a particular instance of EXPL, for example, and knowing when to use Bayes's rule or some other rule still. But this is not to say that broad applicability is not a boon for update rules. If an update rule matches a large range of contexts, or at least matches it better than all other known update rules, then that offers agents a good opportunity to become proficient users of that rule and lets them benefit in new contexts from the experience that they have gained with the rule in earlier contexts.

Another variation one could consider is to bring in the type of meta-inductive reasoners canvassed by Schurz, who keep track of the success rates of other reasoners and use that information to inform their own updating (see section 6.2). Schurz (2019, ch. 6) introduces a variety of meta-inductive strategies and discusses what difference they may make to predictive performance. It would be relatively straightforward to implement and compare those strategies in the context of our evolutionary computations.

Finally, although the simulations involved groups of agents, these agents went about their business all on their own and at no point were in touch with others. That is an important limitation. Alvin Goldman (1999) compellingly argued that much of traditional epistemology had been misguided in ignoring social interactions among people; that our flourishing qua epistemic agents, but also more broadly, can be understood only from a social perspective by recognizing the importance of our being embedded in a community of interacting agents collectively invested in the pursuit of truth. As Goldman pointed out, our knowledge would be a fraction of what it is had we had to explore the world all on our own. In a similar vein, Schurz (2019, p. 193) remarks that individual learning is far more costly than social learning and that "[m]any unsuccessful trial-and-error steps are involved in individual learning that can be avoided by just being informed about the results of these steps."[11] In this view, the traditional restriction to individual thinkers was bound to yield a distorted picture of human epistemology.[12] Therefore, the

11. On the importance of social learning, see also Boyd and Richerson (1985), Caporael (2001), Brewer (2004), Boyd, Richerson, and Henrich (2011), de Ridder (2014), and Oaksford and Hall (2016).

12. The importance of social learning in relation especially to the ecological conception of rationality has also been repeatedly emphasized by Gigerenzer (e.g., in his 2000).

next chapter again compares Bayes's rule and the various probabilistic versions of abduction that we have examined, now in a social setting in which agents' updates can be influenced not only by the evidence they receive directly from the world but also by the views of other agents in their community.

7

The View from Social Epistemology

We ended the previous chapter by noting that there are distinctively social aspects to learning. Not only do we learn *from* others, as when we learn from our parents or our teachers, but we often can acquire new knowledge only if we join forces *with* others. This is true not only for everyday knowledge but also and perhaps in particular for scientific knowledge. Indeed, scientists who might want to work in relative isolation—in the manner of Newton or, more extremely, of the exceedingly reclusive Henry Cavendish—would today stand little chance of making any appreciable progress; modern science has become a community effort (Gribbin, 2002, p. 359). While this has been long recognized especially in the works of sociologists of science (e.g., Kuhn, 1962; Merton, 1973), the more general lesson that the communal aspects of knowledge acquisition call for a rethinking of *epistemology* sank in only after the publication of Goldman's seminal 1999 book on social epistemology.

Once it is appreciated to what extent that we, qua epistemic subjects, are dependent on our functioning as members of groups of agents bent on the same epistemic goal, it becomes natural to ask how we might best organize such groups, supposing that we want to optimize their performance. This is a broad question, of course. Even if we consider a well-delineated case, such as a group of academic researchers employed by the same laboratory, the question could concern topics ranging from strategies for organizing the workplace to after-work socializing or decisions about how to allocate grant money and other resources within a group (Kitcher, 1993). This chapter focuses instead on some of the most fundamental *epistemic* mechanisms that groups might decide to put into place or might be encouraged to adopt, for how their members ought to interact. In particular, it focuses on mechanisms for how group members should update on new evidence while also being informed about the changing belief states of others in the group, whom they know to acquire evidence flowing from a common source (all group members study, for example, the same economic phenomenon, or the same virus, or the same genetic mutation). To nobody's surprise, we are especially interested in comparing Bayesian and abductive versions of social learning.

The notion of *optimizing* group performance will also be understood in an epistemic sense, as relating to the question of how to get at the *truth* of whatever issue the group is working on. Whereas, as stated in chapter4, truth is generally regarded as the overarching epistemic goal, we are often forced to make a trade-off between speed and accuracy: we do want to get at the truth, but we also want to get there reasonably fast, which may mean that we should be content if we can quickly become *highly confident* in the truth or quickly get *close* to the truth.

Thus, the question to be investigated is how members of a group of agents should update on the receipt of new evidence in a setting in which they also have access to relevant beliefs of other group members, supposing that the group wants to strike the best balance between speed (getting at the truth fast) and accuracy (minimizing error rates). The main methodological tool to be used is again that of computational agent-based modeling, but now we also want to explicitly take into account the socio-epistemic interactions. To that end, we build on a model developed by Rainer Hegselmann and Ulrich Krause for studying opinion dynamics in groups of interacting agents focused on a common research question (Hegselmann & Krause, 2002, 2005, 2006, 2009; for related models, see Deffuant et al., 2000; Dittmer, 2001; Weisbuch et al., 2002; Pluchino, Latora, & Rapisarda, 2006; Semeshenko, Gordon, & Nadal, 2008; De Langhe & Greiff, 2010). In this Hegselmann–Krause (HK) model, as it is now commonly known, each agent holds at any point in time only a single opinion on an issue that is the same throughout the process. Later work has generalized the model to those in which agents could at each point in time have opinions on multiple issues (e.g., Lorenz, 2003, 2008; Riegler & Douven, 2009; Wenmackers, Vanpoucke, & Douven, 2012, 2014). In recent joint work with Sylvia Wenmackers (Douven & Wenmackers, 2017), I have presented an extension of the HK model in which each agent at each moment assigns *probabilities* to all members of a set of relevant possibilities.

Douven and Wenmackers's (2017) aim was to compare different update rules within a social setting. Specifically, they compared Bayes's rule with various instances of EXPL. One way in which I want to go beyond Douven and

Wenmackers's work is by considering, next to EXPL, the other formalizations of explanatory reasoning that we examined in the previous chapter. Another generalization concerns the way that an agent determines which of her group members she takes to be her *epistemic peers*, who in Douven and Wenmackers (2017) are those members whose probabilities she takes into account in computing her new probabilities.[1] Whereas Douven and Wenmackers used one particular metric for defining epistemic peerhood, in this chapter I want to consider others as well. Furthermore, I compare different ways in which group members may accommodate the information about their peers' probabilities, thereby going beyond previous research, Douven and Wenmackers having considered only the linear pooling of probabilities (Stone, 1961; for details, see the following).

Whereas the main conclusion from Douven and Wenmackers (2017) was that in a social setting explanatory reasoning can be superior to Bayesian updating both in terms of speed and in terms of accuracy, the present chapter aims at showing how we may be able to find an optimal combination of update rule, rule for determining peerhood, and rule for taking into account one's peers' probabilities, where optimality is still a matter of balancing speed and accuracy. I am again using evolutionary computing by letting combinations of the aforementioned kind compete for survival and reproduction, where in a first batch of simulations their success depends directly on how fast and accurate they are, and in a second batch, more aligned with the simulations reported in the previous chapter, on survival probabilities and so more indirectly on speed and accuracy.

While my main motivation continues to be a comparison of Bayesian and abductive updating, this chapter has a secondary motivation as well. In recent years, much has been made of the so-called wisdom of crowds effect, the often-reproduced finding that averaged group estimates of some parameter value tend to be closer to the truth not only than most of the individual estimates but even than expert estimates (Surowiecki, 2004; Page, 2007). At the same time, Jan Lorenz et al. (2011) report evidence showing that if group members

1. The intuitive idea behind distinguishing peers from non-peers is that the former are and the latter are not considered competent enough to make it worth taking their opinions into account. For background on the notion of epistemic peer, see, for instance, Christensen (2007, 2011, 2016), Douven (2010), Goldman (2010), Ballantyne and Coffman (2011, 2012), De Cruz and De Smedt (2013), De Langhe (2013), Vallinder and Olsson (2013), and De Cruz (2017).

are made aware of their fellow members' estimates, that will not only lead members to adjust their own estimates, but the various resulting adjustments also will tend to have a negative effect on the accuracy of the averaged group estimate. This might seem to give a reason for trying to *prevent* the kind of interactions among group members that were experimentally explored by Lorenz and coauthors. In a similar vein, Kevin Zollman (2007) found that while increasing the connectivity in an epistemic network tends to increase the speed with which the network converges on an opinion, it at the same time tends to *decrease* the chances that the network will arrive at the truth (see also Kummerfeld & Zollman, 2016, and Hahn, Hansen, & Olsson, 2020; but cf. Mason & Watts, 2012, and Rosenstock, Bruner, & O'Connor, 2017).

However, while these authors found a potential negative effect of epistemic interactions among the members of research groups, it is to be noticed that such interactions are not *bound* to be detrimental. To the contrary, as Douven and Wenmackers's (2017) results made manifest, whether the influence will have a positive or a negative effect on the accuracy of the group estimate may all depend on *how* the individual group members' beliefs are influenced by those of their peers. There are update procedures that take into account both new information about the parameter value at issue *and* information regarding the beliefs about that value held by others in the group, and which greatly increase not only the accuracy of the group estimate but also the speed with which that estimate converges to the true value, supposing that group members are repeatedly offered new evidence as well as new information about the beliefs of their peers. Groups might decide to adopt those update procedures to counteract the kind of effects of social influence discovered by Lorenz et al. (2011) and others. Naturally, that presupposes that such procedures have been identified. In the following, I describe in some detail how this could be accomplished.

7.1 The Hegselmann–Krause Model and Beyond

We look at groups of agents, each of whom repeatedly updates her belief state upon the receipt of new information. The new information is of two kinds: on one hand, agents receive new information directly about the subject matter of their research, while on the other hand, they receive new information about the belief states of others in their group after *they* received new "direct" information about that subject matter. All agents always first receive a piece of direct information—which may be different for different agents, even though the information comes from the same source—and then after each of them has updated on that, they receive information about the new belief states of their colleagues, which may occasion a further adjustment of an agent's own belief state. For the first part of this process, I resort to the same update rules that figured in the simulations reported in the previous chapter: Bayes's rule and instances of EXPL, Good's rule, and Popper's rule. The second part involves a variant of the previously mentioned HK model.

Computational agent-based modeling has become a mainstay in the social sciences, used to study a great variety of phenomena and processes—the emergence of social norms, racial segregation, voting behavior, and so on—that are too complex to be investigated using strictly analytical tools. The technique is also rapidly gaining popularity in the study of social aspects of knowledge and belief, fueled by the observation hinted at previously, to wit, that our being socially embedded in one or more communities of truth seekers is crucial to our epistemic success.[2]

Arguably, the most widely used computational agent-based model in this field is the HK model. Hegselmann and Krause themselves used the model mainly to address *descriptive* issues, such as the conditions under which initially diverging opinions in a community of interacting agents tend to converge or when they tend to polarize. More recently, the model has also been applied to a variety of *normative* issues, such as questions concerning reliabilism and truthful assertion (Olsson, 2008), peer disagreement (De Langhe, 2013), and efficient truth approximation in science (Douven & Kelp, 2011).

In the version of the HK model that is most of interest to us (though not in the original, most basic version of the model), communities of agents seek to determine the value τ of some unspecified parameter, where it is given that $\tau \in (0, 1)$. The agents update their estimates of τ repeatedly at discrete time steps, where each agent updates on the basis of (1) information that she receives about τ, and (2) the estimates of τ of those agents who are within her bounded confidence interval (BCI), meaning that their estimate is within some fixed distance ε of the agent's own estimate. More formally, the estimate

2. For some more background on computational agent-based models, see Douven (2019c).

of agent x_i after the $(u + 1)$-st update is given by

$$x_i(u+1) \quad = \quad \alpha \frac{1}{|X_i(u)|} \sum_{j \in X_i(u)} x_j(u) \; + \; (1-\alpha)\tau,$$

with $x_j(u)$ the estimate of agent x_j after the uth update,

$$X_i(u) \quad := \quad \{j : |x_i(u) - x_j(u)| \leqslant \varepsilon\}$$

the set of agents within agent x_i's BCI after the uth update, and $\alpha \in [0, 1]$ a parameter determining the weight that the agent gives to the "social" part of the updating relative to the "evidential" part.[3]

One reason why the HK model is so popular is its great flexibility. In general, it takes little effort to extend the model or to adapt it to a researcher's specific needs. For example, researchers have used simple extensions of the model to study situations in which agents receive evidence beset by some degree of measurement error (e.g., Douven, 2010; Douven & Riegler, 2010; De Langhe, 2013). Matthew Crosscombe and Jonathan Lawry (2016) present another extension to study agents with (partially) vague beliefs. The model has also been extended to situations in which agents are equipped with richer belief states in that they hold opinions on various related or unrelated matters (see, e.g., Lorenz, 2003, 2008; Riegler & Douven, 2009; Wenmackers, Vanpoucke, & Douven, 2012, 2014) and to ones admitting different types of agents, some of which may exhibit epistemically irresponsible behavior (Hegselmann & Krause, 2015; Douven & Hegselmann, 2021; Glass & Glass, 2021).

We are interested mostly in an extension of the HK model proposed in Douven and Wenmackers (2017). In this extension, agents' belief states at a given time are not single real numbers (as in the original model) but probability functions, defined on a set of self-consistent, mutually exclusive, and jointly exhaustive hypotheses $\Psi = \{\psi_i\}_{1 \leqslant i \leqslant n}$. It is still the case, however, that the agents take into account both evidence directly pertaining to the members of Ψ and the belief states (i.e., probability functions) of those in their BCI.

3. In this model, and also in Douven and Wenmackers's extension of it, subsequently discussed, all agents in the BCI are treated on a par. If information about the credibility or expertise of these agents were available, one might plausibly wish to weigh them differently (see, e.g., Tamargo et al., 2014, and Gottifredi et al., 2018). This can easily be accomplished in the framework used in this chapter (see footnote 4 in this chapter).

Naturally, in this model the BCI has a definition different from the previously stated one. Here, two agents are in each other's BCI precisely if the sum of the absolute differences in the probabilities that they assign to the hypotheses is below a given threshold value, or formally, agent i with belief state $\Pr_i$ is within the BCI of agent j with belief state $\Pr_j$ precisely if $\Delta(\Pr_i, \Pr_j) \leqslant \varepsilon$, where

$$\Delta(\Pr_i, \Pr_j) \; := \; \sum_{k=1}^{n} \left|\Pr_i(\psi_k) - \Pr_j(\psi_k)\right|.$$

Note that, given this definition, an agent is always in her own BCI (as is also the case in the original HK model and in the extensions mentioned previously).

The actual updating in Douven and Wenmackers's model then proceeds as follows: In a first, "evidential," step, agent j updates her probabilities on a new piece of evidence she receives, using either Bayes's rule or EXPL; that is, where $\Pr_j^u$ is her probability function after the uth update, she first forms $\Pr_j^{u_{\text{evid}}}$ via either Bayes's rule or EXPL. Then, in a second, "averaging," step, the agent averages over the probability functions of the agents in her resulting BCI, $X\left(\Pr_j^{u_{\text{evid}}}\right)$, which yields $\Pr_j^{u_{\text{av}}}$. Finally, she sets $\Pr_j^{u+1} = \Pr_j^{u_{\text{av}}}$.

The averaging takes place by a method now known as "linear pooling" (Stone, 1961), discussion of which goes back at least to Pierre-Simon Laplace's *Essai Philosophique sur la Probabilité* from 1814 (Bacharach, 1979). In this method, agents average by taking the *straight arithmetic average* of the probabilities of the agents within their BCI. Formally, for all $\psi \in \Psi$ and all $\Pr_k^{u_{\text{evid}}} \in X\left(\Pr_j^{u_{\text{evid}}}\right)$, it holds that

$$\Pr_j^{u_{\text{av}}}(\psi) \quad = \quad \frac{\sum_k \Pr_k^{u_{\text{evid}}}(\psi)}{\left|X\left(\Pr_j^{u_{\text{evid}}}\right)\right|}.$$

The resulting function again satisfies the axioms of probability theory (Stone, 1961).[4]

Douven and Wenmackers's (2017) explicit aim was to compare Bayes's rule with EXPL in a social setting. Their starting point was the finding from Douven (2013), discussed in chapter4, that in an individual setting (without

4. There exist weighted versions of this and the other pooling methods subsequently discussed (see, e.g., Genest & Zidek, 1986), which could be used in the kind of situation in which some agents are known to be more credible or more expert than others (see the previous footnote).

any averaging of probability functions taking place) EXPL tends to outperform Bayes's rule in terms of speed, but Bayes's rule tends to outperform EXPL in terms of accuracy. One could conclude from this that it is a matter of context which of these rules we ought to use (supposing that the choice is between only those two rules), given that typically we care about both speed and accuracy but sometimes more about one and sometimes more about the other.

But Douven and Wenmackers wondered whether a speed–accuracy tradeoff would be needed if agents were to change their probabilities not only on the basis of direct evidence but also as a result of social influence. To address this question, they ran computer simulations using the extension of the HK model just described, with groups of simultaneously and repeatedly updating agents who received evidence that although from the same source (i.e., the same probability distribution) for all agents, was randomly generated per agent. Douven and Wenmackers systematically studied different communities of agents, which differed along two dimensions, to wit, by the value for c that they substituted in EXPL (with agents setting c equal to 0 *de facto* amounting to Bayesians), and by the value of ε, that is, by their being more or less liberal in considering others as their epistemic peers.

The agents in Douven and Wenmackers's simulations all receive binomial data (i.e., independent and identically distributed binary data), the data points coming in one at a time, and their goal is to estimate on that basis the probability of "success." For the sake of concreteness, one may imagine the data to consist of the outcomes of repeated tosses of a coin with unknown bias, where this bias (for example, the probability of heads) can take eleven possible values, ranging from 0 (certain to land tails) to 1 (certain to land heads) in increments of .1. Initially, all these values are equally likely to indicate the true bias of the coin. Thus, at the start of the updating process, it is reasonable for the agents to invest equal confidence in all given bias possibilities. Importantly, while each agent in the community receives data from a different coin, it is known to every agent that all coins have the *same* bias. The community of agents is tasked to estimate this bias.

As said, the relevant dimensions of evaluation were speed and accuracy. The former was operationalized in terms of the percentage of the community that assigns a high probability to the truth (where "high probability" was understood as a probability above .9), and the latter in terms of the Brier score, which we already encountered a number of times.

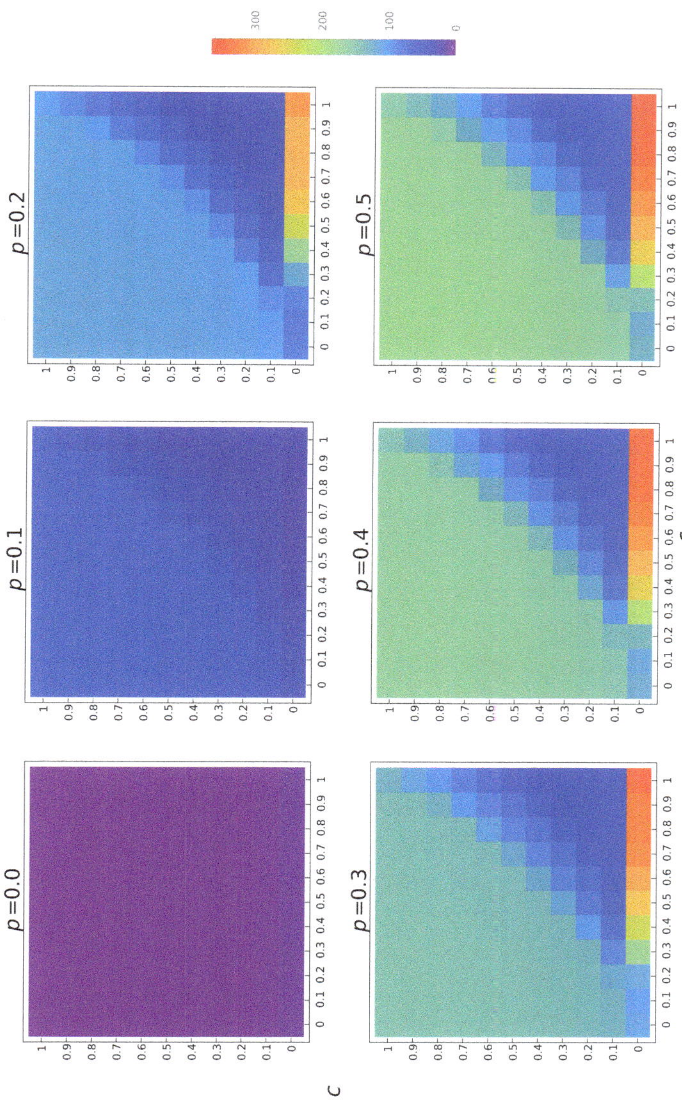

Figure 7.1: Average total Brier scores for six bias possibilities; $p = x$ indicates that the chance for each of the coins to come up heads equals x. (*Source*: Douven & Wenmackers, 2017.)

Douven and Wenmackers's simulations can be thought of as effectively comparing 121 (= 11×11) update procedures with each other, one for each combination of c and ε, with both in $\{0, 0.1, 0.2, \ldots, 1.0\}$. The communities that these authors considered all consisted of fifty agents, each of whom received per simulation the outcomes from 500 tosses. Given that as previous research had shown, results might depend on the specific bias of the coin, simulations were run for biases $p \in \{0, .1, .2, .3, .4, .5\}$.[5] For each of these bias possibilities, fifty simulations were run.

In all simulations, average Brier scores per agent were calculated after each update. Per simulation, these averages were then summed over all 500 tosses, and finally the average over the 50 simulations was calculated. The summarized results are shown in figure 7.1, which is reproduced from Douven and Wenmackers (2017). The bottom row of each of the six plots represents the communities of Bayesian updaters, which differ only by their willingness to count others as their epistemic peers; the square in the bottom-left corner of each plot represents communities of "pure" Bayesians, who attach no value to explanatory strength and are unwilling to be influenced by anyone else in their group. As Douven and Wenmackers observe, among the Bayesian communities the pure ones do best on the count of Brier scores: as clearly visible, Bayesians who *are* willing to be influenced by others incur higher Brier penalties on average. More significantly, *non*-Bayesian updaters who are willing to be influenced by others do better still than pure Bayesians.

With regard to speed, it had already been found that on average, and in the statistical model considered in the simulations, EXPL leads agents to assign a high probability to the truth faster than does Bayes's rule. This result was reproduced in the kind of social setting assumed by Douven and Wenmackers. Figure 7.2 shows graphically for different combinations of c and ε the average proportions of agents (averaged over the simulations) that assign a high probability to the truth, for the relevant bias possibilities, and after each toss (compare figure 6 in Douven & Wenmackers, 2017).

Douven and Wenmackers (2017) were interested in comparing Bayes's rule and EXPL in a social setting along the dimensions of speed and accuracy. They found that although in an individual setting the two rules appeared to have different strengths and weaknesses—Bayes's rule was more accurate

5. For symmetry reasons (see ch. 4, footnote 11), it would have been superfluous to run simulations for the other bias possibilities.

than EXPL, but EXPL was faster than Bayes's rule—this picture changed in a social setting, where EXPL appeared to be both faster and more accurate than Bayes's rule.[6]

It was known that in the kind of individual setting studied in Douven (2013), EXPL users typically had a lower Brier penalty than Bayesians but still tended to have a slightly higher *average* Brier score, because *when* they incurred a higher Brier score than Bayesians it tended to be *much* higher. The cause was that EXPL users tend to pick up the lead in the evidence faster than Bayesians do, which ultimately has the downside that they are more prone to follow *misleading* evidence. As Douven and Wenmackers (2017) showed, in a social setting the risk of being pushed off track by misleading evidence is much diminished due to the averaging mechanism that is called at each step. Specifically, if an agent receives misleading evidence, then the averaging part still brings about a kind of regression to the mean, thereby preventing the agent from straying too far from the true bias probability.[7]

6. Trpin and Pellert (2019) show that these results generalize to cases in which the evidence is uncertain. They effectively compare Jeffrey conditionalization (see section 3.2.1) with an explanation-based counterpart of that rule.

7. Bayesians might object that instead of averaging over the probability functions of their peers, agents should reason back from how those functions are updated to the evidence that the others received and then update their own probability function on the pooled evidence (this would amount to a supra Bayesian approach, in the sense of Keeney & Raiffa, 1976). As pointed out in Douven and Wenmackers (2017, section 5), however, this objection misses the purpose of the statistical toy model used in their paper and also here, which is meant to set up a simulation and not as the ultimate case of interest. In particular, we have no special interest in cases that are analytically tractable, in which all the evidence is at least in principle available to all the agents, and in which the only epistemic goal is to minimize expected Brier penalty. We are interested in learning rules that apply also to situations in which the model is known to be at best an idealization, in which, for instance, information regarding other agents' evidence is or may be incomplete and mediated by probability assignments, in which we have to estimate on how much independent data the assignment of each agent is based, and so on. More practically speaking, and also entirely in line with the view on rationality taken in the previous chapter, it is worth stressing that even in a context that fits the toy model perfectly, processing constraints and in particular working memory constraints may make the "reasoning back" strategy infeasible. This is most evidently true when we assume the updating to take place under time pressure, as is for instance eminently plausible in the kind of situation considered in section 7.4. Bayesians might respond that in such a situation, averaging over others' probability functions is equally problematic. That is not so, however. It has been long known that people can accurately estimate averages without carrying out the actual computations (Spencer, 1961, 1963; Beach & Swenson, 1966; Bulger, Hiles, & Lowe, 1969; Lovie & Lovie, 1976; Malmi & Samson, 1983; Lindskog, Winman, & Juslin, 2013a, 2013b).

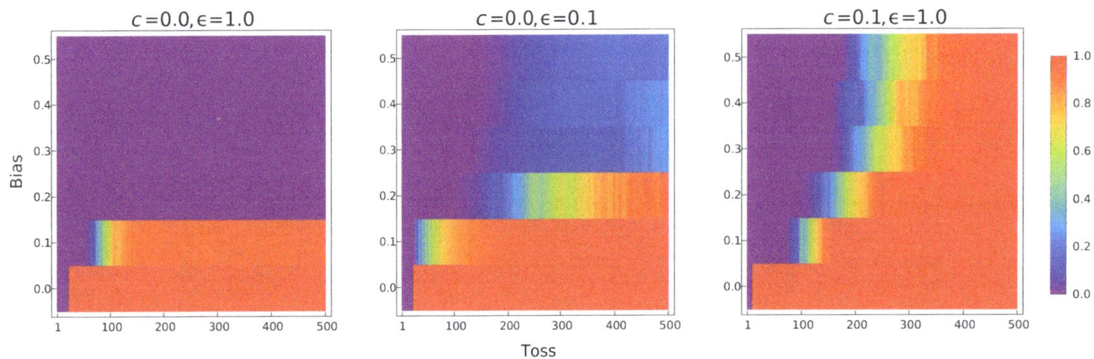

Figure 7.2: Proportions per toss and per bias of agents who assign a high probability to the truth, for specific combinations of c and ε; results are averaged over all simulations.

7.2 Further Generalizing Group Learning

We want to compare update rules in the context of determining how to epistemically organize a research group in order for that group to strike the best balance between being fast and being accurate. Part of the recommendation to a group might be to adopt Bayes's rule, or EXPL, as their members' way of accommodating direct evidence. First, as discussed in section 6.3, there exist other formalizations of abductive reasoning than EXPL, and these should certainly also be considered in a comparison. Second, Douven and Wenmackers (2017) let the agents in their simulations average via linear pooling, but one can think of other ways of merging probability functions, some of which might offer advantages in striking the best balance between speed and accuracy. Finally, we measured distances between agents' probability functions by adding up proposition-wise absolute differences between probabilities; this is the so-called Manhattan distance. Here, too, there are variants that could be tried and that if adopted by a group of agents to determine epistemic peerhood, might help that group to do better epistemically.

While the linear pooling of probability functions assumed by Douven and Wenmackers (2017) has been advocated by various authors (e.g., Aczél & Wagner, 1980; Lehrer & Wagner, 1981), others have expressed a preference for *geometric pooling* (e.g., Genest, 1984), in which one takes the geometric instead of the arithmetic average of probabilities (and then renormalizes, which becomes necessary).[8] Formally, the geometric average of numbers $x_1, x_2, \ldots, x_n$ is $\sqrt[n]{x_1 x_2 \cdots x_n}$. So, instead of setting for any update u and with $\Pr_k^{u_{\text{evid}}} \in X\left(\Pr_j^{u_{\text{evid}}}\right)$ and $\Psi = \{\psi_i\}_{1 \leqslant i \leqslant n}$ a set of jointly exhaustive and mutually exclusive self-consistent propositions,

$$\Pr_j^{u_{\text{av}}}(\psi_i) = \frac{\sum_k \Pr_k^{u_{\text{evid}}}(\psi_i)}{\left|X\left(\Pr_j^{u_{\text{evid}}}\right)\right|}, \qquad \text{for all } i: 1 \leqslant i \leqslant n,$$

8. In fairness, while there is strong empirical support for the thought that people can accurately estimate *arithmetic* averages (see footnote 7 in this chapter), I am not aware of any evidence indicating that they are good at estimating geometric or harmonic averages (the latter to be introduced shortly). So, what follows may be more of interest to, for example, roboticists studying opinion pooling (e.g., Lawry & Lee, 2020; Lee, 2020) than to cognitive scientists and social epistemologists (at least to those social epistemologists who share my conviction that we should occupy ourselves with developing theories of rationality for real people rather than elegant formal models that describe the behavior of agents bearing an at best vague semblance to ordinary mortals).

the proposal is to set

$$\mathrm{Pr}_j^{u_{\mathrm{av}}}(\psi_i) \quad = \quad a\,\sqrt[m]{\prod\nolimits_k \mathrm{Pr}_k^{u_{\mathrm{evid}}}(\psi_i)}, \qquad \text{for all } i:\ 1 \leqslant i \leqslant n,$$

with $m = \left|X\left(\mathrm{Pr}_j^{u_{\mathrm{evid}}}\right)\right|$ and with $a = \left(\sum_{i=1}^{n} \sqrt[m]{\prod_k \mathrm{Pr}_k^{u_{\mathrm{evid}}}(\psi_i)}\right)^{-1}$, to ensure that the probabilities assigned to the members of Ψ sum to 1 again.

For numbers $x_1, x_2, \ldots, x_n$, their harmonic average equals

$$\left(\frac{x_1^{-1} + x_2^{-1} + \cdots + x_n^{-1}}{n}\right)^{-1}.$$

I am not aware of any recommendations in the literature to aggregate probability functions via the harmonic average—"harmonic pooling," one might call this—nor can I think of good a priori reasons for using this procedure. However, neither can I rule out a priori that for the purpose of making group learning fast and accurate, using the harmonic average has an edge over using either the arithmetic or the geometric average. Thus, I also consider the following definition of an average probability function:

$$\mathrm{Pr}_j^{u_{\mathrm{av}}}(\psi_i) \quad = \quad a\left(\frac{\sum_k \mathrm{Pr}_k^{u_{\mathrm{evid}}}(\psi_i)^{-1}}{\left|X\left(\mathrm{Pr}_j^{u_{\mathrm{evid}}}\right)\right|}\right)^{-1}, \qquad \text{for all } i:\ 1 \leqslant i \leqslant n,$$

with a ensuring that the probabilities sum to 1.

With regard to measuring distances between probability functions—in order to determine who is an epistemic peer and who is not—the Manhattan distance used by Douven and Wenmackers (2017) is but one possibility. There are many alternatives. Here, we focus on just two, but readers are invited to use the computer code in the Supplementary Materials to experiment with other distances (see the following for details). The two considered are the Euclidean metric, according to which

$$\Delta(\mathrm{Pr}_i, \mathrm{Pr}_j) \quad := \quad \sum_{k=1}^{n} \left(\mathrm{Pr}_i(\psi_k) - \mathrm{Pr}_j(\psi_k)\right)^2,$$

and the Kullback–Leibler (KL) divergence, according to which

$$\Delta(\mathrm{Pr}_i, \mathrm{Pr}_j) \quad := \quad \sum_{k=1}^{n} \mathrm{Pr}_i(\psi_k) \ln\left(\frac{\mathrm{Pr}_i(\psi_k)}{\mathrm{Pr}_j(\psi_k)}\right).$$

Note that, supposing the KL divergence, because this is not symmetric it will not generally hold that $\Delta(\mathrm{Pr}_i, \mathrm{Pr}_j)$ is equal to $\Delta(\mathrm{Pr}_j, \mathrm{Pr}_i)$. However, in the present context, the asymmetry may actually be an advantage, given that in reality, respecting someone as one's epistemic peer is not always reciprocated.

We can treat the extension of the HK model from Douven and Wenmackers (2017) as a template, into which one rule from each of three groups can be plugged in, viz., one of four evidential update rules: Bayes's rule, EXPL, Good's rule, and Popper's rule; one of three averaging rules: linear pooling, geometric pooling, and harmonic pooling; and one of three rules for determining epistemic peerhood: the Manhattan metric, the Euclidean metric, and the KL divergence. Each of the resulting combinations of rules yields a schema, with slots for c and ε still to be filled in. Does it matter which combination of rules we choose and what values we set for c and ε?

It will probably come as no surprise that the answer is positive. This is easy to verify by rerunning the simulations reported in Douven and Wenmackers (2017) for other instances of the template. By way of illustration, consider figure 7.3, in which the plots summarize results from simulations in which linear pooling was replaced by geometric pooling, but which were otherwise the same as the simulations underlying figure 7.1. We see that this seemingly small difference leads to a *completely* different picture. Now, Bayesians perform very well on the count of accuracy, supposing that they are open to social influence, although on the condition of a greater openness to such influence (even if only slightly greater), EXPL users who assign *some* (not-too-large) bonus for explanatory goodness do well, too. By way of further illustration, figure 7.4 shows that we obtain a somewhat different picture if harmonic pooling is assumed and we replace the Manhattan metric by the Euclidean one.[9]

But accuracy, here measured by the Brier score, is only one dimension of comparison, speed being the other one, and the goal ultimately is to strike the best balance between them. More importantly, we are considering only an illustration here, which by no means provides a systematic answer to our question of how best to organize a group epistemically (which update rule to recommend to the group and which principles to recommend for regulating social influence on belief states).

9. The color scaling was set separately for the plots in figures 7.1, 7.3, and 7.4. Using the same scale in the three figures would have rendered the differences in figures 7.3 and 7.4 difficult to discern.

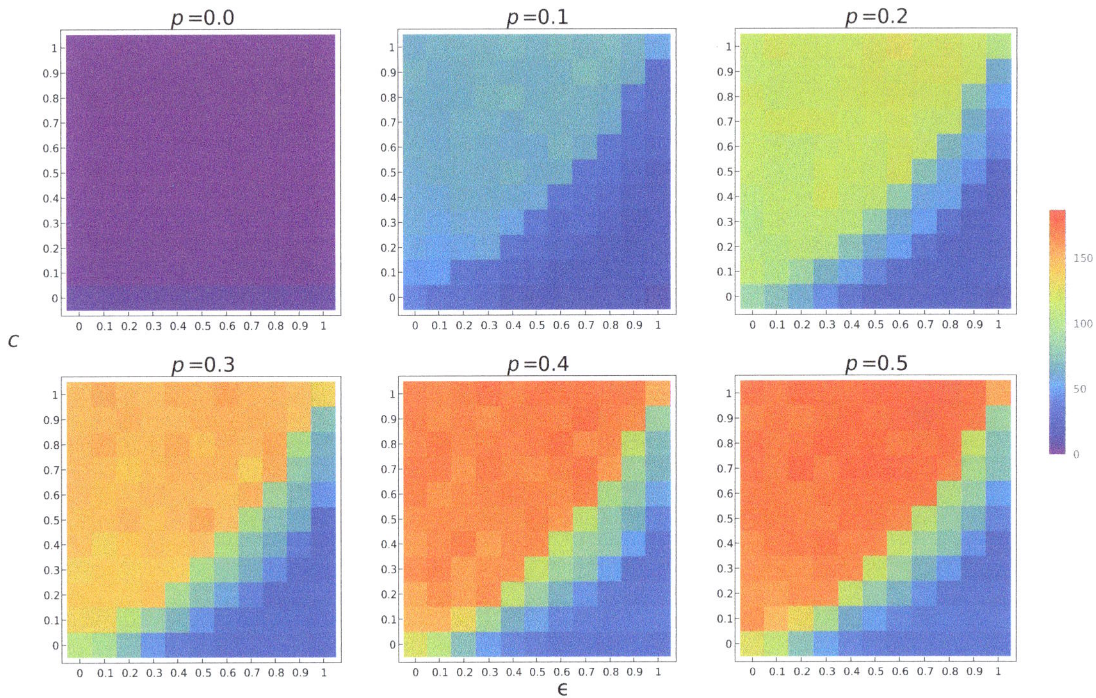

Figure 7.3: Average total Brier scores for six bias possibilities for the combination of EXPL, the Manhattan metric, and geometric pooling; $p = x$ indicates that the chance for each of the coins to come up heads equals x.

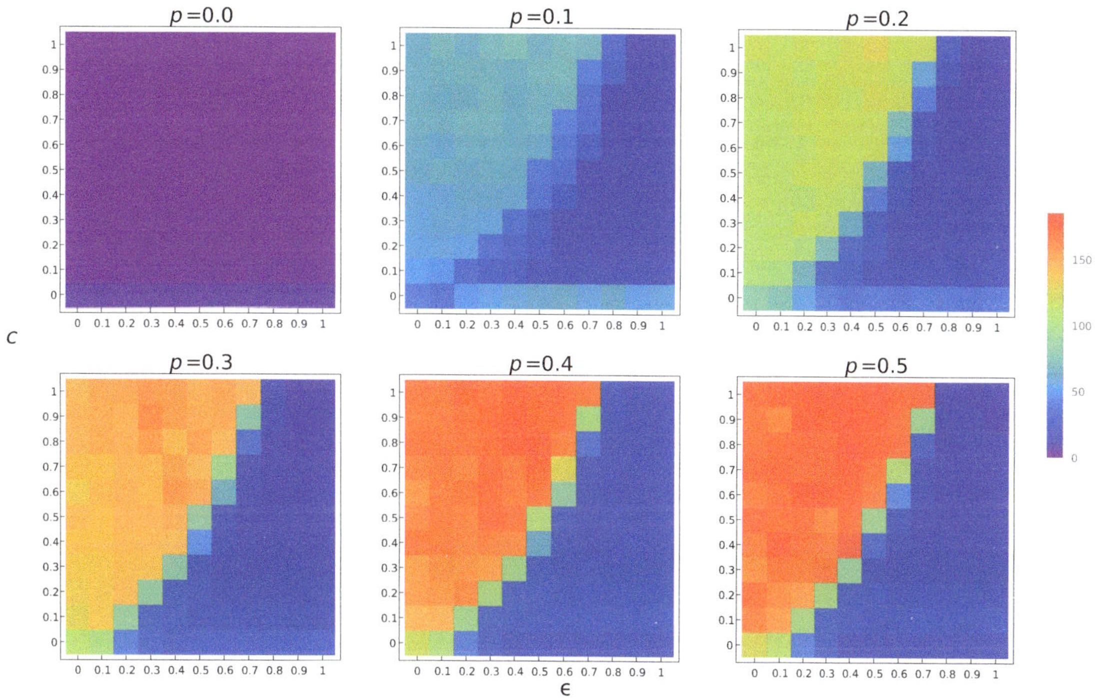

Figure 7.4: Average total Brier scores for six bias possibilities for the combination of EXPL, the Euclidean metric, and harmonic pooling; $p = x$ indicates that the chance for each of the coins to come up heads equals x.

To forestall misunderstanding, it is to be noted that, in answering this question, we do not have to solve a multiobjective optimization problem, with speed and accuracy as objectives (Deb, 2001; Qian, Yu, & Zhou, 2013; Rakshit & Konar, 2015). As mentioned repeatedly, we are thinking of striking the best *balance* between speed and accuracy as our objective. In doing so, we assume that what counts as "best" depends to an *equal* degree on speed and accuracy. This will not always be the case, of course, but for now it serves as a useful simplifying assumption; we come back to this in the final section of this chapter. Although by thinking of the task before us as a single-objective optimization problem we are avoiding some complications, it remains the case that the search space is mixed, with both discrete (update rule, distance, and pooling method) and continuous variables (values for c and ε). This on its own makes analytical methods difficult to apply. A posteriori methods, specifically computational methods, have been applied with greater success to such problems. In the next section, we again make use of the method of evolutionary computing that was introduced in the previous chapter.

7.3 Evolutionary Computing and Optimal Group Learning

The optimization technique used in the following is again a form of agent-based optimization. Recall from chapter 6 that in this technique, different types of agents are thought of as representing different solutions to a given problem and that these solutions are scored with respect to the relevant objective or objectives. From those scores we derive the agents' fitness levels, on the basis of which, in a next step, we select agents either deterministically or stochastically in order to retain them or allow them to reproduce (or both) to form a new population, which then enters into a new competition for survival and reproduction. As previously mentioned, this process of competition, selection, and reproduction can be repeated several times until some predetermined criterion is met.

The evolutionary computing used is slightly more complicated than that used in chapter 6. Rather than individual agents, we consider *groups* of agents, where these groups differ by having adopted different epistemic principles of the kind examined in the previous sections. One could consider groups whose members use (partly) different epistemic principles in order to investigate whether such epistemic in-group diversity might carry evolutionary advantages, but the groups that I looked at consisted of agents who all assumed the same set of epistemic principles.

Sometimes in evolutionary computing, the predetermined stopping criterion is a minimally acceptable fitness level to be attained by some agent (or more generally, by some solution to the target problem). For our present concerns, there does not appear to be such a minimal fitness level that could be set a priori. We would *like* groups to get to the truth immediately and unexceptionally, but that is an unrealistic goal to attempt. What *is* a realistic goal is not clear from the outset but rather something that one would hope to learn from the simulations. So I decided to use the most straightforward stopping criterion to be found in the evolutionary computing literature, to wit, stopping after a fixed number of generations. However, the number I chose—fifty—was established on the basis of some pilot runs that suggested that this number of generations would suffice to achieve satisfactory convergence, in that solutions in the fiftieth generation would deviate only slightly from each other. (This was confirmed by the actual runs carried out, as subsequently described.)

Although all evolutionary algorithms build on the same ideas of variation and selection, they can differ greatly in their details, most notably concerning the extent to which they rely on the basic evolutionary operations of crossover, mutation, and selection as well as the extent to which such operations are applied stochastically or deterministically. In the previous chapter there was no mutation involved, and there was no crossover involved either, in the sense that there was no "mixing" of "genetic" material. In place of crossover, there was a simple copying operation: the 50 percent best doctors were selected in each round and allowed to move on to the next generation, together with an identical copy of them. The algorithm I used for the simulations reported in this section involved selection and crossover, detailed subsequently, and to a lesser extent, mutation. In this algorithm, selection proceeded deterministically but crossover and mutation proceeded stochastically. Selection was done on the basis of speed and accuracy, the latter operationalized in terms of the average Brier score incurred by a group and the former in terms of the first time a majority of the group assigned a high probability (understood as a probability above .9) to the true hypothesis.[10] Both criteria factored equally in determining fitness values.

The simulations that I ran were similar to those reported in Douven and Wenmackers in that the agents in any one group received binomial data from a common source, where the data one agent received were independent from those received by any other agent. In this situation each agent repeatedly flips her own coin even though all coins have the same (initially unknown) bias, and the shared task of the group is to find out what that bias is. For concreteness, I describe the data in the following as the outcomes of coin flips.

More exactly, the algorithm used consists of the following five steps:

Step 1. Create a starting population of groups each consisting of fifty agents, such that all combinations of update rules (Bayes's rule, EXPL, Good's rule, and Popper's rule), distances (Euclidean, Manhattan, and KL divergence), and pooling methods (linear pooling, geometric pooling, and harmonic pooling) are represented by exactly one group, yielding a total of thirty-six groups; for the groups consisting of non-Bayesian updaters (users of EXPL, Good's rule, or Popper's rule), choose randomly per group an explanatory bonus value from the standard uniform distribution; and for each group, choose randomly a value for ε (i.e., the "peer radius") also from the standard uniform distribution.

Step 2. For every agent in every group, generate a sequence of 500 coin flips, where all coins have the same randomly chosen bias (the choices limited in the same exact way as in Douven & Wenmackers, 2017, so that all possible biases are multiples of 0.1). Let agents update their probabilities after each toss according to whichever combination of rules their group has adopted, using the generalized HK model described previously, and after each update, calculate Brier scores for all agents.[11] Also, register after each update the percentage of agents

10. I also experimented with different scoring rules and different threshold values. These experiments did not lead to any noteworthy qualitative differences. Comparing forms of abductive reasoning with Bayesian reasoning using different stopping rules (see footnote 10 in ch. 4) is left as an avenue for future research.

11. To forestall misunderstanding, it is to be noted that the agents in the model all belong to the same group, so that the algorithm enables only intragroup communication and not intergroup communication.

who assign a probability greater than .9 to whichever the true bias hypothesis is in this run. For each group, repeat this scoring process twenty-five times; so choose anew a bias for the coins, let the agents update on 500 coin flips, and score each agent with respect to accuracy (Brier score) and speed (probability greater than .9 assigned to the truth), and keep doing this until all agents have gone through twenty-five such rounds of scoring. Then calculate the group's average Brier score (i.e., the total Brier score incurred by the group's members averaged over the twenty-five runs) and also the average toss number after which a majority of agents in the group assign a probability greater than .9 to the truth (the average being over the twenty-five runs).

Step 3. Rank the thirty-six groups in the population twice, once according to their average Brier score and once according to their average speed (expressed in terms of steps needed to achieve truthcloseness), with lower ranks being better in both cases. Then add up the ranks and select the eighteen groups with the lowest resulting sums (following, e.g., Baker, 1985). Draw (with replacement) from these "fittest" groups eighteen pairs, and then select randomly and independently from each pair an update rule, a distance, and a pooling method; draw a value for the explanation bonus c from a normal distribution truncated between 0 and 1 inclusive with as mean the mean of the c values of the members of the pair and as standard deviation the standard deviation of the c values of the whole previous generation; and determine the value for ε in exactly the same way. Put the results of these selections together to obtain eighteen new "offspring" groups.

Step 4. Form the next generation of groups by conjoining the eighteen best groups from the parent generation and their offspring (another eighteen groups) formed as explained in step 3.

Step 5. Iterate steps 2–4.

Figure 7.5 shows a schematic representation of the given steps.

I have two comments on this algorithm. First, as mentioned, selection occurs deterministically while variation occurs stochastically. Given the way the values of c and ε are picked for the offspring, an element of random

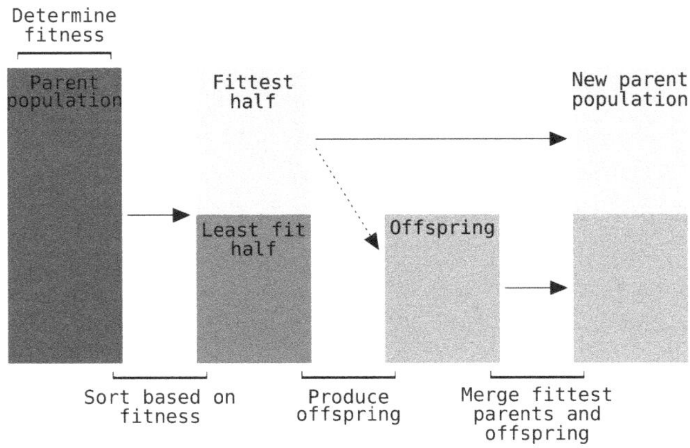

Figure 7.5: Schematic representation of how the evolutionary algorithm creates a new generation of groups. (Solid arrows suggest that groups move from one stage to the next, and the dotted arrow suggests that although the groups themselves do not move to a next stage, their genetic material does.)

mutation is built into the variation part. In the piloting stage of the present research, I experimented with stochastic selection as well, in particular with an algorithm in which a group's rank on the basis of the scoring determined its chances of being selected for producing offspring. Because this did not lead to results qualitatively different from those obtained with deterministic selection but did slow down the computations quite a bit, I went for deterministic selection in the full simulations.

Second, these simulations, too, were conducted using the Julia language. Appendix E explains how to obtain and use the code for the simulations. Readers are encouraged to go through the code in the notebook and to experiment with it; doing so should further enhance understanding of the simulations. The code will also allow readers to try out variant update rules, distances, and aggregation functions. For example, the distances that I used are from the `Distances.jl` package for Julia, which currently supports a total of twenty-seven distances, a number of which are certainly plausible options for determining peerhood (e.g., the Jensen–Shannon divergence and the Hellinger distance).

As said, one run consisted of an evolutionary sequence of fifty generations in which except for the starting generation, each generation was the result of the process described in steps 2–4. The plan was to start by conducting fifteen runs and to continue with additional runs depending on how much variance was found in the first fifteen.

Turning now to an evaluation of the results that were obtained, one can get a first impression of them from Figures 7.6 and 7.7. These figures show the outcomes from a single, randomly chosen run. Figure 7.6 consists of three series of fifty stacked histograms, each histogram showing the percentage of tokens of each update rule type (top plot), distance type (middle plot), and pooling type (bottom plot) present in the corresponding generation. It is clear from this figure that after nineteen generations, all groups agree on a best combination of update rule, distance, and pooling method, to wit, EXPL, the Euclidean metric, and harmonic pooling. Figure 7.7 shows that there is great agreement among the groups about the best values for c and ε, which are both in the higher range. Figure 7.8 shows that how the population of groups evolves has a noticeable impact on how accurate they are on average, as well as on how fast they are on average.

Whereas the aforementioned figures show the results of a randomly selected run, very similar results were obtained in the other runs. Notably, there appeared to be a strong convergence not only in that all runs showed all groups to endorse, after twenty to thirty generations, a combination of EXPL, the Euclidean metric, and the harmonic pooling method, but also in that in each run all groups came to hold values for both c and ε typically in the vicinity of .9. More specifically, pooling the results from all runs, I found, averaging over all last-generation groups, a c value of .94 (± 0.04), and an ε value of .91 (± 0.05). These values hardly differ from the average values of the top groups in the last generations from the fifteen runs: .95 (± 0.03) for c, and .89 (± 0.07) for ε. In view of these results, I refrained from conducting runs in addition to the fifteen with which I started.

To see the effect that the evolutionary process had on the criteria of interest, note that the average total Brier score incurred by the groups in the first generation was 6,468.19 (± 730.07) whereas that incurred by the groups in the last generation was 616.08 (± 114.59). This amounts to an average Brier score *per agent* of 129.36 for first-generation groups, and 12.32 for last-generation groups. Similarly, the first toss number at which a majority of agents in a group assigned a high probability to the truth was, on average, 153.60 (± 6.20) for groups in the first generation, and 31.47 (± 1.16) for groups in the last generation.

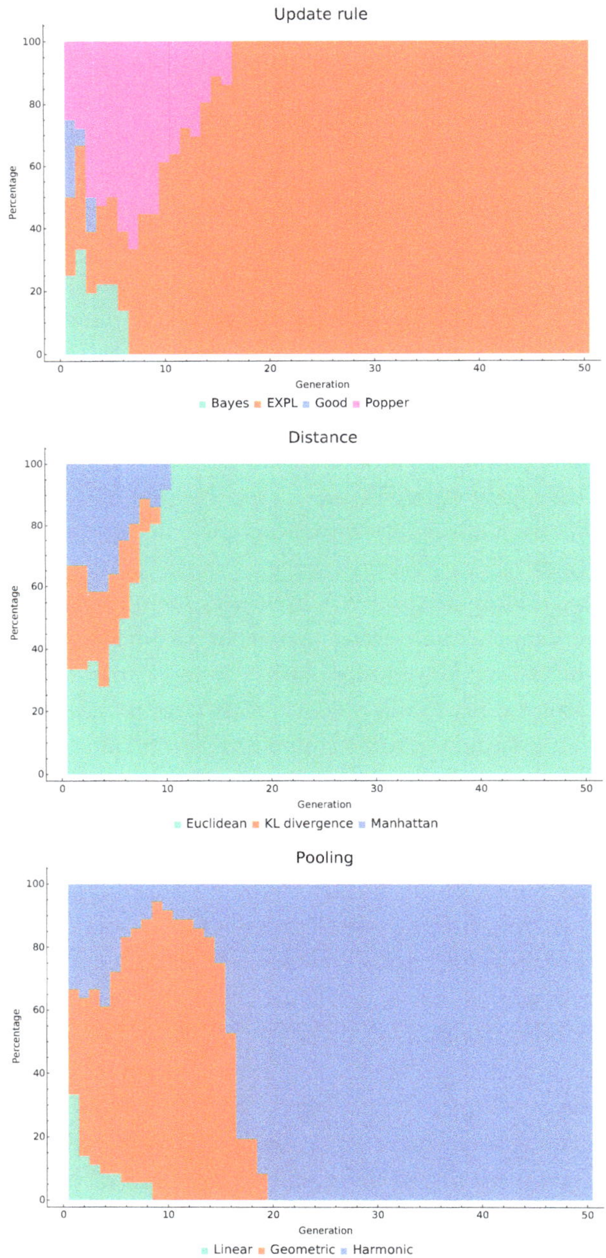

Figure 7.6: For a randomly chosen run, per-generation percentages of rule types (*top*), distance types (*middle*), and pooling types (*bottom*).

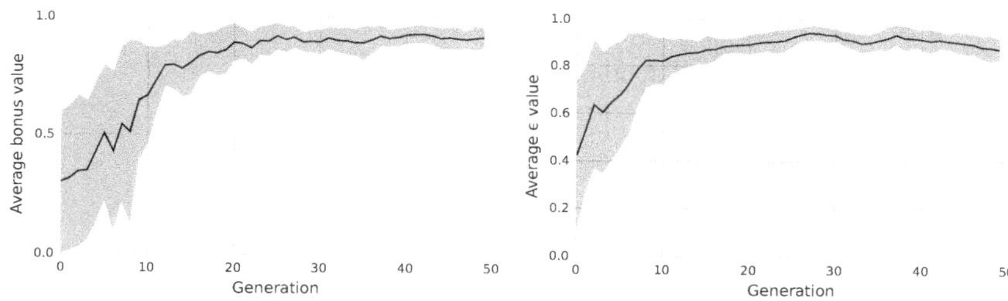

Figure 7.7: Average bonus and ε values for the fifty generations in the same run that generated the data shown in figure 7.6; shaded bands indicate one standard deviation from the mean.

To bring the significance of the overall outcome into further relief, one can run the fitness-level determining procedure used in the evolutionary computations on a group of *noninteracting* Bayesians (agents updating via Bayes's rule but not letting themselves be influenced by the opinions of their peers) and, for comparison, on a group of agents that has adopted the setting that came out best in the foregoing, so a combination of EXPL as an update rule, with $c = 0.95$; the Euclidean metric for determining peerhood, with $\varepsilon = .89$ as determining the criterion for peerhood; and the harmonic pooling function for aggregating probability functions. For either group, (1) draw a random

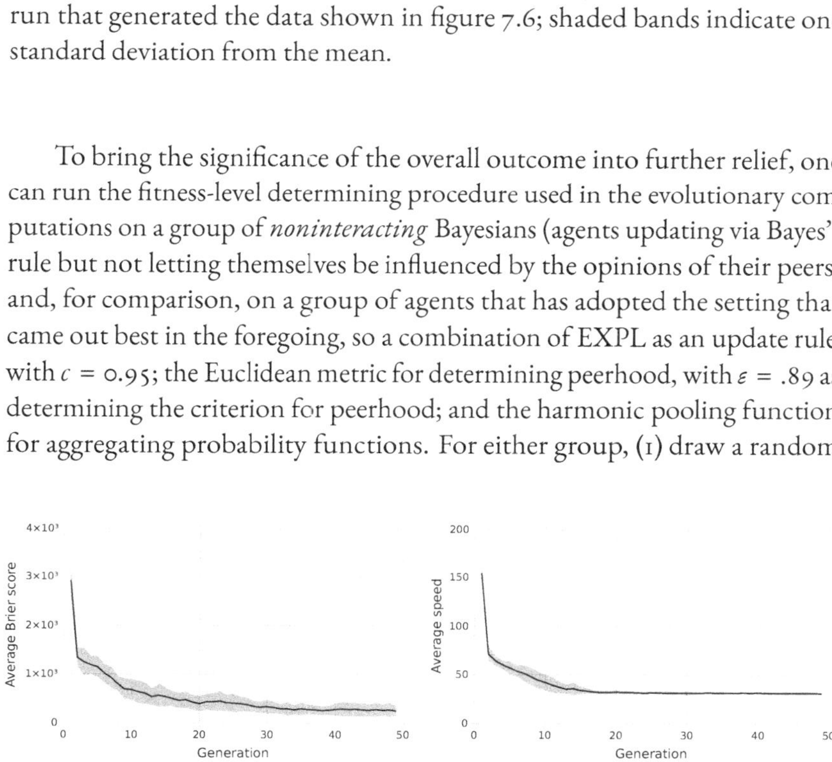

Figure 7.8: Average total Brier scores and average first toss after which the majority assigns a high probability to the truth, for the fifty generations in the same run that generated the data shown in figures 7.6 and 7.7; shaded bands indicate one standard deviation from the mean.

bias hypothesis; (2) generate for each agent her own sequence of 500 tosses with the drawn bias; (3) let each agent consecutively update on those tosses and, for those in the second group, also take into account her peers' opinions in the way prescribed by the said epistemic principles; (4) register the Brier scores for each agent after each update; (5) register the first toss after which a majority of the group assigns a probability above .9 to whichever bias hypothesis was drawn in step (1); and (6) repeat the foregoing five steps twenty-five times just as in the evolutionary computations. I ran these simulations ten times, in which the average total Brier score was 3,798.82 (± 347.43) for the Bayesians and 524.91 (± 59.57) for the group using the "best" settings and in which the average toss at which a majority assigned a probability above .9 to the truth was 118.58 (± 10.00) for the Bayesians and 29.90 (± 4.12) for the other group. In other words, by using the settings that the evolutionary computations showed to be optimal, a group can, in the type of situation considered here, on average achieve a more than sevenfold increase in accuracy and at the same time an almost fourfold increase in speed relative to a group of strict Bayesian updaters.

7.4 Optimal Group Learning in Action

There are a great many variations one could consider in the framework developed in the foregoing. In this section, I want to look at a variation that puts the simulations reported in the previous chapter in a social perspective. Instead of *single* doctors, consider *teams* of doctors, each team being, for example, fifty strong. All doctors in a team are independently gathering evidence about one and the same patient whose life they are jointly trying to save. Suppose that a team decides to intervene precisely if a majority of its members assign a probability above .9 to the same hypothesis. Teams are evaluated in the same way individual doctors were evaluated in the simulations presented in chapter 6. In particular, they are evaluated on the basis of treating one hundred patients, where the relevant score is the average probability that a patient they treat will live. Here, I look only at the kind of environment in which that probability comes from a Weibull distribution. Interested readers may wish to use the code in the Jupyter notebook belonging to this section to explore a Gamma environment (or other environments, by modeling probability of death via still other probability distributions).

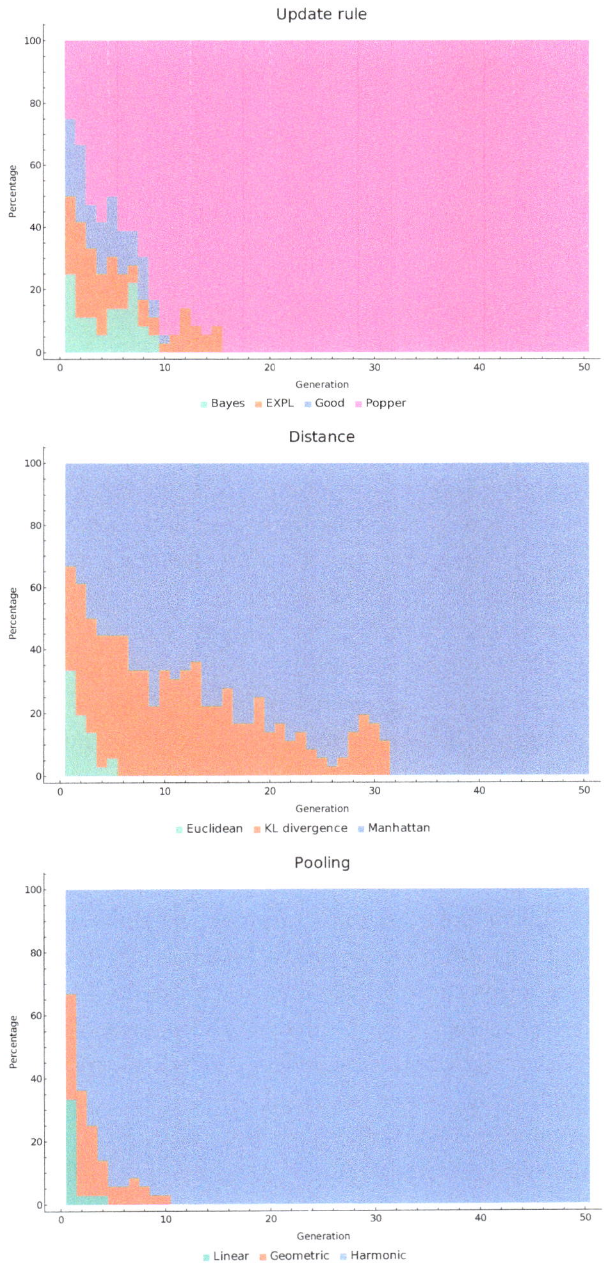

Figure 7.9: For a randomly chosen run, per-generation percentages of rule types (*top*), distance types (*middle*), and pooling types (*bottom*).

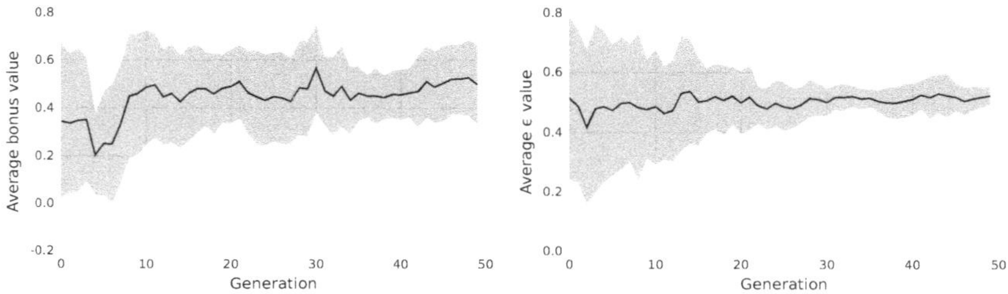

Figure 7.10: Average bonus and ε values for the fifty generations in the same run that generated the data shown in figure 7.9; shaded bands indicate one standard deviation from the mean.

To see whether we can make epistemic recommendations to such teams, we rerun the evolutionary procedure described in the previous section. We keep everything the same, except, as said, how the teams are scored (average survival probabilities in place of speed and accuracy). Figures 7.9 and 7.10 show the outcomes of a randomly selected simulation. This simulation is representative of the lot at least insofar as Popper's rule is again the overall winner: Of the $36 \times 15 = 540$ groups in the last generations, 276 consisted of Popperians, 193 of EXPL users, and 71 of Bayesians, groups using Good's rule never making it until the end. This is in line with what we found in chapter 6 for individual doctors in a Weibull environment. And identical to what we found in the previous section, all groups in all simulations came to use the harmonic pooling method. As for distances, the outcome is different now: whereas in the previous simulations all last-generation groups used the Euclidean metric, in the new simulations this is so for only 180 of the last-generation groups, the other groups using either the KL divergence (252) or the Manhattan metric (108). Taken over all last-generation groups in all simulations, the average bonus value assigned by last-generation groups is .49 (± 0.16) and the average ε value is .52 (± 0.03).

To emphasize once again the importance of *social* aspects to learning, I mention that averaged over all simulations, the probability that a patient would live went up from .78 (± 0.01) in the first generation to .95 (± 0.01) in the last. Compare this with the fact that in the simulations presented in the previous chapter, in which patients were treated by individual doctors, the average probability for the last generation that a patient would live was only .87 (± 0.02). In real life, there is a reasonable expectation that a patient's chances of survival will go up when they are looked at by several doctors instead of just by one. But that is at least partly because we would expect each of the doctors to bring her or his own expertise, complementing that doctor's colleagues' expertise. Nothing of this kind is involved in our simulations. That we can nevertheless expect a better outcome from a group of doctors can be explained only by the fact that they learn in part by taking the opinions of their colleagues into account.

Further details can be obtained from the Supplementary Materials. For now, I note that the new results buttress the main message of this book that abduction is not a rule that is to be followed mechanically but rather designates a mode of reasoning that needs to be filled in depending on the situation, where it may require skill to find the best fill-in for any given case. We see here in particular that it may matter to how we ought to reason abductively whether we are acting completely on our own or are part of a group and revise our opinions not only on the basis of evidence that we receive directly from the world but also in response to the opinions of other group members.

7.5 The Upshot

This chapter continued the comparison among update rules begun in chapter 6, specifically pitting Bayes's rule against various probabilistic versions of abduction. It did so as part of a broader research question, to wit, whether we can recommend best epistemic practices to groups of agents who independently receive evidence from a common source. To address this question, we again used the method of agent-based optimization, and more specifically an evolutionary algorithm, although now the units of selection were groups of agents, where the agents within any one group could communicate with each other and thus impact each other's belief states. Assuming that a group is interested in striking the best balance between speed and accuracy, and making certain assumptions about the evidence-generating process, it was found that one can make a clear recommendation in the form of a unique combination of update rule, distance measure, and pooling method, together with an indication of the optimal range for the c and ε values, the former indicating the explanatory bonus and the latter willingness to count others as peers.

This result was reached by assuming that speed and accuracy mattered equally. As said, however, this assumption is not guaranteed to hold. In some situations, we may be willing to sacrifice quite a bit of accuracy for the benefit of greater speed, whereas in other situations the opposite may hold. The code that was used for the simulations and that is in the Supplementary Materials allows interested readers to verify that giving different weights to speed and accuracy can lead to vastly different outcomes. For instance, rerunning the simulations with all the weight given to accuracy consistently led to a combination of Popper's rule, the Euclidean metric, and geometric pooling as the optimal combination, with values for c and ε in the mid-to-high range. Apart from the relative importance of speed and accuracy, the evidence in the simulations also had properties that do not hold generally—for instance, if the data really concern coin tosses, one might be more likely to encounter biases close to .5 than ones more to either extreme—and varying the evidence-generating process can affect the outcomes as well. All this is underpinned by the simulations reported in the previous section, in which as in the simulations presented in chapter 6, we focused not on speed and accuracy directly but rather on survival probabilities, and we let those determine, indirectly, which trade-off between speed and accuracy to choose. This led to results concerning best combinations of epistemic principles that were different from those obtained in section 7.3.

As discussed in the previous chapter, cognitive psychologists working on theories of ecological rationality have generally rejected as an illusion the idea, cherished by philosophers, that rationality can be captured by a number of universally valid principles and have argued that we can understand the notion of rationality only if we take into account (1) the various ways—biological and cognitive—in which humans are limited; (2) the environments in which humans are to operate and to interact; and (3) the goals that humans have in those environments. Biological, cognitive, and environmental constraints can all vary greatly from one individual to another and even from one moment to another. As explained earlier, this led Elqayam, Gigerenzer, and others to conclude that whether someone reasons or acts rationally depends on whether the reasoning or behavior helps the person to achieve whatever goal she pursues in whichever environment she is in, given the cognitive resources available to her in that environment. These researchers were concerned with the rationality of individuals, not of groups, but the results from this chapter suggest that their proposal makes sense at the group level as well.

The upshot is that we *can* make recommendations for how to organize research groups epistemically, but these recommendations will depend on what type of phenomenon or process the group is studying and on what the aim of a group is. In view of the aforementioned work on ecological rationality, a stronger conclusion could not have been expected. And as for the overarching topic of this book, we have seen further evidence that it can make a lot of sense to urge the members of groups to reason abductively (in some form or other), and more sense than to recommend Bayes's rule. Again, this is not to discredit Bayes's rule, but rather to push back against the claim that violations of that rule necessarily indicate epistemically irresponsible behavior.

8

An Abductive Response to the Skeptic

> We should be careful to get out of an experience only the wisdom that is in it—and stop there; lest we be like the cat that sits down on a hot stove-lid. She will never sit down on a hot stove-lid again—and that is well; but also she will never sit down on a cold one anymore.
>
> —Mark Twain, *Following the Equator*

8.1 The Skeptical Challenge

Explanationists faced the challenge of showing that reasoning abductively in a way that conflicts with Bayesian reasoning could *ever* be rational. In the previous chapters, I have addressed that challenge at some length by showing arguments that allegedly demonstrate the irrationality of abduction to be seriously flawed and, on the positive side, by showing that abduction, although perhaps not universally recommendable, is sometimes an ecologically rational mode of reasoning.

There are still some open issues, however. We demonstrated the ecological rationality of abduction in computer simulations. What about abduction in the real world? In section 3.5, we saw that the participants in the experiments reported in Douven and Schupbach (2015a) tended to be more accurate as they gave more weight, in updating, to explanatory considerations. That is *some* evidence for the real-world value of abduction. We also discussed some successes of abductive reasoning in section 2.1. Showing that abduction is a reliable mode of inference in a broad range of contexts, in that judgments of explanatory superiority reliably point in the direction of the truth in those contexts and hence that giving weight to such judgments often tends to make us more accurate, might just be a matter of going beyond anecdotal evidence of the kind examined in section 2.1 and launch a systematic empirical investigation. For instance, it would probably not be very complicated to see whether the extent to which medical doctors rely on explanatory considerations in their work correlates with their success rates, a project that could show the real-world significance of the findings reported in sections 6.3 and 7.4.

There is an additional hurdle, mentioned in section 6.2, which is especially pressing because it concerns the question of whether it is rational to reason abductively in contexts in which philosophers have hoped that abduction could make a decisive dialectical difference. I refer to contexts in which abduction is supposed to break a deadlock due to underdetermination by the evidence, as explained in section 2.3. To revisit the hurdle in these contexts, consider that while scientific realists like to claim that Thomson was led to the discovery of the electron via abductive reasoning—the existence of the electron best explained the phenomena he had observed in his cathode tube experiments—antirealists deny realists the right to speak of a discovery here. And antirealists will not be impressed by our argument for the ecological rationality of abduction, because as they would rightly point out, that argument did not concern the kind of abduction relevant to the realism debate. Indeed, it did not concern any debate in which abduction is supposed to justify existence claims going beyond what at least one party to the debate holds that we have direct epistemic access to.

In this chapter, I aim to show how we can defend abduction in a context in which such a defense is arguably hardest to accomplish, to wit, in the debate about Cartesian skepticism, according to which we must remain agnostic about the existence of an external world. After all, in that debate one faces an opponent who is unwilling to grant one almost anything, unlike, for instance, in the debate with the scientific antirealist, in which it is typically possible to find some common ground that one might then hope to leverage in an attempt to convert one's opponent.

More specifically, Cartesian skeptics (henceforth simply "skeptics") are typically opposed to admitting as evidence anything that goes beyond the purely phenomenal, and equally typically, they disown the use of any inferential principles, including any form of abduction, that might enable one to move from premises about the phenomenal alone to a conclusion about the external world. Not all nonskeptics may find this troublesome. Although the aforementioned facts may stand in the way of a *resolution* of the skepticism debate, which I understand to require that the debate be settled in a way that is acceptable to both parties involved, those facts need not bar an *argument* against skepticism. Supposing a less restrictive conception of evidence than the skeptic's or stronger rules of inference than she is willing to countenance may enable one to argue on empirical grounds for the external world hypothesis. Such an argument may leave the skeptic cold and therefore may not bring any nearer a resolution of the debate, but if we nonskeptics can convince or reassure ourselves of the tenability of our own position, then what further good is there to be gained by convincing *others* of it? As Psillos observed (see section 6.2), in philosophy the goal is not always to convince one's opponent and can instead be to justify one's position to one's own satisfaction. Philosophy is not politics.[1]

The chief aim of this chapter is to show that a resolution of the debate becomes possible once it is recognized that the skeptic embraces overly restrictive and ill-motivated positions vis-à-vis both evidence and inference, and that more reasonable alternatives are available and that embracing those alternatives will crucially, and inter alia, allow us to commit to abductive reasoning. Naturally, for this aim it will be imperative to make clear that these alternatives should appear more reasonable from the skeptic's point of view. A secondary aim is to show that the main aim is likely to be more important than may appear to those who believe that all the nonskeptic should care about is an argument against skepticism; as I argue, whether or not nonskeptics *need* no more than such an argument, it is quite unclear that they are in a better position to give one than to give a resolution of the skepticism debate.

Focusing on the latter aim first, I begin in section 8.2 by considering two well-known arguments against skepticism that rather explicitly do not seek to resolve the debate, and by arguing that one of them assumes a conception of evidence and the other a conception of inference that the skeptic can reasonably reject, and that should seem problematic even from the nonskeptic's perspective. In section 8.3, I argue that the skeptic's views on evidence and inference are problematic as well, however, and I propose alternative views that differ from both the skeptic's and those supposed by the aforementioned

1. For other examples of this line of thinking see Pryor (2000) and Byrne (2004). In Pryor's terms, it is enough for nonskeptics to pursue the "modest anti-skeptical project," which requires that we "establish to our satisfaction that we can justifiably believe and know such things as that there is a hand" (Pryor, 2000, p. 517).

arguments but that should by anyone's lights appear more reasonable than either. Finally, section 8.4 presents a convergence result showing that given those more reasonable views, it is at least in principle possible to resolve the skepticism debate.

8.2 Mooreans and Russellians against the Skeptic

To see that an antiskeptical argument is not automatically pointless if it presupposes a more liberal conception of evidence than the skeptic's, recall G. E. Moore's (1962) famous "proof of an external world." A common initial reaction to this argument is that in it Moore is begging the question by assuming that he has perceptual evidence of his having two hands. But note that one may admit this yet insist, as some argue Moore himself does, that by making the said assumption he is no more begging the question against the skeptic than the skeptic is begging the question against her opponent by assuming that our evidence is limited to the purely phenomenal. Suppose that this can be successfully maintained. Then the least Moore's argument would seem to yield is a kind of standoff between the skeptic and the nonskeptic. And given that to many the skeptic so clearly seems to have a winning hand—undeservedly, it is typically added—it would be a remarkable achievement to show that the skeptic in any case does not win the debate (even if he does not lose it).[2]

Some nonskeptics might want to go further and claim that considerations of question-begging should not detain us here. All that matters, they might say, is that the premises and presuppositions of Moore's argument are acceptable to the nonskeptic; whether Moore is begging any skeptical questions or whether, if he does, this is offset by the skeptic's begging nonskeptical questions in turn, is neither here nor there. To these nonskeptics' eyes, Moore has offered us a successful argument against skepticism, and it is none the worse for failing to persuade the skeptic.

---

2. Epistemic conservatives could even argue that inasmuch as the one position is as good (or bad) as the other, dialectically speaking, the sensible thing for us to do is not to abandon what we believe anyway—namely, that we do have lots of interesting knowledge, including knowledge of an external world—in favor of a position that no one has ever seriously held (as opposed to temporarily entertained for philosophical purposes). But epistemic conservatism is controversial; see, e.g., Christensen (1994) and Vahid (2004).

Much the same could be said about the Russellian response to skepticism, which argues for the external world hypothesis on the basis of this hypothesis's explanatory power.[3] The best explanation of our perceptual experiences and of the complex patterns and regularities manifest in those experiences is that the *apparent* causes of our experiences—mostly external objects—typically are their *real* causes. Therefore, we are justified in believing that there is an external world. The response assumes, patently, that the external-world hypothesis *is* the best explanation of our experiences. Some authors seem to think of this assumption as immediately compelling. Others, most notably Laurence BonJour and Jonathan Vogel, have argued for it extensively—and quite ingeniously, to my mind.[4] While this is not to say that the assumption is entirely undisputed, I am willing to grant it here.[5] Note that this is far from enough to settle the skepticism debate. The Russellian response also relies on some qualitative version of abduction, and why should skeptics want to sanction abduction in *any* of its forms? Even so, the Russellian response is not necessarily a failure. It, too, might achieve a standoff between the skeptic and her opponent, and for reasons previously mentioned, that would be of considerable significance.

Some might even think that whether the Russellian response achieves a standoff is immaterial; what is important is simply that the argument is acceptable from a nonskeptical perspective, however many skeptical issues it may beg and however few nonskeptical issues the skeptic may have to beg in order to defend herself against it (if she has to beg any at all).

But it is not at all obvious that the Moorean and the Russellian responses to the skeptic *are* successful in either of the senses just explained, that is, whether they yield standoffs with the skeptic or at least arguments that are acceptable from a nonskeptic's viewpoint. In fact, the following considerations give some grounds for thinking not merely that the skeptic has more right to her position regarding evidence than the Moorean has to hers, as well as more right to reject certain forms of abduction, or indeed any form, than the Russellian has to endorse abduction generally—which seems to preclude that

3. I call it "Russellian" because the first response of this type (or at least the first I know of) is to be found in Russell (1912, ch. 2). More elaborate and refined versions of it have been propounded since, however; see, for instance, Harman (1973, chs. 8, 11), Goldman (1988, ch. 9), Vogel (1990, 2005), and chapter 5, written by BonJour, of BonJour and Sosa (2003).

4. See Vogel (1990, 2005) and BonJour and Sosa (2003, ch. 5).

5. Sosa (1997, pp. 413–414), for one, disputes it.

the foregoing arguments yield standoffs—but also that the Moorean and Russellian positions, and concomitantly the Moorean and Russellian responses to skepticism, should, upon reflection, be unappealing to the nonskeptic herself.

Whereas the skeptic may grant that we have an instinctive trust in our senses, she will deny that our instincts are worthy of trust on this point, and she can seemingly do so without having to commit herself to any implausibly strong assumptions. In particular, her denial need not involve anything as strong as the principle that we ought to have good reasons for trusting our senses, or perhaps even know that they are trustworthy, before we can trust them. To defend her claim, the skeptic could make do with what appears to be a rather minimal and unproblematic assumption, namely, that it is perfectly all right for us to trust our senses, *provided* that we have no reasons for *doubting* their trustworthiness. After all, she will have no difficulty pointing to such reasons. For example, we probably all have had conflicting experiences of the sort in which what at one moment appeared to be a bent stick the next moment appeared to be a straight one, or in which what at first appeared to be a sheep later appeared to be a rock. Such experiences fall far short of *establishing* that our senses are untrustworthy: trustworthiness does not require infallibility. Yet even the Moorean will have to concede that they do give us *some* reason to doubt our senses' trustworthiness and concomitantly the admissability of sentences such as "I have two hands" as possible evidence reports.

Note, that the skeptic can thus grant a lot to nonskeptics. For instance, nothing seems to prevent her from concurring with Crispin Wright (1991) that barring evidence against it, we have an unearned warrant to believe the hypothesis that sensory perception is reliable. Also, she can treat that hypothesis "liberally" in Jim Pryor's (2004, p. 354) sense, meaning that the hypothesis can be among the conditions that justify one defeasibly in believing a given proposition A even if one lacks antecedent justification for the hypothesis, although, importantly, evidence against the hypothesis would undermine one's defeasible justification for A. So, it seems that a skeptic who argues along the lines suggested in the text is not threatened by antiskeptical responses—such as Wright's and Pryor's but also Alex Byrne's (2004)—which proceed on the assumption that the skeptic must be assuming that if we are to know anything via perception then we must antecedently know, or at least be antecedently justified in believing, that perception is reliable. As may be noted, I am assuming that Wright and Pryor would count the reasons for doubting the trustworthiness of our senses as (possibly inconclusive) evidence against the hypothesis that our senses are trustworthy. Although they do not discuss in any detail the matter of possible counterevidence, they would seem hard-pressed to give a generally acceptable account of evidence on which reasons of the kind mentioned in the text do *not* count as evidence against the said hypothesis. (On the popular Bayesian conception of evidence, which we also assume in the following sections, it would seem near to obvious that those reasons count as evidence against that hypothesis.)

Again, basically the same that was just said about the Moorean response can be said with respect to the Russellian response. It has happened more than once that I have assumed something because it furnished the best explanation for things that I knew to be the case only to find out later that the assumption was false. I am not an exception in this respect, I am sure. In fact, similar failings of abduction are known from the history of science. In section 2.1, we discussed that although from the viewpoint of nineteenth-century astronomy the small deviation from Newtonian predictions of Mercury's orbit around the sun was best explained by assuming the existence of a yet-to-be-discovered planet, that planet never was discovered, and the deviation is now generally thought to be explained by the General Theory of Relativity as due specifically to the curvature of space-time. Could Russell or anyone else be said to have as much right to endorse a mode of inference whose reliability has come to appear at least somewhat questionable as the skeptic has a right to disavow that rule? In order to back up a negative answer to this question, all the skeptic would have to assume is a principle according to which it is all right to reason abductively, as long as no reasons for doubting abduction have arisen.

At this juncture, some might want to reply that there is an inconsistency in the skeptic's position because she can raise her concerns about the reliability of sensory perception and abduction only by presuming sensory perception to be reliable. ("How else could she claim with any confidence that our senses as well as abduction at least sometimes fail us?") This would be wrong, however. As to sensory perception, note that one can register to have had conflicting perceptions without being able to sort the veridical ones (if such there be) from the nonveridical ones; that there were conflicts between them is enough to aver that at a minimum, they cannot all have been veridical. As to abduction, suppose that sensory perception is *not* reliable. Then that itself is a failure of abduction. For if, as the Russellian assumes, our perceptual experiences are best explained by assuming that their apparent causes are their real causes, then abduction justifies us in believing that sensory perception is reliable (that is precisely what the reliability of our senses amounts to—that typically the apparent causes of our perceptual experiences are their real causes). So, the reliability of abduction is called into question either way, whether it is assumed that sensory perception is reliable or whether it is assumed that sensory perception is not reliable.

A more sensible reply would point to the fact that, as we have briefly noted, occasional failures are compatible with reliability (just not with infallibility). Thus, to start with abduction, it might be that, its registered failures notwithstanding, this rule is reliable. Some externalists might hold that if abduction is in fact reliable, then inferring to the best explanation is *ipso facto* a legitimate procedure even in the face of doubts with regard to the rule's reliability. I am not sure whether this hard-nosed kind of externalism has any actual sympathizers. If it does have them, then I would expect this to be due to the fact that some believe it to be our only hope for resisting skepticism. As it follows from this chapter that this is not so, I set aside the designated externalist position without a hearing. That is not necessarily to embrace internalism; for all I need to assume, it may be epistemically unassailable to infer to the best explanation even lacking any reason whatever to believe that abduction is reliable. Nor is it necessarily to say that once failures of the rule have come to our attention we can nevermore utilize it, or at least not legitimately so. What I do think is that on any plausible epistemology, be it internalist or externalist, once such failures have come to our attention, we face the task of showing, or at least making plausible, that these failures are mere accidents and not symptoms of a systematic defect before we can utilize abduction or utilize it again. But how to do this?

It seems that unless we somehow *define* truth in terms of (best) explanation—an option that we have already rejected—the only way to show that abduction is reliable, if it is, is by means of an empirical argument. At first glance, such an argument may not seem too hard to come by. Are we not often enough in a position to verify a conclusion that we had previously reached on the basis of explanatory considerations? Newton was *right* about the Dutch having beaten the British. Le Verrier was *right* about what caused the perturbations in Uranus's orbit. I and doubtless many others were *right* about the head and the headless body found in Jackson belonging together. These conclusions had all been reached abductively. Most of the pertinent evidence may in fact be of a more humdrum type. Suppose that you conclude that your neighbors are back from vacation, as that would best explain the presence of their car in the driveway. If soon afterward you see them busy in the garden, it is very tempting to classify this as a success of abduction, lending some credibility to the hypothesis that abduction is a reliable mode of inference. Still further support for that hypothesis could easily accrue from a more systematic examination of applications of abduction.

There is a problem, however, that we already noted in the introduction to this chapter. As we can now put the problem, it is that an inductive approach as just sketched relies on a Moorean conception of evidence. And so far it is still an open question whether the nonskeptic can embrace that conception with as much right as the skeptic can reject it.

Some may think that an empirical argument for abduction need not rely on a Moorean conception of evidence. Indeed, it seems that we could restrict our evidence to sense data (or impressions or appearances or something else) and use abduction to arrive at predictions about future sense data. For instance, given that it appears to me that my neighbors' car is in their driveway, I predict that it will soon appear to me that I am seeing my neighbors. If time and again such predictions are borne out, then would that not be enough to warrant the conclusion that abduction is reliable? The problem is that this would at most warrant the conclusion that abduction is reliable when it is applied to sense data statements for obtaining justification in further sense data statements and not that abduction is reliable when it is applied to sense data statements for obtaining justification in external world statements. And eventually we want the latter.

To return, then, to the question of whether the nonskeptic has as much right to rely on a Moorean conception of evidence as the skeptic has to reject it, I expect that with respect to our senses' reliability, too, all but perhaps the most radical externalists will agree that once doubts about that reliability have arisen, we face the task of showing, or at least making plausible, that our senses are in fact reliable before we can rely on them or rely on them again. Naturalized epistemologists have long been urging that on this matter we turn to psychologists for assistance, and it seems that so far no one has been able to come up with a more promising idea. But in trying to answer the question of whether sensory perception is reliable, psychologists make claims of the following sort (and now I am exaggerating but not essentially so): "I put up my hands ten times before the subject's eyes and in all cases it seemed to the person that there were hands before her." That research practice is perfectly warranted from their viewpoint because they are not engaged in a debate with skeptics. From our present philosophical viewpoint, however, the practice is *not* warranted: by judging that we put up our hands before the subject we are unarguably relying on our senses, which is to violate the previously stated principle that in the face of doubts about their reliability we should suspend such reliance. That is to say that by admitting the said kind of claims as evidence we are begging the question, not just against the skeptic but against all except perhaps advocates of the variety of externalism that I have set aside for our purposes.

The foregoing considerations exhibit a seemingly telling asymmetry between the skeptic and the Moorean and also an asymmetry between the skeptic and the Russellian. They show that the skeptic can raise doubts about the Moorean conception of evidence as well as about the Russellian conception of inference, that her opponents will have to acknowledge as such and that, for all they have hitherto shown, cannot be removed in a non-question-begging way; and for all that we have been told so far, the Moorean and Russellian are not in a similar position to discredit the skeptic's positions regarding evidence and inference. That seems enough to conclude that the skeptic has a greater right to her positions vis-à-vis evidence and inference than the Moorean and Russellian have to theirs—on *any* reasonable understanding of what it is for a party to have greater right to a position than another party has to a contrary position. As a result, it seems hard to maintain that the Moorean and Russellian responses to skepticism succeed in realizing a standoff with the skeptic. Moreover, given that, as we have said, anyone except (perhaps) the most extreme externalists may be expected to concur that once doubts about the reliability of our senses or of abduction have arisen they must first be removed before we can rely (or rely again) on these faculties, and given that so far no one has been able to think of a way for removing such doubts that does not assume the Moorean or Russellian conceptions of evidence and inference, respectively, it seems that these conceptions of evidence and inference, and thereby the Moorean and Russellian responses to skepticism, should appear highly problematic even from the perspective of the nonskeptic.

For all we presently know, what goes for the Moorean and Russellian conceptions of evidence and inference goes for every other nonskeptical conception of evidence or inference that anyone might ever seriously consider. As a result, it is far from clear whether there could be *any* conceptions of evidence and inference that in the final analysis the nonskeptic, but not the skeptic would be comfortable to adopt. It thus seems that nonskeptics cannot be quite so confident as some of them appear to be that there exist any successful arguments against skepticism that do not at the same time resolve the debate.

What *is* clear is that if both skeptics and nonskeptics are left with the skeptical conceptions of evidence and inference as their only options, then we must all end up being skeptics. For it has been long known that starting from these conceptions, one just has too slender a basis for the construction of *any* successful argument for the external world. As I further argue, however, other conceptions of evidence and inference *are* available—conceptions that should seem preferable to all hands, including the skeptic.

8.3 Evidence, Inference, and Expert Functions

If the registered fallibility of our senses should be of epistemic significance and should discomfit the Moorean position vis-à-vis evidence, one still wonders how it could motivate the skeptic's drastic retreat to a sense data conception of evidence. In fact, the Moorean and the skeptical responses to recorded failures of our senses both seem disproportionate. Once such failures have come to our attention, we cannot, contra the Moorean, simply ignore them and go on trusting our senses undeterred. But by distrusting our senses altogether or at least by effectively treating them as though they are to be distrusted, skeptics seem to err in the opposite direction; they seem to overreact. Speaking figuratively, paraphrasing this chapter's epigraph, it seems that the skeptic, unlike the Moorean, is willing to get out of the experienced failures of our senses the wisdom that is in them. Unfortunately, she does not stop there; rather she is like the cat who once sat on a hot stove-lid and now will never again sit even on a cold one.

By way of analogy, consider how ordinary people react when they realize that their memories have betrayed them, that something they seemed to remember with great clarity was actually false, such as when they were certain that they had attended some party but then later discovered that they simply could not have been there. I think that no reasonable person ever concluded on the basis of such failures that she would better remain agnostic about everything she seems to recollect from memory. Nor, however, is it likely that after she has been confronted with its fallibility, such a person would go on trusting her memory in exactly the same way or to exactly the same extent as she did before that event. What such failures *are* likely to occasion in her, I believe, is a sense of caution vis-à-vis her memories, a kind of metacognitive awareness that her recollections are not fully dependable. I, for one, do trust my memory, but due to experiences of the aforementioned kind I do not trust it 100 percent but only about 95 percent. That does not make me an agnostic about my recollections, although it does mean that for me they no longer have the status of infallibility that I once thought could be accorded to them.

The skeptic may be expected to respond with something like the following: "Evidence ought to be absolutely infallible: anything accepted as evidence should be guaranteed to be true. This requires a sense data conception of evidence, for our senses may betray us but we cannot go wrong in judging how things appear to us. So I'm not overreacting: cats that want to be sure that they will never sit on a hot stove-lid again do well not to sit on *any* stove-lids." To the best of my knowledge, however, no skeptic has ever succeeded in producing a convincing argument for this absolute infallibility requirement. It might be asserted that the requirement is plausible on intuitive grounds alone, so that no argument for it is needed. Yet the intuition (if it is one) seems hardly decisive on its own.

For consider that another intuition that we have about evidence is that evidence ought to be informative and useful to at least some extent. Why invest the huge amounts of time, effort, and money in gathering evidence that we do invest, when we get nothing useful in return? This means trouble for the skeptic if Hans Reichenbach is right that the requirement that evidence be informative (however minimally) is at odds with the requirement that evidence be infallible. As most other logical empiricists, Reichenbach had long thought that the only sentences that we could not wrongly judge to be true, and therefore the only sentences that could serve to report evidence, were of the sort of "I have the impression of there being a table in front of me." But in *Experience and Prediction* (Reichenbach, 1938) he came to reconsider this view, and in chapter 3 of that book he argues convincingly that it is *false* that we cannot be mistaken about such sentences and that we can make our "basic statements"—the sentences that we accept as evidence reports—totally secure only by making them devoid of any content and thereby depriving them entirely of any use they might otherwise have. At bottom, the problem as Reichenbach describes it is that in making the judgment that I have the impression of there being a table in front of me, I am relying on my memory in classifying the impression as being that of a table. I am thereby implicitly comparing it with previous impressions I have had. But my memory may be confused, and the type of impression that I am now having might yesterday have made me judge that I was having the impression of there being a chair, or whatever, in front of me. As my judgment that my memory is not right now confused in this way is fallible, so is my judgment that I am currently having the impression of a table. To circumvent this problem, Reichenbach points out, one would have to stop classifying one's impressions in ways which rely on one's memory, but in so doing one's evidence would become utterly uninformative.[6]

Skeptics could retort that whereas the preceding may show that the requirement of infallibility cannot be maintained unimpaired, we can still insist that evidence should be as secure as is compatible with the requirement that it should not be entirely useless. She could then claim that because, for instance, "It appears to me that there is a table in front of me" is not entirely empty of content but is more secure than "There is a table in front of me," the attenuated requirement still favors the skeptical position regarding evidence over the Moorean one.

However, conceding that a trade-off will have to be made between certitude and informativeness almost automatically invites the question whether we should not be willing to sacrifice a little more certitude for some additional informativeness. The intuitively plausible answer might seem to be *yes*, but then the problem is that we will somehow have to balance the different values involved, and that it is unclear how to do this in a principled way.

This much does seem clear: if we were fully convinced that our senses are reliable, then it would be eminently reasonable for us to admit "I have two hands" and similar sentences as possible evidence reports, just as the Moorean does, because thereby we would seem to be making an excellent deal in which a little loss of certitude is traded for a significant increase in informativeness. If by contrast we were fully convinced that our senses are *un*reliable, then it would be eminently reasonable for us to restrict the possible evidence sentences to sense data statements or some such, just as the skeptic does, because thereby we would seem to get a lot of certitude at the cost of no significant loss of informativeness.

As intimated previously, however, we must be honest enough in our response to the skeptic to admit that there are reasons to doubt the reliability

6. For a more recent and more general argument for the unreliability of our introspective judgments, see Schwitzgebel (2008).

of our senses. Acknowledging as much seems incompatible with being fully convinced that our senses are reliable—at least as long as we have not been able to neutralize those reasons. On the other hand, whereas the registered failures of our senses may be sufficient to shake our instinctive trust in our senses, they hardly suffice to *establish* that our senses are unreliable. As a result, at the present stage of inquiry it would be as unreasonable to be fully convinced of the reliability of our senses as it would be to be fully convinced of their unreliability. What to do in this situation?

Someone might conceivably be able to devise a method that would allow us to choose in a principled manner between, for instance, "I have two hands" and "It appears to me that I have two hands" as possible evidence reports, depending, perhaps among other things, on our degree of conviction that our senses are reliable. Although it is, before the actual specification of such a method, hard to say anything definitive about it, I suspect that it could never be wholly satisfactory because however we should opt according to the method, it would always seem that in going along with the skeptic we would be according the registered failures of our senses too much weight, whereas in going along with the Moorean we would be according them too little. Be this as it may, a satisfactory solution to our problem does seem to be available. Here, I propose an account of evidence that allows us to assume a sort of middle position between the skeptic's and Moore's, and thereby to get out of the registered fallibility of our senses only the wisdom that is in it and stop there.

The proposal makes use of the notion of uncertain evidence. Let E be some sentence that according to the Moorean can report evidence, such as the sentence "The sun is out." Then e is the sentence "It appears that E," that is, in the present case the sentence "It appears that the sun is out." Now suppose that it comes to appear to you that the sun is out. Then, presumably, if you were a Moorean, you would take E to report your evidence. If instead you were a skeptic, you presumably would take e to report your evidence. But how should you express the evidence that you are receiving if you think—as, we said, would be reasonable to do—that the Moorean is not being quite sensitive enough to the fallibility of our senses but the skeptic is being oversensitive in this respect? The following would seem a natural suggestion in that case: "I am uncertain about what exactly the evidence is I am receiving. Were I a Moorean, I would say that it is that the sun is out, but were I a skeptic, I would say that it is that it *appears* that the sun is out." Abbreviate this "uncertain evidence report" (as we may call it), or any other to the same effect, by "E/e." The important question now is how such a report could be useful and, more specifically, how one could learn anything from it.

The proposed answer to this question draws on the machinery of Bayesian epistemology. As discussed, Bayesians hold that we ought to accommodate new evidence via the rule of conditionalization, which dictates that a person's new probability for a given sentence A after she has come to learn another sentence B (and nothing else) be equal to her earlier probability for A conditional on B. The rule is silent about how to proceed with the kind of uncertain evidence considered here. However, two relatively recent additions to Bayesian epistemology help us on this point.[7]

The first is the theory of expert functions as developed by, among others, Haim Gaifman (1986) and van Fraassen (1989). In van Fraassen's presentation of the theory, the central definitions are the following:

Definition 8.1 *Let* Pr *be person P's probability function. Then q is said to be an* expert function *for P concerning family $\mathcal{F}$ of sentences iff, for all $A \in \mathcal{F}$,* $\Pr(A \mid q(A) = x) = x$.

Notice that it follows from this and the standard ratio definition of conditional probability that if q is an expert function for P concerning $\mathcal{F}$, then for all $A, B \in \mathcal{F}$: $\Pr(A \mid B) = q(A \mid B)$.

Definition 8.2 *Let persons $P_1, \dots, P_n$ have probability functions* $\Pr_1, \dots, \Pr_n$. *Then $P_1, \dots, P_n$ constitute a* panel of experts *for person P on $\mathcal{F}$ iff there are functions q and w such that (1) q is an expert function for P in the sense of*

7. Because considerations similar to those that led us to doubt both the Moorean and the skeptical conceptions of evidence led Jeffrey to the generalized rule of conditionalization that now goes by his name (see section 3.2), one might think that Jeffrey's rule could be of help in solving the problem at hand. Perhaps it can; see the first appendix in Douven (2005). However, the theory of expert functions subsequently appealed to allows us to state a rule for updating on uncertain evidence that *is* mathematically very similar to Jeffrey's rule but has a bit more structure; and the extra structure enables us to relate an update on uncertain evidence directly and in a seemingly natural way to the probability that is most directly related to the uncertainty of the evidence, namely, the probability that sensory perception is reliable. Another reason for preferring the theory of expert functions to Jeffrey's rule has to do with unity of presentation. As shown further on, the theory of expert functions can be utilized to model one's accounting for explanatory considerations in learning from evidence when one is uncertain about how much weight to give to such considerations; I do not see how Jeffrey's rule can be used for that.

definition 8.1, and (2) w is a weight function such that $q(A) = \sum_{i=1}^{n} w(P_i)\,\mathrm{Pr}_i(A)$ *for all* $A \in \mathcal{F}$.

To say that w is a weight function is to say, first, that $w(P_i) \geqslant 0$ for $i = 1, \ldots, n$, and, second, that $\sum_{i=1}^{n} w(P_i) = 1$. In principle, this function could stand for anything, although it is most naturally interpreted as measuring person P's confidence that the respective expert is right about the sentences in $\mathcal{F}$; for instance, if $w(P_i) = w(P_j)/2$, then she trusts in expert P_i only half as much as she trusts in expert P_j.

A consequence of definition 8.2 is that whereas P's probability for A conditional on B (for arbitrary $A, B \in \mathcal{F}$) will be some average of the various $\mathrm{Pr}_i(A \mid B)$'s, it will in general not be the average as weighted by the weight function w but one weighted by some other weight function w_B, which may differ from w, as B may provide evidence for or against the expertise of the various members. (Naturally, the functions *need* not differ: B may be irrelevant to the question of the members' expertise.)

The second addition to Bayesian epistemology that we need consists of so-called counterfactual probabilities, which were proposed by philosophers of science in the 1980s in the context of what is now commonly known as "the problem of old evidence." Roughly, the problem is that as a simple consequence of the probability calculus, no data that are already known at the time that a theory is conceived can confirm that theory, which seems intuitively wrong. To do justice to our intuitions, various authors have proposed that confirmation by known data is to be measured relative not to our actual probabilities but to the counterfactual probabilities that we would have had if the data had been unknown to us.[8] To see what role counterfactual probabilities play in our account, suppose that it appears to you that the sun is out. Even though you are unsure, we still assume, how to express the evidence you are receiving, it seems not difficult at all to determine your probability for a given A conditional on your present evidence E/e on the counterfactual supposition that you endorse a Moorean conception of evidence nor to determine your probability for A conditional on E/e on the counterfactual supposition that you share the skeptic's standpoint regarding evidence. Let us use "$\mathfrak{m}(A \mid E/e)$" to designate the former, and "$\mathfrak{s}(A \mid E/e)$" to designate the latter. Because you share neither the Moorean nor the skeptical standpoint

8. See, e.g., Horwich (1982) and Howson (1984); also Sprenger and Hartmann (2019, ch. 5).

regarding evidence, these are not actual probabilities but counterfactual ones. In a sense, this is like having at hand expert opinions from two experts who hold radically different views about which sentences are suitable for reporting evidence. The obvious idea now is to apply the theory of expert functions to these counterfactual probabilities and make your conditional probability $\Pr(A \mid E/e)$ for A given E/e a weighted average of $\mathfrak{m}(A \mid E/e)$ and $\mathfrak{s}(A \mid E/e)$.

The next question is how the weights are to be determined. A plausible suggestion is that the weights should reflect your confidence in the reliability of sensory perception. Specifically, it seems that to the extent that you believe sensory perception to be reliable, you will agree with the Moorean on what can be accepted as evidence and concomitantly think it correct to set $\Pr(A \mid E/e) = \mathfrak{m}(A \mid E/e)$. Equally, to the extent that you believe sensory perception to be unreliable, you will agree with the skeptic that evidence is restricted to the phenomenal and concomitantly believe it to be correct to set $\Pr(A \mid E/e) = \mathfrak{s}(A \mid E/e)$. So, if we call the hypothesis that sensory perception is reliable "RSP," the natural thought is that the weights should be determined by your probability for RSP. If we want to follow strictly the theory of expert functions, we cannot simply let the weights be Pr(RSP) and Pr(¬RSP): according to this theory we should be sensitive to the fact that new evidence may alter the confidence that you have in the "experts." This requirement is satisfied by letting $\Pr(\text{RSP} \mid E/e)$ and $\Pr(\neg\text{RSP} \mid E/e)$ be the weights. Thus, on the current proposal conditional probabilities on uncertain evidence of the kind that we are considering are to be determined by the following formula:

$$(8.1)\quad \Pr(A \mid E/e) = \Pr(\text{RSP} \mid E/e)\,\mathfrak{m}(A \mid E/e) + \Pr(\neg\text{RSP} \mid E/e)\,\mathfrak{s}(A \mid E/e).$$

Once it is clear how to determine probabilities conditional on uncertain evidence, we can conditionalize on such evidence.

Naturally, (8.1) cannot be fully general, for it cannot provide a way for determining one's probability for RSP conditional on some piece of uncertain evidence. This may seem to be a problem, given that we need such conditional probabilities in order to apply (8.1). But the problem is only apparent. We proposed to formulate uncertain evidence reports in the way that we did because we wanted to have a type of evidence that does not presuppose an answer, one way or the other, to the question of the correctness of RSP. It seems that precisely because of this deliberately built-in neutrality as regards

RSP, pieces of uncertain evidence cannot be telling either for or against that hypothesis. That is to say, it seems that we can set Pr(RSP | *E*/*e*) equal to Pr(RSP) for all uncertain evidence *E*/*e* (and thus also set Pr(¬RSP | *E*/*e*) equal to Pr(¬RSP)). I should add that even if this easy solution were unavailable, (8.1)'s not being entirely general could not be held against it, at least not if we accept the theory of expert functions. For we cannot defer to experts for *all* our opinions; eventually we will have to provide answers ourselves to such questions as how we should weigh the various experts relative to one another and whether it would not be better to consult an additional expert. But this could not be a valid objection to the theory: that we are not always in a position to consult experts does not make expert advice less valuable when it can be had.

Because the preceding account of learning from uncertain evidence may seem fairly abstract, I briefly illustrate it with the aid of an example which also prepares for the use that we subsequently make of (8.1).

Example 8.1 Call the hypothesis that abduction is a reliable mode of inference "RABD." I now show how uncertain evidence may affect one's probability for RABD, provided that one accepts (8.1). To that end, let Pr be a given person's probability function, where this person accepts (8.1) and where $\mathfrak{m}$ and $\mathfrak{s}$ are the probability functions of her imaginary "Moorean" and "skeptical" experts, respectively; just for purposes of illustration, say that Pr(RSP) = Pr(RABD) = .5, $\mathfrak{m}$(RABD) = .7, and $\mathfrak{s}$(RABD) = .3. Furthermore, let *E*/*e* be the following evidence report: "Were I a Moorean, I would express the evidence that I am presently receiving as 'My neighbors are home'; were I a skeptic instead, I would express it as 'It appears to me that my neighbors are home'." Now suppose that our person had earlier deemed that the presence of the neighbors' car in their driveway, or at least the seeming presence of the seeming neighbors' seeming car in the seeming driveway, was best explained by the supposition that they were home. Then, we may assume, at least from her imaginary Moorean expert's viewpoint, *E*/*e* counts as a success of abduction and thus as supporting RABD: application of the rule has led, or would have led, the person to conclude that her neighbors are home, and now, from the said viewpoint at least, the evidence vindicates that conclusion. Say that $\mathfrak{m}$(RABD | *E*/*e*) = .8. From the viewpoint of the imaginary skeptical expert, on the other hand, *E*/*e* will at least not be negatively relevant to RABD: that her neighbors *appear* to be home may not confirm what the person had earlier judged to be the best explanation of the seeming presence of their car in the driveway—namely, that the neighbors *are* home—but the evidence certainly does not show that conclusion to be false nor, we may assume, does it render the conclusion less likely. For simplicity, say that, from the skeptical expert's viewpoint, the evidence is irrelevant to RABD, that is, $\mathfrak{s}(\text{RABD} \mid E/e) = \mathfrak{s}(\text{RABD})$. Finally, for the reason given in the preceding paragraph, we may set $\Pr(\text{RSP} \mid E/e)$ equal to $\Pr(\text{RSP})$. Thus, if our person comes to accept E/e and conditionalizes on it, then her probability for RABD will go up from .5 to .55, for $\Pr(\text{RABD} \mid E/e) = \Pr(\text{RSP} \mid E/e)\,\mathfrak{m}(\text{RABD} \mid E/e) + \Pr(\neg\text{RSP} \mid E/e)\,\mathfrak{s}(\text{RABD} \mid E/e) = .5 \times .8 + .5 \times .3 = .55$. ♦

Next let us turn to the question of the confirmation-theoretic status of abduction. Just as conceding to the skeptic that the Moorean conception of evidence cannot be maintained is not automatically to concede that only sense data can qualify as evidence, so conceding to her that the Russellian position vis-à-vis abduction cannot be maintained is not to concede that explanatory factors can play no role whatever in our inferential practices. With respect to the role of such factors, a middle position would seem to be desirable as well. To endorse abduction categorically would, in light of its registered failures, seem reckless. Yet to reject it wholesale would seem an ill-motivated move, for in light of much that has been discussed in previous chapters it is hard to escape the impression that explanatory considerations can guide us toward the truth frequently enough. It appears that the tools that we used in the preceding also allow us to state the desired middle position with respect to the question of which confirmation-theoretic role we are to grant to abduction.

In chapter 4, we saw that, contrary to what most Bayesians hold, updating probabilities by dint of conditionalization is compatible with assigning bonus points for explanatory reasons. The trick, we explained, is to account for explanatory relations between sentences when one determines one's conditional probabilities; by doing so, one can coherently (i.e., non-Dutch-bookably) give weight to explanatory considerations in updating probabilities. In chapter 4, this point was illustrated by means of EXPL, but in the meantime we have encountered other probabilistic versions of abduction as well. We also saw that even if abduction came with Dutch-bookability, it still would not follow that abductive reasoning makes us irrational. That would be so only if there could be no compensating advantages. And as was amply shown, there *can* be such advantages.

Here, I will not make any assumptions about *how* weight is given to explanatory considerations. Indeed, if you believe that there is a meaningful way to bring explanatory considerations to bear on the determination of priors—contrary to what was suggested in section 1.2.2—then you could assume that *that* is how explanatory considerations are taken into account in the following.

That we *can* rationally, even coherently, give weight to explanatory considerations does not mean that we *must* or *will* give weight to such considerations. Realistically speaking, whether we will want to assign a role to explanatory considerations will depend on how confident we are that such considerations guide us toward the truth. Someone convinced that they do point us to the truth will presumably want to assign bonus points on the basis of explanatory considerations (how big a bonus she will want to assign will depend, among other factors perhaps, on how strong in her view the connection between truth and explanation is). On the other hand, someone who thinks that although they may perhaps be of pragmatic concern, explanatory considerations are unrelated to truth, will in determining probabilities not want to take them into account at all. But what is one to do if one is convinced only to some intermediate degree that explanatory considerations and truth are related in a way that would make giving the former a confirmation-theoretic role a prudent epistemic policy?

Suppose that you want to determine your probability for a hypothesis H conditional on evidence E, where you believe that H explains E better than any rival hypothesis. Now try to imagine what probabilities you would assign to H given E were you to be fully convinced that explanatory considerations reliably guide us toward the truth, respectively were you to be fully convinced that such considerations are at best of pragmatic interest. It is not too unrealistic to suppose that you can come up with the requisite counterfactual probabilities; designate them by "$\mathfrak{R}(H \mid E)$" and "$\mathfrak{S}(H \mid E)$," respectively. We can again think of these as constituting a panel of experts in the sense of definition 8.2, consisting of a "Russellian" expert who in determining probabilities gives full weight to explanatory factors, and a "skeptical" expert who in doing the same disregards them completely. Unsurprisingly, then, the proposal is that you make your actual probability for H conditional on E a weighted average of the counterfactual conditional probabilities $\mathfrak{R}(H \mid E)$ and $\mathfrak{S}(H \mid E)$, where for obvious reasons this time the weights are your probabilities for the reliability of abduction and its negation, respectively, both conditional on E (for the reason explained in the comment on definition 8.2). Thus, on the current proposal your conditional probability for H given E is given by

$$(8.2)\quad \Pr(H \mid E) = \Pr(\text{RABD} \mid E)\,\mathfrak{R}(H \mid E) + \Pr(\neg\text{RABD} \mid E)\,\mathfrak{S}(H \mid E).$$

For basically the same reason that (8.1) could not be entirely general, (8.2) cannot be entirely general either. In the case of (8.1), it seemed plausible to think that $\Pr(\text{RSP} \mid E/e) = \Pr(\text{RSP})$ for all E/e. A parallel assumption for RABD would seem unwarranted, though: (8.2) is not restricted to evidence that is formulated in a special way so as to guarantee neutrality regarding RABD's truth. Sometimes it will be clear, or can be readily assumed, that there is no explanatory connection between RABD and the evidence; this will be so for our applications of (8.2) in the following. But if not, then in determining the probability of RABD conditional on the evidence you are left to your own devices. For reasons already pointed to, that is no defect of (8.2); not all our opinions can come from experts anyway.

By way of further comment on (8.2), note that the question of the status of this rule is orthogonal to the question of whether we admit only sense data statements as evidence, or also uncertain evidence reports, or even statements such as "I have two hands." Whether one thinks that only sense data statements are admissible or whether one also wants to admit "I have two hands" and its ilk, (8.2) will permit one to give weight to any possible explanatory connections between one's evidence and the hypothesis or hypotheses one is interested in, where the exact weight reflects one's confidence that explanatory power is a mark of truth; in the former case, the variable E will merely have a narrower range than in the latter.[9]

9. But how is one to proceed in case there is an explanatory connection between a hypothesis and some uncertain piece of evidence, which would seem to require the involvement not only of an imaginary Russellian expert and an imaginary expert skeptical about the confirmation-theoretic significance of explanatory considerations, but also of a Moorean expert and one skeptical about the reliability of sensory perception? Neither (8.1) nor (8.2) provides for the possibility of a consultation of all these experts together. However, in Douven (2005, p. 276) it is shown that (8.1) and (8.2) can be plausibly combined in a single rule, which does provide for the said possibility. Yet for our present purposes such a (more complicated) combined rule can be dispensed with, as we are interested only in explanatory connections between a certain hypothesis—RSP, as it happens—and certain sense data statements. To account for these connections, (8.2) suffices.

To end this section, let me also briefly illustrate by means of an example how (8.2) works in practice.

Example 8.2 Consider a person who accepts (8.2). Say that this person has probability function Pr with the associated expert functions $\mathfrak{R}$ and $\mathfrak{S}$, and say that Pr(RSP) = Pr(RABD) = .5, $\mathfrak{R}$(RSP) = .7, and $\mathfrak{S}$(RSP) = .3. Let E report some pattern in the person's sense data; for instance, let it be the sentence "I am currently having sensorimotor and tactile experiences as if I am typing on a keyboard, auditory experiences of soft clicks as if keys are being struck, and visual experiences as if letters pop up on my computer screen."[10] This is a sentence that both the Moorean and the skeptic may be supposed to admit as a possible evidence report. Further, the person takes E to be best explained by the hypothesis that her senses reliably inform her about her surroundings. Indeed, what better (simpler, more straightforward) explanation could there be for the occurrence of the pattern than that things actually are as they appear to be, that is, that the person is really typing on a keyboard?[11] That means nothing for the person's imaginary expert skeptical about explanatory considerations; so, $\mathfrak{S}(\text{RSP} \mid E) = \mathfrak{S}(\text{RSP})$. But things look quite different from the perspective of her imaginary Russellian expert, who is fully convinced of the confirmation-theoretic significance of explanatory power. Given that RSP best explains E, the expert's probability for RSP conditional on E exceeds her unconditional probability for that hypothesis, specifically $\mathfrak{R}$(RSP) = .8. Finally, the person deems E irrelevant to RABD, that is, $\Pr(\text{RABD} \mid E) = \Pr(\text{RABD})$. Then, similar to what we had in example 8.1, we find that $\Pr(\text{RSP} \mid E) = \Pr(\text{RABD} \mid E)\,\mathfrak{R}(\text{RSP} \mid E) + \Pr(\neg\text{RABD} \mid E)\,\mathfrak{S}(\text{RSP} \mid E) = .55$. Hence, if the person comes to learn E and subsequently conditionalizes on it, her probability for RSP will increase from .5 to .55. ♦

10. Many similar examples can be distilled from Vogel's 1990 paper as well as from BonJour's rather extensive catalog of characteristics of sense data that seem best explained by RSP (in BonJour & Sosa, 2003, pp. 88–91).

11. Are these experiences not already explained by assuming that her senses are *now* correctly informing her about her surroundings? They are, but we are after the *best* explanation, and potential explanations are to be assessed, among other things, on the basis of their scope or generality. On this basis, the hypothesis that her senses *typically* inform her correctly about her surroundings is to be preferred to the much less general hypothesis that her senses *now* inform her correctly about her surroundings.

8.4 How to Resolve the Skepticism Debate?

I will now show that if we accept the middle positions regarding evidence and abduction presented previously, then, given some very weak further assumptions that no party to the skepticism debate should find disputable, a resolution of this debate becomes possible, at least in principle.

The hypotheses in which we are interested are RSP, which states that our senses are reliable, and RABD, which states that abductive reasoning is reliable. The main thing to be shown is that someone whose probability function satisfies certain (weak) constraints and who accepts both (8.1) and (8.2) may come to assign a high probability to both RSP and RABD. Example 8.1 shows how, assuming (8.1), RABD may be confirmed by uncertain evidence of a particular type, and example 8.2 shows how, assuming (8.2), RSP may be confirmed by evidence reporting certain patterns in a person's sense experiences. What may not be immediately obvious from those examples and what the following will make clear is that, given the use of both (8.1) and (8.2) and supposing that there will be "enough" evidence of the kinds encountered in the examples for us to arrive at probabilistic conclusions about RSP and RABD that are strong enough to allow for the refutation of skepticism, the initial probabilistic assumptions about evidence and hypotheses (and their interrelations) can be exceedingly weak.

The person that we consider has an initial probability function Pr with associated imaginary expert functions $\mathfrak{m}$, $\mathfrak{s}$, $\mathfrak{R}$, and $\mathfrak{S}$, as introduced in the previous section. Our first assumption is that the person herself as well as all her imaginary experts assign positive initial probabilities to both RSP and RABD; in other words, for all $p \in \{\mathrm{Pr}, \mathfrak{m}, \mathfrak{s}, \mathfrak{R}, \mathfrak{S}\}$ and $H \in \{\mathrm{RSP}, \mathrm{RABD}\}$:

$$(8.3) \qquad p(H) > 0.$$

Note that this is compatible with the person's finding initially vastly more probable that our senses are not reliable than that they are, and also with her finding initially vastly more probable that abductive reasoning is not reliable than that it is; she may in fact be nearly certain of the unreliability of both (but not entirely certain). I should also note that neither (8.3) nor any further assumption to be made is incompatible with the person's (and her imaginary experts') assigning a very high prior probability to, for instance, "We are brains in a vat," and a very low one to the external-world hypothesis.[12]

The second assumption concerns the various pieces of evidence our person comes to know in the course of her investigations.

(8.4) Insofar as the evidence that becomes available to our person is relevant to the hypotheses at issue, it can be represented as a sequence σ each of whose elements is either a piece of uncertain evidence that is relevant to RABD—using our earlier notation, we designate these elements by "E_i/e_i" ($i \in \mathbb{N}$)—or a sense data statement that is relevant to RSP—these elements are denoted by "E_i'" ($i \in \mathbb{N}$).

Informally, the idea is that the pieces of uncertain evidence all qualify as successes of abduction at least from the perspective of a Moorean; so among them could be the uncertain evidence report called "E/e" in example 8.1. The E_i', on the other hand, all report patterns in our sense experiences that are best explained by supposing that our senses reliably inform us about our surroundings; among them could be, for instance, the evidence sentence called "E" in example 8.2. If $i < j$, then E_i/e_i comes before E_j/e_j in σ, and similarly for E_i' and E_j'. Apart from these, no assumptions need to be made about the ordering of the elements of σ.

The remaining assumptions make more precise how the elements of σ bear on RSP and RABD relative to Pr and the various expert functions. To state these assumptions, I adopt the notational convention that if among the first $m + n$ elements of σ there are m evidence reports of the sort that we indicate by "E_i'" and n of the sort that we indicate by "E_i/e_i," then, with $p \in \{\text{Pr}, \mathfrak{m}, \mathfrak{s}, \mathfrak{R}, \mathfrak{S}\}$, p_n^m is p updated, consecutively, on the first $m + n$ elements of σ, in the order in which they appear in the sequence. So if, for example, σ should begin

$$\langle E_1/e_1, E_2/e_2, E_1', E_3/e_3, E_4/e_4, E_2', E_3', E_4', \ldots \rangle,$$

then p_2^0 is the result of updating p consecutively on the first two elements of σ, and p_4^3 the result of updating it consecutively on the first seven elements. To give a slightly more concrete example (supposing still the same beginning

12. However, as can easily be verified, given assumptions (8.3), (8.4), and (8.5), she could not coherently assign probability 0 to the external-world hypothesis (nor could her imaginary experts).

of σ), $\mathfrak{S}_4^3(\text{RSP} \mid E_4')$ is the probability assigned by the person's imaginary expert skeptical about explanation to RSP conditional on E_4' after the person has come to know the first seven elements of σ.

The third assumption is that no element of σ is from the perspective of any of the imaginary experts negatively relevant to either RSP or RABD, that is to say, for $p \in \{\mathfrak{m}, \mathfrak{s}, \mathfrak{R}, \mathfrak{S}\}$, $H \in \{\text{RSP}, \text{RABD}\}$, and all E_m', $E_n/e_n \in \sigma$

$$\text{(8.5)} \qquad \begin{aligned} p_n^{m-1}(H \mid E_m') &\geqslant p_n^{m-1}(H); \\ p_{n-1}^m(H \mid E_n/e_n) &\geqslant p_{n-1}^m(H). \end{aligned}$$

Given that, informally, we think of the E_i' as reporting experiences that are best explained by RSP, it would seem plausible to assume that at least $\mathfrak{R}_n^{m-1}(\text{RSP} \mid E_m') > \mathfrak{R}_n^{m-1}(\text{RSP})$ for all E_m'. Similarly, because we think of the E_i/e_i as successes of abduction from the standpoint of the Moorean, it would seem plausible to assume that $\mathfrak{m}_{n-1}^m(\text{RABD} \mid E_n/e_n) > \mathfrak{m}_{n-1}^m(\text{RABD})$ for all E_n/e_n. Nevertheless, for the result we are aiming at, the weaker (8.5) is enough.

Next, we have that for all E_m', E_n/e_n

$$\text{(8.6)} \qquad \begin{aligned} \mathfrak{R}_n^{m-1}(\text{RSP} \mid E_m') &\geqslant \mathfrak{S}_n^{m-1}(\text{RSP} \mid E_m'); \\ \mathfrak{m}_{n-1}^m(\text{RABD} \mid E_n/e_n) &\geqslant \mathfrak{s}_{n-1}^m(\text{RABD} \mid E_n/e_n). \end{aligned}$$

This will be immediately obvious from our description of the elements of σ in relation to RSP and RABD, as perceived by the various experts; that description in effect seems to warrant $>$'s instead of the $\geqslant$'s, but again the weaker assumption suffices.

A further assumption is that for all E_m', E_n/e_n,

$$\text{(8.7)} \qquad \begin{aligned} \Pr_{n-1}^m(\text{RSP} \mid E_n/e_n) &\geqslant \Pr_{n-1}^m(\text{RSP}); \\ \Pr_n^{m-1}(\text{RABD} \mid E_m') &\geqslant \Pr_n^{m-1}(\text{RABD}). \end{aligned}$$

It is reasonable to suppose, as previously stated, that pieces of uncertain evidence are probabilistically irrelevant to RSP; the $\geqslant$ in the first part of (8.7) is merely to make the assumption as weak as possible. As to the second part, note that it is perfectly conceivable that evidence that is at least from a Russellian perspective explanatorily relevant to RSP, such as the E_i''s in σ are supposed to be, is from *any* perspective completely irrelevant to RABD; in any event there is no reason to suppose that such evidence, or at least some such pieces of evidence, should be negatively relevant to RABD. Take our example, "I am currently having sensorimotor and tactile experiences as if I am typing on a keyboard, auditory experiences of soft clicks as if keys are being struck, and visual experiences as if letters pop up on my computer screen." One would seem hard-pressed to give an argument to the effect that this sentence, or any sentence like it, is *bound* to be negatively relevant to RABD, or at least that by supposing otherwise we would be begging skeptical issues.

The last two assumptions are

$$(8.8) \qquad \lim_{m,n\to\infty} \mathfrak{R}_n^{m-1}(\text{RSP} \mid E'_m) = 1$$

and

$$(8.9) \qquad \lim_{m,n\to\infty} \mathfrak{m}_{n-1}^{m}(\text{RABD} \mid E_n/e_n) = 1.$$

That is, we assume that, as the person comes to know more and more elements of σ, her imaginary Russellian expert's probability for RSP converges to 1 and her imaginary Moorean expert's probability for RABD converges to 1 also. Because these assumptions are clearly more substantive than the previous ones, I want to emphasize that the skeptic should have no difficulty granting them, for they are still exceedingly weak. To see just *how* weak, note that, for instance, (8.9) assumes that *in the infinite limit* (and not necessarily sooner) after our person has come to know infinitely many pieces of evidence *all* of which (and not merely some, or most, of which) report *from a Moorean perspective* (and not necessarily from anyone else's perspective) successes of abduction and *none* of which report from *any* (and not just some) perspective failures of abduction, the imaginary *Moorean* expert will be certain of RABD conditional on any further such piece of evidence (for all we are supposing, the imaginary *skeptical* expert's corresponding probability may well be no higher than her initial probability for RABD). What could one be denying, by making this assumption, that the skeptic might hold dear? Similar remarks apply to (8.8).[13]

The link between these assumptions and the skepticism issue is established by the following theorem, which is proved in appendix D:

13. I would expect the Humean skeptic to find both (8.8) and (8.9) disputable. She presumably would think that any Bayesian ought to choose an initial probability distribution guaranteeing that no claim not entailed by any finite conjunction of evidence sentences can have its probability altered in the course of her (the Bayesian's) coming to know more evidence. A fortiori, it would from a Humean perspective not be reasonable for the Russellian to become

Theorem 8.1 *If (A1)–(8.9) hold, then*

$$\lim_{m,n\to\infty} \Pr_n^m(\text{RSP}) \;=\; \lim_{m,n\to\infty} \Pr_n^m(\text{RABD}) \;=\; 1.$$

From an epistemological point of view, what happens in the infinite limit may seem to be of academic interest at best. Note, though, that by the definition of a mathematical limit, it follows from theorem 8.1 that

Corollary 8.1 *If (8.3)–(8.9) hold, then for any c such that* $0 \leqslant c < 1$

(1) *there are* E'_k, $E_l/e_l \in \sigma$ *such that for all* $E_n/e_n \in \sigma$ *with* $n \geqslant l$ *and all* $E'_m \in \sigma$ *with* $m \geqslant k$ *it holds that* $\Pr_n^m(\text{RSP}) > c$, *and*
(2) *there are* $E'_{k'}$, $E_{l'}/e_{l'} \in \sigma$ *such that for all* $E_{n'}/e_{n'} \in \sigma$ *with* $n' \geqslant l'$ *and all* $E'_{m'} \in \sigma$ *with* $m' \geqslant k'$ *it holds that* $\Pr_{n'}^{m'}(\text{RABD}) > c$.

So, given some threshold value such that probabilities exceeding it qualify as high, it follows from theorem 8.1 that if (8.3)–(8.9) hold, then after the person has come to know some *finite* initial segment of σ it will be, and remain, highly probable to her that sensory perception is reliable and also highly probable that abduction is a reliable mode of inference.

How does this result bear on skepticism? Suppose, for now, that RSP does become highly probable to me and that that justifies me in believing RSP. Then, assuming that I am justified in believing how things appear to me, I am justified in believing that (1) my senses reliably inform me about my surroundings, and hence things generally are as they appear to be in my perceptual experiences, and (2) most or even all of my perceptual experiences appear to be of an external world. Together beliefs (1) and (2) justify my belief that there exists an external world. The same conclusion is warranted on the supposition that I become justified in RABD, still assuming the explanatory superiority of the external-world hypothesis. Thus, supposing the conditions of theorem 8.1 to be satisfied, we can become justified in believing that there is an external world, along two different routes, in effect.

---

increasingly confident in RSP as she comes to know more and more elements of σ, nor would it be reasonable for the Moorean to become increasingly confident in RABD under the same circumstances. So, by assuming that the external-world skeptic is willing to grant (8.8) and (8.9), and thus to grant that learning from experience is possible at least for the Russellian and the Moorean, I am effectively assuming that the external-world skeptic is not also a Humean skeptic. Again, I take this assumption to be justified in light of Schurz's work on meta-induction (Schurz, 2019).

I assumed that high probability would be enough for justification. According to some authors, that is indeed the case.[14] But most nowadays take Henry Kyburg's (1961) so-called lottery paradox to show that the connection between probability and justification cannot be so straightforward. Many of these authors believe that high probability is only *nearly* sufficient for justification and that some defeater has to be added to the condition of high probability to keep the lottery paradox at bay (e.g., Korb, 1992; Pollock, 1995; Ryan, 1996; Douven, 2002b; Olin, 2003). There is no agreement among them about what this defeater should come to, but they do generally agree that it should apply as selectively as possible to the type of propositions that engender the lottery paradox (such as, most typically, propositions to the effect that a given ticket in a particular lottery is a loser). As neither RSP nor RABD is of this type, these authors would probably grant that our result shows that we can become justified in believing these hypotheses.[15]

As mentioned in section 2.2.2, the connection between probability and justification may actually be still more complicated than philosophers have so far believed. But even in the worst case, in which we cannot infer anything about the justificatory status of a statement however probable it is (barring, I suppose, its being certain), the nonskeptic would still seem to have gained the upper hand if RSP and RABD had been shown to be highly probable. What is left of the pull of skepticism if we are in a position to claim that, in all likelihood, there is an external world?

I end with a number of comments on the previous result. First, I have tried to be as minimalistic as possible in choosing the inferential machinery to be used in the debate against the skeptic. Specifically, I assumed Bayes's rule, the theory of expert functions, and counterfactual probabilities. Might

---

14. See, among others, Kyburg (1961), Klein (1985), and Foley (1992).

15. Depending on one's account of knowledge, this may suffice to show that we can come to know that in all likelihood there is an external world (supposing that there *is*). I will not argue for this claim here, however, as it should be enough to persuade the skeptic to show that, starting from probabilistic assumptions she should find unobjectionable, we can become *justified in believing* that there exists an external world (perhaps it is already enough to show that this can become highly probable to us). And according to Wright (1991), it is justified belief in, rather than knowledge of, an external world that is at stake in the skepticism debate. Specifically, he maintains that "[k]nowledge is not really the proper central concern of epistemologico-sceptical enquiry. . . . We can live with the concession that we do not, strictly, *know* some of the things we believe ourselves to know, provided we can retain the thought that we are fully justified in accepting them" (p. 88). In the same vein, see Russell (1912, ch. 2), but see also Nagel (2014, ch. 2) for a dissenting opinion.

the skeptic still object that I have thereby been loading the dice against her? I do not see how. Of course, the *Humean* skeptic might complain about the use of Bayes's rule in particular, but the foregoing addressed the *Cartesian* skeptic, who, we may assume, is perfectly comfortable using Bayes's rule to make inferences about future sensory experiences.[16] Similarly for the theory of expert functions and for the concept of counterfactual probability: I can think of no reason why either should be objectionable from the skeptic's viewpoint or why adopting one or both of them would, in itself, tip the scale in favor of the nonskeptic. These elements were needed for formalizing views of evidence and inference that are different from those held by skeptics and that the skeptic *might* want to reject. But I have done my best to argue that the new proposals are more reasonable than the skeptic's views on these matters and should appear as such also to the skeptic.

Second, it cannot be stressed enough that we do not have a resolution of the skepticism debate here. We have no more than *stipulated* the existence of the sequence σ with its specific connections to the person's probability function. One may wonder how sensible it is to suppose that if we were to do the empirical research required for a resolution of the skepticism debate, the data would turn out as favorably for the nonskeptic as is required for our result to apply. It may be said that while perhaps it is sensible to suppose that by far most of our experiences are best explained by assuming that our senses are reliable and also that a systematic investigation of the success rate of abduction would reveal that this mode of reasoning mostly serves us well, σ contains only nonnegative evidence for our hypotheses RSP and RABD and that this is too much to assume. First, however, note that the preceding is compatible with there being finitely many pieces of evidence *negatively* relevant to either RSP or RABD (or both), as long as it is assumed that after the last such piece of evidence has become known to her, our person and her imaginary experts still assign some positive probability to both RSP and RABD. Second, I strongly suspect that in particular (8.5) can be relaxed so as to permit occasional negative evidence to pop up even in the long run while still allowing us to derive the requisite convergence result. Third, and most importantly, it should not be forgotten that the claim that I am making in this section is very much an in-principle claim: it is in principle possible to resolve the skepticism debate empirically, that is, to settle it in a way that should be acceptable to all parties.

16. As for the Humean skeptic, I would once more refer her to Schurz (2019).

To the extent that at the beginning of this chapter such a resolution seemed *im*possible in principle, the result obtained here should still count as a step forward.

Third, while the convergence result may very well *describe* how we nonskeptics actually come to rely on our senses as well as on abductive reasoning and thereby come to be nonskeptics, the result is in the first instance meant to have *normative* implications. In sections 8.2 and 8.3, reasons were given for preferring the middle positions regarding evidence and inference to the traditional "extreme" positions regarding these matters (both the skeptical and the nonskeptical positions). Further, supposing these middle positions and a number of very weak assumptions which, I have argued in this section, should seem reasonable to all parties involved, and supposing "favorable" data, a convergence result could be stated to the effect that accumulating evidence will justify us in believing that our senses are reliable as well as in believing that abduction is a reliable mode of inference. And that, as was also argued, is enough to justify us in believing in the existence of the external world. Hence, if the data turn out favorably, then it follows from the result that we *ought* to be non-skeptics regarding the external world.

Fourth, Wright (1985) and others have argued against Moore's proof of an external world that although Moore does have justification for "I have two hands," this justification does not transmit to "There exists an external world," even if we assume the entailment of the latter sentence by the former to be known. The reason for this transmission failure is that, to these authors' eyes, Moore must be antecedently justified in the conclusion of this argument if he is to have justification for "I have two hands." If they are right, then even if justification is closed under known entailment, that still does not mean one's justification for certain premises justifies one in believing a conclusion that is known to follow from these premises. Whether they are right is still a matter of controversy.[17] Be this as it may, Wright's and others' appraisal of Moore's argument is irrelevant to the two arguments that, I claimed, justify us in believing that there is an external world if assumptions (8.3)–(8.9) hold. Consider that the empirical argument that in that case would justify us in believing RSP and RABD merely assumes that we assign a positive probability, *which may otherwise be as small as one likes*, to the external-world hypothesis. It would certainly be wrong to say that the external world hypothesis has

17. See, for instance, Pryor (2004) for some incisive criticisms.

been *assumed* in an argument if, for all the argument must suppose, that hypothesis's probability may be vanishingly small.

Finally, and most relevantly to our overarching concern, it has been seen how, from minimal assumptions, we can defend abduction in a way that makes it useful in the kind of context in which philosophers had *hoped* it would be useful. The skepticism debate may well provide the most challenging of these contexts, given that (as said) there is really not much one can start from without begging any issues. In particular, while the skeptic might be happy to admit abduction to make inferences about sensory impressions, she might object to the use of abduction in order to infer the existence of objects and processes causing, but not reducible to, such impressions. That would seem to make the prospects of an empirical defense of abduction bleak. But taking on board conceptions of evidence and inference that, I have argued, should make sense to all parties, it was seen how we can more or less simultaneously build confidence in the reliability of our senses and in applications of abductive reasoning that let us infer the existence of real objects and processes.

This same construction can be reapplied in the scientific realism debate, which provides an only slightly less challenging context than that of the skepticism debate. Scientific antirealists are typically willing to grant the existence of observables—anything that can be seen with the naked eye—but regard any claim postulating unobservables as a case of epistemic overreach. For instance, van Fraassen (1985) has raised skeptical concerns about realist readings of micrographs, in particular about taking such micrographs as evidence for the existence of bacteria, viruses, molecules, and other such things. According to him, there is no non-question-begging evidence showing that our current microscopes are veridical. But few realists have been swayed by his arguments. The following excerpt from an interview with Ronald Giere expresses a widely shared realist sentiment:

> An overall kind of nonrealism or antirealism I find very hard to even take seriously. I know people do it. But look at electron microscope pictures of chromosomes, of DNA—we have pretty nice pictures of DNA! The overall idea that there is an observational realm and that we should not go beyond that I think is just not tenable. (Giere in Callebaut, 1993, p. 171)

However, even though most people with some background in science will be sympathetic to Giere's view, it would seem prudent to be at least initially somewhat more cautious in our interpretations of micrographs. In the case of perception, we all have first-hand evidence of its fallibility, evidence which, we said, should make us wary of going along with Moore's proof of the external world even if it should not make us side with the skeptic. Although not as generally known, however, there is also evidence showing that modern microscopes are not completely trustworthy and that, for example, procedures used for preparing specimens can produce various sorts of artifacts, some of which can go unrecognized for quite some time, perhaps even forever (Rastogi et al., 2013). So here, too, a kind of intermediate position between van Fraassen's and Giere's would seem the more reasonable one, at least as a starting point.

Then, in order to test empirically whether abduction is reliable also when we infer, or become more confident in, the existence of certain unobservable kinds of entities postulated by best-explaining theories, we can imagine having at hand an expert deeming microscopes at least somewhat trustworthy, next to an expert who instead shares van Fraassen's view on microscopes. And also parallel to what we had before, we have our pair of experts holding different views on the reliability of abduction as applied to unobservables. The rest then unfolds as it did previously. Suppose that using some type of microscope, we "discover" a virus that had earlier been postulated on explanatory grounds. That would constitute evidence for the reliability of abduction from the perspective of the expert who believes microscopes to be at least somewhat trustworthy. Similarly, Hacking (1981, pp. 144–145) points out that when we study blood cells both by means of a low-resolution electron microscope and by means of a fluorescent microscope—types of microscopes that exploit different physical principles—the visual outputs we get from these microscopes tend to be very similar. As he rightly remarks, that would seem best explained by assuming that these types of microscopes are veridical and thus would constitute evidence for that assumption at least from the perspective of the expert who has some confidence in the reliability of abduction.[18]

18. For further details, see Douven (2005).

Epilogue

I have aimed to rectify what I see as persistent misunderstandings about, as well as a general underappreciation of, a mode of reasoning that, evidence indicates, forms an important part of our cognitive tool box: abduction, which gives special weight to explanatory relations, as when, for instance, the appreciation that our evidence makes perfect sense in light of one hypothesis but not so much in light of its rivals raises our confidence in the former at the expense of our confidence in each of the latter.

That does not mean that there is literally a single rule of abduction that people can apply wherever and whenever they are able to make judgments of explanation quality. I have argued that critics are basically right when they contend that abduction is a slogan but wrong to deem this a shortcoming. Deductive logic offers a set of domain-general and topic-neutral rules, and perhaps this made many think that the rules guiding our nondeductive reasoning must be of that kind as well. On the other hand, chapter 6 looked at a psychologically oriented theory of rationality that makes it from the start reasonable to expect that the rules for nondeductive reasoning will *not* hold uniformly. Even more generally, the *Encyclopædia Britannica* (2006) circumscribes intelligence as the "ability to adapt effectively to the environment." While this is only one among many definitions of intelligence around, most would agree that the said kind of adaptability is at least a key characteristic of human intelligence. And this, too, should make it unsurprising if people are found to reason in ways that can be fine tuned depending on the environment so as to more readily pick up the inductive cues provided by that environment. I looked at rules of inference that appear to have that feature, and those rules were seen to do better than Bayes's rule in environments that were nothing out of the ordinary. The rules, while context specific, were tied together by the slogan that explanation guides inference.

If you had heard about abduction before you read this book, what you heard was probably mostly critical. Current thinking about how we ought to respond to new information, how we should revise our beliefs or degrees of belief in light of new evidence, is so strongly dominated by Bayesian confirmation theory that the idea that we are better served, even if perhaps only sometimes, by any mode of reasoning that could go against Bayesian tenets may be taboo to many. Bayesians have nothing against abduction as a kind of assistive tool, a helper function for determining priors perhaps, or (sometimes) likelihoods, but to the extent that abduction conflicts with Bayesian dogma, we should avoid it like the plague.

Bayesians have been quite explicit about the doom that awaits abductive reasoners: they make themselves vulnerable to exploitation, and they make their degrees of belief unnecessarily inaccurate. Worse yet, they do so willingly, or at least they could have figured out a priori that those things were going to happen! If that is not a token of irrationality, what is?

At first blush, the arguments for these claims can easily seem unassailable, supported even by *proofs*. On closer scrutiny, however, those proofs do not appear to connect so well to Bayesian allegations of irrationality on the part of, for instance, abductive reasoners. That is because, in arguing for the broader claims, Bayesians sneaked in some assumptions that the friends of abduction need not sign up to. We saw that if we think of update rules as packaged together with other epistemic principles and also with decision-theoretic principles, then there are ways to reason abductively free of charge, meaning that, for instance, you can assign bonuses for best explanations and still avoid Dutch books. That such additional principles are irrelevant and thus can be ignored, or that there must be something amiss with the ones that make for a non-Dutch-bookable combination with abduction, has been assumed by Bayesians without argument. Similarly, their inaccuracy-minimization argument assumes, again without any support, that the one and only sense of inaccuracy minimization that should matter to questions of epistemic rationality is that of expected next-step inaccuracy minimization. Moreover, Bayesians present the latter argument as being purely epistemic, assuming, again without argument, that scoring is completely free of pragmatic encroachment. As if that were not enough, we found that, for independently plausible ways of measuring inaccuracy, the claim is false to begin with.

Apart from all that, the dynamic Dutch book and inaccuracy-minimization arguments would still have been successful only had they shown that using EXPL or any other non-Bayesian update rule incurs costs to no benefit. But the most that they have shown is the "incurs costs" part and not the "to no benefit" part. As said, I strongly doubt that the dynamic Dutch book argument shows much of interest at all. However, cutting the Bayesian some slack (e.g., by granting that the Brier score is the only correct way of measuring inaccuracy), at least the inaccuracy-minimization argument could be said to accomplish *something*, viz., shifting the burden of proof to those wanting to maintain the rationality of forms of non-Bayesian updating, making it incumbent on them to show that there *are*, or at least *can* be, benefits to such updating procedures.

That is what I have sought to do, using various probabilistic instances of abduction. The main point of chapters 6 and 7 was to show that in some contexts the use of such rules has clear advantages over the use of Bayes's rule. I very much doubt that what I have shown generalizes to each and every context. Indeed, we showed in section 7.4 that sometimes Bayes's rule did best in a context of social learning, where teams of doctors were evaluated on the basis of their ability to save their patients' lives. From my perspective, that is perfectly fine. My goal has emphatically *not* been to show that the widespread popularity that Bayesianism enjoys is completely undeserved; adopting Bayesian norms may be the right thing to do in certain contexts. As I put it in section 1.2.1, what is misguided is Bayesian *imperialism*, which deems any deviation of those norms a token of irrationality. In chapter 6, I embraced insights from Elqayam, Gigerenzer, and others, who have argued against a universalist conception of rationality and in favor of a more local and contextual notion that makes room for a plurality of normative criteria. Rationality, in this view, requires a match between, on one hand, a subject's epistemic or practical behavior and, on the other, her goals, her cognitive capacities, and the environment in which she finds herself.

A caveat about the probabilistic versions of abduction that have featured in the foregoing: no claim has been made to the effect that these rules are realistic in any way, that they are actually being employed by people, or that they are at least idealizations of rules that we actually employ. Some of the rules—in particular Good's and Popper's rules—were loosely inspired by the findings from Douven and Schupbach (2015b), which showed that Good's and Popper's measures of explanatory strength came close to capturing human judgments of explanatory strength. It does not follow from this, however, that people use Good's or Popper's rule to accommodate new evidence, which in turn does not imply that these rules cannot at least sometimes serve as useful approximations of actual human reasoning. It would certainly be worth conducting further experimental work to investigate the extent to which these rules are able to accurately predict people's deviations from Bayesian reasoning. That work might suggest versions of explanation-based update rules that are more realistic than EXPL, Good's rule, or Popper's rule. If these to-be-discovered rules can be formalized—admittedly, a big *if*—we might be able to investigate their normative status much in the way in which we put EXPL and the other rules to the test in chapters 6 and 7. But, as intimated at the start (section 1.2.1), I am committing neither to the descriptive adequacy nor to the normative correctness of EXPL or any of the other explications of abduction that featured in the foregoing. In the present work, these abductive rules serve only dialectical purposes. In particular, they were introduced to help make two main points: (1) that abductive reasoning can in some contexts outperform Bayes's rule in ways that are of immediate importance to us; and (2) that the exact form of abductive reasoning on which we should rely may differ from one context to the next, so that there can be no single best formalization of abduction to which we can turn irrespective of the circumstances. The choice for EXPL was primarily motivated by the fact that it had been used in one of the main *attacks* on abduction, making it seemingly safe to choose the same explication in a *defense* of abduction. Good's and Popper's rule were, as said, introduced as variants of EXPL to make the point that different contexts may call for different explications of abduction; those rules were inspired by empirical findings about how real people update their probabilities, but without the ambition to capture real updating behavior.[1]

For those interested in pursuing follow-up research aimed at providing more realistic versions of abduction, I note here that there is no shortage of alternative explications of abductive reasoning that prima facie merit some atten-

1. Thus I take Climenhaga's (2017b) critique that EXPL and the other probabilistic versions of abduction that we considered may lead to trouble when simultaneously applied at different levels of explanation to be misguided, not only because, as said (and as pointed out in various previous publications), I consider none of those formulations a definite statement of a possible realistic instance of abduction, but also because, even more importantly, different levels of explanation may create different contexts, and there is no reason to hold that whichever explication of abduction is best suited for one level of explanation must also be best suited for any other level of explanation. (Climenhaga does say at various junctures that he believes that it is preferable to have only one version of abduction, without however providing an argument for this claim.)

tion. One could not only consider different measures of explanatory goodness but also related measures, notably measures of coherence (see section 1.2.1) and measures of the degree of competition among hypotheses (Schupbach & Glass, 2017; Glass, 2021a). Moreover, contributions to the debate about explanatory reasoning tend to be noncommittal regarding theories of explanation—as this book has also been—and in general there has been little contact between the debate about rational updating and the debate about explanation. It might be worth investigating whether the various proposals concerning explanation that have been made in the philosophy of science can be compared in the kind of settings developed in chapters 6 and 7. For instance, might a suitably formalized causal model of explanation (perhaps building on Pearl, 2000) do better in the sort of simulations reported in those chapters than an equally suitably formalized unificationist model (perhaps building on the aforementioned measures of coherence)? It would be an interesting finding if, for example, inferring to the best causal explanation turned out to be generally more successful, or in some contexts more successful, than inferring to the best unifying account.

Jeffrey (1992) famously argued that Bayesian epistemology should not be conceived as providing a set of rules that apply mechanically and universally but rather as providing tools whose proper handling requires experience and skill. Indeed, he thought of Bayesian reasoning as constituting an *art*: "the art of judgment." The view on abduction that I have advocated in this work is very much in the same spirit except that it presents the art as being even more encompassing: we also have to be able to pick the, or a, right updating rule for any given context. That may sometimes be Bayes's rule, but it can also be some version of abduction.

Scientific progress has to an important part been due to the fact that we have been able to devise better ways of obtaining data about the world. But, to repeat from chapter 1, it has also been due to the fact that we have learned more about what we can *infer* from the data that we have obtained: that we have learned more about how to learn. We have not just expanded our knowledge; we have improved how knowledge is being expanded. To my eyes, becoming clearer about our inferential practices, in particular about the confidence that these practices permit us to have in our beliefs, constitutes progress—*scientific* progress, even if perhaps not of the most glamorous kind. I have tried to contribute to this progress by, I hope, removing some misconceptions of abductive reasoning and by showing that, in the right hands and when used judiciously, it can be a powerful tool.

My main concern has been to make a positive case for abduction as a mode of nondeductive reasoning in its own right, not subservient to this or that confirmation theory. Given the dominance of Bayesianism, however, this book may well be seen as a critique of that position. So, in closing, I want to be still a bit clearer about how, in my view, the two accounts relate to each other and what I would regard as helpful responses from Bayesian colleagues and what as unhelpful responses.

Although not concerned with abduction, Giere (1988) voiced serious misgivings about Bayesianism, and he did so largely on the same grounds that motivated the advocates of ecological rationality. Most centrally, he argued that people cannot possibly obey Bayesian norms given their limited capacity to gather, memorize, and process information. This consideration led Giere to reject Bayesianism altogether and to adopt in its stead a Simonian satisficing account of rationality.

It has always appeared to me, however, that there is a tension between Giere's unqualified rejection of Bayesianism and the version of the so-called semantic view of theories that he offers in the same book, a version that many still regard as the best on offer. According to the semantic view, scientific theories should not be thought of as axiomatic systems, connected to the world via certain bridge principles, but rather as sets of models. Giere convincingly argues that there is an important element missing from the standard presentation of the semantic view, to wit, the observation that theories also comprise hypotheses postulating similarity relations between the models that they define and certain worldly systems. To present a theory is to present a set of models *and* to point out which parts of those models reflect which parts of reality to what degree.

While I share many of Giere's qualms about Bayesianism—even if these only lead me to reject Bayesian imperialism—I still believe that precisely his version of the semantic view allows for a more favorable assessment of Bayesianism than it receives in Giere (1988). For in spite of all the criticisms, one could still argue that Bayesianism offers a set of models of reasoning—models for synchronic and diachronic belief management and for decision making—that not only allow us to represent with some degree of precision important aspects of how people do behave cognitively but also captures to some extent our pretheoretical ideas about what it is to behave in cognitively appropriate ways (e.g., how to choose wisely).

Clearly, though, this is not to say that there is no room for improvement. If Bayesianism captures important aspects of our reasoning to some extent, but we are able to capture those aspects more accurately or more comprehensively by acknowledging the role that explanatory considerations sometimes play in reasoning, then that is progress. If it turns out that by attending to such considerations we sometimes get more of what we want from our reasoning than if we were pure Bayesian agents, then the same proposal is also closer to describing our pretheoretic normative thinking about rationality. The proposal made in this book, at least insofar as it pertains to graded beliefs, is far from a radical departure from Bayesianism. It emphasizes the importance of explanation to reasoning, but it could be rightly remarked that from a neutral perspective the update mechanisms that we looked at, which are sensitive to explanation quality, are but relatively minor deviations from Bayes's rule. At an abstract level, the endeavor undertaken in the foregoing is perhaps best thought of as an attempted de-idealization (McMullin, 1985) or concretization (Nowak, 1980) of an existing account of rationality, meaning that we try to pass from certain idealizations to less idealized assumptions and thereby to get closer to an account of rationality for real people. That is how I hope to have improved on Bayesianism.

Paul Moser (1990, p. 6) urges specifically with respect to decision theory that "adequate principles of rational decision making should be psychologically realistic relative to actual human decision making." The present work has been written in the conviction that the same holds, *mutatis mutandis*, for theories of rationality generally and that our efforts should be geared toward making such theories more responsive to human psychology, in particular to cognitive limitations. We can acknowledge as much without abandoning the project of conceiving a normative theory of rationality. Indeed, it would seem that *only* a theory of rationality that attends to human psychology can have normative force, inasmuch as the time-honored principle that "ought" implies "can" would be totally empty if "can" were not to refer to what is possible for actual human beings to accomplish.

To be sure, theorizing may become hopelessly difficult if we try to do away with idealizations altogether. But our goal can and should still be to see which of the idealizations that we are making can be replaced by more realistic assumptions while nonetheless retaining a theory that *both* accurately describes people's belief management and decision making *and* reflects our main pretheoretical convictions regarding rationality. This goal is not fostered by taking on board *more* and *crasser* idealizations just to shield one's preferred theory from refutation.[2] In this light, my recommendation to Bayesians wanting to object to the criticisms that this book levels against the dynamic Dutch book and inaccuracy-minimization argument is to refrain from taking on board assumptions that would bring us further away from a theory of rationality for real people, for instance, assumptions that require cognitive capacities entirely unlike any we ordinary mortals possess.

For instance, in section 4.2, I compared Bayes's rule with an instance of EXPL by letting two players compete in a game in which they asserted what they thought was the true hypothesis as soon as that was subjectively highly likely to them. The inspiration for this came from a remark by Frank Jackson about high subjective probability warranting assertion. Bayesians might refuse to go along with that idea and propose instead that we should assert something only if doing so maximizes expected utility. Accordingly, they might say that my comparison was unfair, for a true Bayesian would never have agreed to play by the rules that I was assuming.[3] It seems to me, however, that the just-mentioned Bayesian account of assertion can at best pertain to artificial, computationally unconstrained agents and certainly not to us. The flow of our conversations would be seriously disrupted if we had to constantly perform the computations that the Bayesian account of assertion would demand from us.

Similarly, Bayesians could respond to my critique of their inaccuracy minimization argument, which as I said focuses on next-step inaccuracy, by trying to extend it to cover the accuracy or epistemic utility of *all* future updates. In doing so, they might have to suppose that we can envision all possible data streams that we might receive and all possible decisions that we might have to make after the receipt of each piece of data, that we can estimate how much inaccuracy or epistemic utility will matter to us at any particular

2. Exactly what idealizations we are willing to make may also depend on context. If accepting a couple of extra idealizing assumptions makes your theory much more powerful or elegant, then for some purposes that may be fine. At other times, you may be willing to sacrifice some computational power or elegance for the sake of making your theory more realistic or more predictively accurate.

3. Richard Pettigrew raised this objection in personal communication, for which I thank him.

moment in time, and perhaps more.[4] But why not? Fantasize away—there is a market for philosophical fiction!

Perhaps not all Bayesian colleagues have an interest in developing a theory of rationality for *real* people or see any value in trying to connect with the extant psychological literature on human reasoning. Cognitive limitations are hard to model formally and may be even harder to generalize, given that they can differ from one individual to another. Thus, attending to such limitations could come in the way of an activity that many philosophers like so much, to wit, formal model building. If that comes with the price tag of having to accept some highly unrealistic assumptions, then so be it.

But not all Bayesians are alike. In section 1.2.1, I referred to Lycan's typology of explanationist positions, going from a weak form according to which abduction is *a* (possibly derivative) mode of rational nondeductive reasoning to a very strong form according to which it is ultimately the only rational mode of reasoning. As mentioned there, my favored form is closest to the intermediate position that Lycan calls "Sturdy Explanationism," according to which abduction is a fundamental mode of rational reasoning, but not necessarily excluding the existence of other modes of rational reasoning. Whereas I am not aware of work trying to delineate types of Bayesianism along these same lines, a typology much like the one that Lycan proposed for explanationism would make sense and might be fruitful for Bayesianism as well. In particular, there would seem a clear place for what could be called "Sturdy Bayesianism," according to which Bayesian reasoning is both fundamental and rational but not necessarily the only such form of reasoning. If I am not mistaken, there are already Bayesian philosophers of science who appear to conceive of Bayesianism as providing valuable tools for modeling important parts of scientific reasoning and human reasoning more generally, rather than as a universal theory of rationality (e.g., Sprenger & Hartmann, 2019; Vassend, 2019, in press; Sprenger, 2020).[5] I cannot speak for these philosophers and I

4. For instance, they might suppose this to undercut the first objection to Briggs and Pettigrew's (2020) argument that was raised in chapter 4, footnote 17.

5. Some statisticians in the Bayesian camp definitely seem to qualify as sturdy Bayesians. For instance, Gelman and Shalizi (2012, 2013) have argued that while we are often well advised to analyze our data using Bayesian statistical techniques, in bringing the results of such analyses to bear on our theories we are not committed to Bayesian practices. Indeed, Gelman and Shalizi (2012, 2013) advocate a hypothetico-deductive approach at the level of theory choice. However, as Borsboom and Haig (2013) argue in their commentary on Gelman and Shalizi (2013), this is a place where one might more fruitfully want to invoke abduction. Furthermore,

do not know whether they would want to be identified as sturdy Bayesians, but my impression—and hope—is that they would be prepared to seriously consider abduction as a possible further valuable tool for studying rational reasoning. In section 1.2.2, we saw that Lipton, Psillos, and others argued that explanationism is compatible with Bayesianism. In doing so, however, they were happy to go along with Bayesianism's imperialist aspirations and sought to show how explanationism might be able to effectively serve those. By contrast, Sturdy Bayesianism is evidently compatible with the brand of explanationism defended in this book in a way that does not make one subservient to the other.

This book will have accomplished its mission not if all readers agree with each and every argument I have given—a most unlikely event to happen—but if philosophers and psychologists of reasoning find some motivation in it to start considering abduction as being at least worthy of more attention than it has been given in the last twenty-five years or so. Abduction could certainly benefit from such increased attention. We looked at formal models of abduction, both in an individual and in a group setting, but—to repeat—those were only toy models, serving strictly dialectical purposes. We identified as an important avenue for future research the task of devising models that connect more closely to how actual people, or groups of actual people, update on newly acquired information. To the best of my knowledge, all the relevant data available at the moment was reviewed in chapter 3—and clearly, that was not a whole lot of data. My hope is that we will see further experiments of the kind reported in that chapter and that the outcomes of these experiments will help to refine our models of abduction, which in turn will suggest further experimental work, and so on.

there is nothing that would prevent sturdy Bayesians from taking on board Lipton's idea that abductive reasoning may help us fix our priors or Psillos's idea that Peircean abduction may be invoked to narrow down the number of theories that are candidates for further exploration (for both ideas, see section 1.2.2). Thus, Bayesian statistics may be supplemented by abduction in a variety of ways.

Appendices

APPENDIX A

Proof of Theorem 4.1

In the situation considered in both van Fraassen's and our dynamic Dutch book example in chapter 4, we had a set of mutually exclusive and jointly exhaustive hypotheses $H_1, \ldots, H_n$ such that, for all i, $\Pr(H_i) > 0$. We further had evidence statements $E_1, \ldots, E_m$, which were independent conditional on any of the hypotheses. We compared two methods for calculating the probabilities of conjunctions of those evidence statements, the special multiplication rule

$$\text{(SMR)} \qquad \Pr(E_1 \wedge \cdots \wedge E_n) = \sum_{i=1}^{m} \prod_{j=1}^{n} \Pr(H_i) \Pr(E_j \mid H_i),$$

and the more general method

$$\text{(GM)} \quad \Pr(E_1 \wedge \cdots \wedge E_n) =$$

$$\sum_{i=1}^{m} \Pr(H_i) \Pr(E_1 \mid H_i) \times \prod_{j=2}^{n} \sum_{i=1}^{m} \Pr(H_i \mid E_1 \wedge \cdots \wedge E_{j-1}) \Pr(E_j \mid H_i).$$

The claim to be proven then was that

Theorem 4.1 *SMR and GM are equivalent methods for calculating* $\Pr(E_1 \wedge \cdots \wedge E_k)$*, with* $1 < k \leqslant m$*, if and only if conditional probabilities are calculated according to Bayes's theorem.*

Proof: Sufficiency is easy; we prove necessity. For simplicity, we consider only the case where we have two hypotheses H_1 and H_2 that are assumed to be jointly exhaustive and mutually exclusive, and two outcome events E_1 and E_2 that are independent in the previous sense; the generalization is straightforward.

If we calculate conditional probabilities not according to Bayes's theorem but according to the formula representing some probabilistic updating rule at variance with Bayes's rule (such as EXPL, or Good's rule, or Popper's), then for all $i: 1 \leqslant i \leqslant n$ and $k: 1 \leqslant k \leqslant m$ there is a g such that

$$\Pr(H_i \mid E_k) = \frac{\Pr(H_i) \times \Pr(E_k \mid H_i)}{\sum_{j=1}^{n} \Pr(H_j) \times \Pr(E_k \mid H_j)} + g(H_i, E_k),$$

where $g(H_i, E_k) = -\sum_{j \neq i} g(H_j, E_k)$, and it is not the case that $g(H_i, E_k) = 0$ for all i, k (for then our updating rule would no longer be at variance with Bayes's rule). Suppose, toward a contradiction, that SMR and GM are equivalent. Then in the case that we consider, the following equations must hold:

$$\begin{aligned}
&\Pr(E_1 \mid H_1)\Pr(E_2 \mid H_1)\Pr(H_1) + \Pr(E_1 \mid H_2)\Pr(E_2 \mid H_2)\Pr(H_2) = \\
&= \Big[\Pr(H_1)\Pr(E_1 \mid H_1) + \Pr(H_2)\Pr(E_1 \mid H_2)\Big] \times \\
&\quad \Big[\Pr(H_1|E_1)\Pr(E_2 \mid H_1) + \Pr(H_2 \mid E_1)\Pr(E_2 \mid H_2)\Big] \\
&= \Big[\Pr(H_1)\Pr(E_1 \mid H_1) + \Pr(H_2)\Pr(E_1 \mid H_2)\Big] \times \\
&\quad \Bigg[\left(\frac{\Pr(H_1)\Pr(E_1 \mid H_1)}{\Pr(H_1)\Pr(E_1 \mid H_1) + \Pr(H_2)\Pr(E_1 \mid H_2)} + g(H_1, E_1)\right) \times \Pr(E_2 \mid H_1) + \\
&\quad \left(\frac{\Pr(H_2)\Pr(E_1 \mid H_2)}{\Pr(H_1)\Pr(E_1 \mid H_1) + \Pr(H_2)\Pr(E_1 \mid H_2)} + g(H_2, E_1)\right) \times \Pr(E_2|H_2)\Bigg] \\
&= \left[\frac{\Pr(H_1)\Pr(E_1 \mid H_1)\Pr(E_2 \mid H_1)}{\Pr(H_1)\Pr(E_1 \mid H_1) + \Pr(H_2)\Pr(E_1 \mid H_2)} + g(H_1 \mid E_1)\Pr(E_2 \mid H_1)\right] + \\
&\quad \left[\frac{\Pr(H_2)\Pr(E_1 \mid H_2)\Pr(E_2 \mid H_2)}{\Pr(H_1)\Pr(E_1 \mid H_1) + \Pr(H_2)\Pr(E_1 \mid H_2)} + g(H_2 \mid E_1)\Pr(E_2 \mid H_2)\right] \\
&= \Pr(H_1)\Pr(E_1 \mid H_1)\Pr(E_2 \mid H_1) + \\
&\quad \Big[g(H_1, E_1)\Pr(E_2 \mid H_1)\big(\Pr(H_1)\Pr(E_1 \mid H_1) + \Pr(H_2)\Pr(E_1 \mid H_2)\big)\Big] + \\
&\quad \Pr(H_2)\Pr(E_1 \mid H_2)\Pr(E_2 \mid H_2) + \\
&\quad \Big[g(H_2, E_1)\Pr(E_2 \mid H_2)\big(\Pr(H_1)\Pr(E_1 \mid H_1) + \Pr(H_2)\Pr(E_1 \mid H_2)\big)\Big].
\end{aligned}$$

Because (1) $g(H_1, E_1) = -g(H_2, E_1)$, and (2) $\Pr(E_2 \mid H_1) \neq \Pr(E_2 \mid H_2)$, it must be that $g(H_1, E_1) = g(H_2, E_1) = 0$. That means that Bayes's rule and the non-Bayesian rule do not disagree over $\Pr_{E_1}(H_i)$. However, we can interchange E_1 and E_2 and then it follows by parity of reasoning that they cannot disagree over $\Pr_{E_2}(H_i)$ either. This contradicts our assumption that the rules are at variance with each other. Consequently, SMR and GM are not equivalent. ■

APPENDIX B

Proof of Theorem 5.1

Recall that the weights of a VS rule (as defined in ch. 5) are all positive and add up to 1, and that they are said to reflect truthlikeness in a minimally adequate sense precisely if hypotheses are assigned weights as a function of their distance from the truth, with hypotheses farther from the truth being assigned larger weights than hypotheses closer to the truth.

Theorem 5.1 *Every VS rule whose weights reflect truthlikeness in a minimally adequate sense is improper.*

Proof: Without loss of generality, consider a hypothesis partition of three hypotheses, H_1, H_2, and H_3. Then, for an arbitrary VS rule $\mathcal{V}$ and a given person's probability assignment $\mathbf{p} = (p_1, p_2, p_3)$ to the aforementioned hypotheses, with p_i the probability assigned to H_i, this person's expected $\mathcal{V}$ score for a probability assignment $\mathbf{p}^*$ to the same hypotheses is given by the function

$$\begin{aligned}\mathbb{E}_{\mathbf{p}}[\mathcal{V}(\mathbf{p}^*)] \;=\; & p_1\big(w_{11}(1-p_1^*)^2 + w_{21}(p_2^*)^2 + w_{31}(p_3^*)^2\big)\\ &+ p_2\big(w_{12}(p_1^*)^2 + w_{22}(1-p_2^*)^2 + w_{32}(p_3^*)^2\big)\\ &+ p_3\big(w_{13}(p_1^*)^2 + w_{23}(p_2^*)^2 + w_{33}(1-p_3^*)^2\big).\end{aligned}$$

Also without loss of generality, assume that the hypotheses are ordered by their distances from each other, with H_2 being equally far from H_1 and H_3, and H_1 and H_3 being twice as far from each other as they are from H_2. Then $w_{11} = w_{33}$, $w_{21} = w_{23}$, $w_{31} = w_{13}$, and $w_{12} = w_{32}$, so that we can simplify notation by defining $w_1 := w_{11} = w_{33}$; $w_2 := w_{21} = w_{23}$; $w_3 := w_{31} = w_{13}$; $w_4 := w_{12} = w_{32}$; and $w_5 := w_{22}$. For $\mathcal{V}$ to be proper, it must hold that $\arg\min_{\mathbf{p}^*} \mathbb{E}_{\mathbf{p}}[\mathcal{V}(\mathbf{p}^*)] = \mathbf{p}$, for any distribution $\mathbf{p}$ on $\{H_1, H_2, H_3\}$. To see whether this holds, we use the method of Lagrange multipliers. Specifically, where $f(\mathbf{p}^*) = p_1^* + p_2^* + p_3^*$, we must find values for p_1^*, p_2^*, p_3^*, and λ such that $\nabla\mathbb{E}_{\mathbf{p}}[\mathcal{U}(\mathbf{p}^*)] = \lambda\nabla f(\mathbf{p}^*)$ and $f(\mathbf{p}^*) = 1$. Calculating the first-order partial derivatives of $\mathbb{E}_{\mathbf{p}}[\mathcal{U}(\mathbf{p}^*)]$, we find

$$\begin{aligned}(\partial/\partial p_1^*)\mathbb{E}_{\mathbf{p}}[\mathcal{U}(\mathbf{p}^*)] &= -2w_1p_1(1-p_1^*) + 2w_3p_3p_1^* + 2w_4p_2p_1^*;\\ (\partial/\partial p_2^*)\mathbb{E}_{\mathbf{p}}[\mathcal{U}(\mathbf{p}^*)] &= -2w_5p_2(1-p_2^*) + 2w_2p_1p_2^* + 2w_2p_3p_2^*;\\ (\partial/\partial p_3^*)\mathbb{E}_{\mathbf{p}}[\mathcal{U}(\mathbf{p}^*)] &= -2w_1p_3(1-p_3^*) + 2w_3p_1p_3^* + 2w_4p_2p_3^*.\end{aligned}$$

Because $\nabla f(\mathbf{p}^*) = \mathbf{1}$, we have $(\partial/\partial p_i^*)\mathbb{E}_{\mathbf{p}}[\mathcal{U}(\mathbf{p}^*)] = \lambda$ for all $i \leqslant 3$. So in particular, expanding the partial derivatives in p_1^* and p_3^* and dividing both by 2, we have

$$-w_1p_1 + w_1p_1p_1^* + w_3p_3p_1^* + w_4p_2p_1^* = -w_1p_3 + w_1p_3p_3^* + w_3p_1p_3^* + w_4p_2p_3^*,$$

and hence

$$w_1p_1p_1^* + w_3p_3p_1^* + w_4p_2p_1^* - w_1p_3p_3^* - w_3p_1p_3^* - w_4p_2p_3^* - w_1p_1 + w_1p_3 = 0.$$

Suppose that $\mathcal{U}$ is proper, so that $\mathbb{E}_{\mathbf{p}}[\mathcal{U}(\mathbf{p}^*)]$ reaches its minimum if $p_1 = p_1^*$, $p_2 = p_2^*$, and $p_3 = p_3^*$. Then there must be values for the w_i such that

$$w_1{p_1}^2 + w_3p_3p_1 + w_4p_2p_1 - w_1{p_3}^2 - w_3p_1p_3 - w_4p_2p_3 - w_1p_1 + w_1p_3 = 0.$$

However, factoring the left-hand side yields

$$(p_1 - p_3)(-w_1 + w_1p_1 + w_4p_2 + w_1p_3).$$

This equals 0 if and only if either (1) $p_1 = p_3$ or (2) $w_1 = w_4$, where the latter follows from the fact that the condition that the right-hand factor equals 0 can be rewritten $w_1(1 - p_1 - p_3) = w_4p_2$, in conjunction with the fact that the p_i sum to 1. Because, as said, for $\mathcal{U}$ to be proper, it must hold for all $\mathbf{p}$ that $\arg\min_{\mathbf{p}^*}\mathbb{E}_{\mathbf{p}}[\mathcal{U}(\mathbf{p}^*)] = \mathbf{p}$, we may pick a $\mathbf{p}$ such that $p_1 \neq p_3$, thereby violating condition 1. Regarding condition 2, note that whichever precise values the w_i assume, w_1 must be smaller than 1/3 (given that it is assigned to the supposed truth) and w_4 must be greater than 1/3 (given that it is assigned to the two hypotheses supposed false). Consequently, on the supposition that $\mathcal{U}$ is proper, we can minimize $\mathbb{E}_{\mathbf{p}}[\mathcal{U}(\mathbf{p}^*)]$ subject to the given constraint precisely if the truthlikeness weights assigned by the rule do *not* reflect truthlikeness in a minimally adequate sense. By assumption, the weights do reflect truthlikeness in a minimally adequate sense. Given that we made no further assumptions about $\mathcal{V}$, it follows that every VS rule is improper if it assigns truthlikeness weights in a minimally adequate fashion. ■

Remark: This proof proceeds by constructing a specific counterexample involving three hypotheses that are assumed to stand in specific relations of truthlikeness to each other. To see that this assumption does not undermine the generality of the proof, we note that the said relations are perfectly possible according to all modern measures of truthlikeness. (See p. 140 for references.) As a matter of fact, one can think of the example presented in chapter 5 concerning the possible grades (A, B, or C) a given student may receive as instantiating exactly the relations of truthlikeness that are assumed to hold in the counterexample. It is also to be noted, however, that not *all* known measures of truthlikeness will suffice for the purposes of the proof. Most famously, Pavel Tichý (1974) discovered that on Popper's (1963) measure all false theories are equally far from the truth, contrary to what Popper had hoped to achieve with his measure.

APPENDIX C

Proof of Theorem 5.2

In this appendix we prove

Theorem 5.2 *Let S be the standard unit* $(n-1)$*-simplex, let* $\mathbf{p}$ *and* $\mathbf{p}^*$ *range over vectors in S, and let* $m: S \longrightarrow S$ *be defined as follows:*

$$m(\mathbf{p}) \;:=\; \arg\min_{\mathbf{p}^*} \sum_{i=1}^{n} \sum_{j=1}^{n} p_i w_{ij}(\delta_{ij} - p_j^*)^2$$

with δ_{ij} *the Kronecker delta, and with* $w_{ij} > 0$ *for all* i, j*, and* $\sum_{i=1}^{n} \sum_{j=1}^{n} w_{ij} = 1$*. Then (1) there is a* $\mathbf{p}^+ \in S$ *such that* $m(\mathbf{p}^+) = \mathbf{p}^+$*; and furthermore (2) there is only one such* $\mathbf{p}^+$*, and (3) it depends only on the* w_{ij}*.*

Proof: Clause 1 follows from Brouwer's (1911) fixed-point theorem, which (in one version) states that every continuous function from a simplex onto itself has a fixed point. It does not follow from Brouwer's theorem that the fixed point is unique.

To prove clause 2, then, one first verifies that the function that is being minimized at each step on the way to the fixed point has the Hessian

$$\begin{bmatrix} 2(p_1 w_{11} + \cdots + p_n w_{1n}) & 0 & \cdots & 0 \\ 0 & 2(p_1 w_{21} + \cdots p_n w_{2n}) & \cdots & 0 \\ \vdots & \vdots & \ddots & \vdots \\ 0 & 0 & \cdots & 2(p_1 w_{n1} + \cdots + p_n w_{nn}) \end{bmatrix}$$

This is a diagonal matrix, so its eigenvalues are the diagonal elements, which given the constraints on the p_i and w_{ij}, can be seen to be all necessarily positive. Therefore, the Hessian is positive definite everywhere, and given that a simplex is a convex set, it follows that the function that is minimized is strictly convex, and hence the minimum it reaches is unique. So, at each step toward the fixed point, a unique minimum is reached. As a result, the minimum reached at the fixed point is unique as well.

For clause 3, finally, note that at the fixed point the function that is being minimized is of the form

$$m^+(\mathbf{p}) = \sum_{i=1}^{n}\sum_{j=1}^{n} p_i w_{ij}(\partial_{ij} - p_j)^2.$$

Because the fixed point $\mathbf{p}^+$ is a minimum, it holds that $\nabla m^+(\mathbf{p}^+) = \mathbf{0}$. We obtain a system of n polynomial equations with n variables and with the w_{ij} as coefficients by setting $(\partial/\partial p_i)m^+(\mathbf{p}^+) = 0$, for all $i \leqslant n$. This system has a unique solution (in virtue of the first two clauses), which is bound to be strictly in terms of the coefficients. ∎

APPENDIX D

Proof of Theorem 8.1

In this appendix we prove

Theorem 8.1 *If (8.3)–(8.9) hold, then*

$$\lim_{m,n\to\infty} \Pr_n^m(\mathrm{RSP}) \;=\; \lim_{m,n\to\infty} \Pr_n^m(\mathrm{RABD}) \;=\; 1.$$

Proof: One first verifies by a routine induction on $m + n$ that $p_n^m(H)$ is a monotonously nondecreasing function on σ for $p \in \{\Pr, \mathfrak{m}, \mathfrak{s}, \mathfrak{R}, \mathfrak{S}\}$ and $H \in \{\mathrm{RSP}, \mathrm{RABD}\}$. Note that because all these functions are bounded, they converge, and so their limits exist. Further observe that it is a direct consequence of (8.1), the rule for determining probabilities conditional on uncertain evidence, that, for any A, if (1) $\Pr(\mathrm{RSP} \mid E/e) \geqslant \Pr(\mathrm{RSP})$ and (2) $\mathfrak{m}(A \mid E/e) \geqslant \mathfrak{s}(A \mid E/e)$, then

$$\text{(D.1)} \quad \Pr(A \mid E/e) \;\geqslant\; \Pr(\mathrm{RSP})\,\mathfrak{m}(A \mid E/e) \;+\; \Pr(\neg\mathrm{RSP})\,\mathfrak{s}(A \mid E/e).$$

It follows from (D.1) and assumptions (8.6) and (8.7) that, for all $n, m \in \mathbb{N}$

$$\text{(D.2)} \quad \Pr_{n-1}^m(\mathrm{RABD} \mid E_n/e_n) \;\geqslant\; \Pr_{n-1}^m(\mathrm{RSP})\,\mathfrak{m}_{n-1}^m(\mathrm{RABD} \mid E_n/e_n) + \Pr_{n-1}^m(\neg\mathrm{RSP})\,\mathfrak{s}_{n-1}^m(\mathrm{RABD} \mid E_n/e_n).$$

Next note that, by (8.2), (8.5), and (8.7), we have for some $k \geqslant 1$ that

$$\Pr_{n-1}^m(\mathrm{RSP}) \;\geqslant\; \Pr_{n-k}^{m-1}(\mathrm{RABD} \mid E'_m)\,\mathfrak{R}_{n-k}^{m-1}(\mathrm{RSP} \mid E'_m) + \Pr_{n-k}^{m-1}(\neg\mathrm{RABD} \mid E'_m)\,\mathfrak{S}_{n-k}^{m-1}(\mathrm{RSP} \mid E'_m).$$

Together with (D.2), this yields

$$\begin{aligned}
&\mathrm{Pr}^{m}_{n-1}(\mathrm{RABD} \mid E_n/e_n) \ \geqslant \\
&\Big[\mathrm{Pr}^{m-1}_{n-k}(\mathrm{RABD} \mid E'_m)\,\mathfrak{R}^{m-1}_{n-k}(\mathrm{RSP} \mid E'_m) + \\
&\qquad \mathrm{Pr}^{m-1}_{n-k}(\neg\mathrm{RABD} \mid E'_m)\,\mathfrak{S}^{m-1}_{n-k}(\mathrm{RSP} \mid E'_m)\Big] \times \\
&\qquad \mathfrak{m}^{m}_{n-1}(\mathrm{RABD} \mid E_n/e_n) \ + \ \mathrm{Pr}^{m}_{n-1}(\neg\mathrm{RSP})\,\mathfrak{s}^{m}_{n-1}(\mathrm{RABD} \mid E_n/e_n).
\end{aligned}$$

Expanding yields

$$\begin{aligned}
&\mathrm{Pr}^{m}_{n-1}(\mathrm{RABD} \mid E_n/e_n) \ \geqslant \\
&\quad \mathrm{Pr}^{m-1}_{n-k}(\mathrm{RABD} \mid E'_m)\,\mathfrak{R}^{m-1}_{n-k}(\mathrm{RSP} \mid E'_m)\,\mathfrak{m}^{m}_{n-1}(\mathrm{RABD} \mid E_n/e_n) + \\
&\quad \mathrm{Pr}^{m-1}_{n-k}(\neg\mathrm{RABD} \mid E'_m)\,\mathfrak{S}^{m-1}_{n-k}(\mathrm{RSP} \mid E'_m)\,\mathfrak{m}^{m}_{n-1}(\mathrm{RABD} \mid E_n/e_n) + \\
&\qquad \mathrm{Pr}^{m}_{n-1}(\neg\mathrm{RSP})\,\mathfrak{s}^{m}_{n-1}(\mathrm{RABD} \mid E_n/e_n).
\end{aligned}$$

Rearranging then yields

$$\begin{aligned}
\text{(D.3)}\quad &\mathrm{Pr}^{m}_{n-1}(\mathrm{RABD} \mid E_n/e_n) - \\
&\mathrm{Pr}^{m-1}_{n-k}(\mathrm{RABD} \mid E'_m)\,\mathfrak{R}^{m-1}_{n-k}(\mathrm{RSP} \mid E'_m)\,\mathfrak{m}^{m}_{n-1}(\mathrm{RABD} \mid E_n/e_n) \ \geqslant \\
&\quad \mathrm{Pr}^{m-1}_{n-k}(\neg\mathrm{RABD} \mid E'_m)\,\mathfrak{S}^{m-1}_{n-k}(\mathrm{RSP} \mid E'_m)\,\mathfrak{m}^{m}_{n-1}(\mathrm{RABD} \mid E_n/e_n) + \\
&\qquad \mathrm{Pr}^{m}_{n-1}(\neg\mathrm{RSP})\,\mathfrak{s}^{m}_{n-1}(\mathrm{RABD} \mid E_n/e_n).
\end{aligned}$$

Now note that, by the limit rule for subtraction

$$\begin{aligned}
\text{(D.4)}\quad &\lim_{m,n\to\infty}\Big[\mathrm{Pr}^{m}_{n-1}(\mathrm{RABD} \mid E_n/e_n) - \\
&\mathrm{Pr}^{m-1}_{n-k}(\mathrm{RABD} \mid E'_m)\,\mathfrak{R}^{m-1}_{n-k}(\mathrm{RSP} \mid E'_m)\,\mathfrak{m}^{m}_{n-1}(\mathrm{RABD} \mid E_n/e_n)\Big] \ = \\
&\quad \lim_{m,n\to\infty} \mathrm{Pr}^{m}_{n-1}(\mathrm{RABD} \mid E_n/e_n) \ - \\
&\lim_{m,n\to\infty}\Big[\mathrm{Pr}^{m-1}_{n-k}(\mathrm{RABD} \mid E'_m)\,\mathfrak{R}^{m-1}_{n-k}(\mathrm{RSP} \mid E'_m)\,\mathfrak{m}^{m}_{n-1}(\mathrm{RABD} \mid E_n/e_n)\Big],
\end{aligned}$$

and that, by the limit rule for products

(D.5)

$$\lim_{m,n\to\infty}\left[\Pr^{m-1}_{n-k}(\text{RABD}\mid E'_m)\,\mathfrak{R}^{m-1}_{n-k}(\text{RSP}\mid E'_m)\,\mathfrak{m}^{m}_{n-1}(\text{RABD}\mid E_n/e_n)\right] = \lim_{m,n\to\infty}\Pr^{m-1}_{n-k}(\text{RABD}\mid E'_m)\times\lim_{m,n\to\infty}\mathfrak{R}^{m-1}_{n-k}(\text{RSP}\mid E'_m)\times \lim_{m,n\to\infty}\mathfrak{m}^{m}_{n-1}(\text{RABD}\mid E_n/e_n).$$

Given assumptions (8.8) and (8.9), (D.5) simplifies to

(D.6)

$$\lim_{m,n\to\infty}\left[\Pr^{m-1}_{n-k}(\text{RABD}\mid E'_m)\,\mathfrak{R}^{m-1}_{n-k}(\text{RSP}\mid E'_m)\,\mathfrak{m}^{m}_{n-1}(\text{RABD}\mid E_n/e_n)\right] = \lim_{m,n\to\infty}\Pr^{m-1}_{n-k}(\text{RABD}\mid E'_m).$$

And from equations (D.3)–(D.6) we obtain

(D.7)
$$\lim_{m,n\to\infty}\left[\Pr^{m}_{n-1}(\text{RABD}\mid E_n/e_n)-\Pr^{m-1}_{n-k}(\text{RABD}\mid E'_m)\right] \geqslant \lim_{m,n\to\infty}\left[\Pr^{m-1}_{n-k}(\neg\text{RABD}\mid E'_m)\,\mathfrak{S}^{m-1}_{n-k}(\text{RSP}\mid E'_m)\,\mathfrak{m}^{m}_{n-1}(\text{RABD}\mid E_n/e_n) + \Pr^{m}_{n-1}(\neg\text{RSP})\,\mathfrak{s}^{m}_{n-1}(\text{RABD}\mid E_n/e_n)\right].$$

Because $\Pr^{m}_{n}(\text{RABD})$ converges, we have

(D.8)
$$\lim_{m,n\to\infty}\left[\Pr^{m}_{n-1}(\text{RABD}\mid E_n/e_n)-\Pr^{m-1}_{n-k}(\text{RABD}\mid E'_m)\right] = 0.$$

From (D.7) and (D.8) it follows that

$$\lim_{m,n\to\infty}\left[\Pr^{m-1}_{n-k}(\neg\text{RABD}\mid E'_m)\,\mathfrak{S}^{m-1}_{n-k}(\text{RSP}\mid E'_m)\,\mathfrak{m}^{m}_{n-1}(\text{RABD}\mid E_n/e_n) + \Pr^{m}_{n-1}(\neg\text{RSP})\,\mathfrak{s}^{m}_{n-1}(\text{RABD}\mid E_n/e_n)\right] = 0.$$

Hence it must be that both

(D.9)

$$\lim_{m,n\to\infty}\left[\Pr^{m-1}_{n-k}(\neg\text{RABD}\mid E'_m)\,\mathfrak{S}^{m-1}_{n-k}(\text{RSP}\mid E'_m)\,\mathfrak{m}^{m}_{n-1}(\text{RABD}\mid E_n/e_n)\right] = 0$$

and

$$\text{(D.10)} \qquad \lim_{m,n\to\infty} \left[\Pr^m_{n-1}(\neg\text{RSP})\, \mathfrak{s}^m_{n-1}(\text{RABD} \mid E_n/e_n) \right] = 0.$$

Because, first, by (8.9) it holds that $\lim_{m,n\to\infty} \mathfrak{m}^m_{n-1}(\text{RABD} \mid E_n/e_n) = 1$; second, $\mathfrak{S}^{m-1}_{n-k}(\text{RSP} \mid E'_m)$ is monotonously nondecreasing; and third, by (8.3) it holds that $\mathfrak{S}^0_0(\text{RSP}) > 0$, we have $\lim_{m,n\to\infty} \mathfrak{S}^{m-1}_{n-k}(\text{RSP} \mid E'_m) \neq 0$. Therefore, (D.9) can only be the case if

$$\lim_{m,n\to\infty} \Pr^{m-1}_{n-k}(\neg\text{RABD} \mid E'_m) = 0$$

and so

$$\lim_{m,n\to\infty} \Pr^{m-1}_{n-k}(\text{RABD} \mid E'_m) = 1,$$

that is, because $\Pr^m_n(\text{RABD})$ is monotonously nondecreasing,

$$\text{(D.11)} \qquad \lim_{m,n\to\infty} \Pr^m_n(\text{RABD}) = 1.$$

And it follows from (D.10) and that as $\mathfrak{s}^m_{n-1}(\text{RABD} \mid E_n/e_n)$ is monotonously nondecreasing and $\mathfrak{s}^0_0(\text{RABD}) > 0$, $\lim_{m,n\to\infty} \mathfrak{s}^m_{n-1}(\text{RABD} \mid E_n/e_n) \neq 0$, that

$$\lim_{m,n\to\infty} \Pr^m_{n-1}(\neg\text{RSP}) = 0$$

and hence that

$$\lim_{m,n\to\infty} \Pr^m_{n-1}(\text{RSP}) = 1,$$

that is,

$$\text{(D.12)} \qquad \lim_{m,n\to\infty} \Pr^m_n(\text{RSP}) = 1.$$

Finally, from (D.11) and (D.12) we get

$$\lim_{m,n\to\infty} \Pr^m_n(\text{RABD}) = \lim_{m,n\to\infty} \Pr^m_n(\text{RSP}) = 1,$$

which had to be proved. ■

APPENDIX E

Using Julia

Chapters 6 and 7 relied heavily on computer simulations. All those simulations were conducted in Julia, a new, free, and open-source computer language, distributed under the MIT license. Julia is a general-purpose language geared toward high-performance computing. It was originally developed at MIT by Alan Edelman with a team of colleagues, post docs, and doctoral students; now it has contributors from around the globe. Three of the main developers were awarded the James H. Wilkerson Prize for Numerical Software for their contributions to the creation of Julia. Importantly, Julia is written almost entirely in Julia, so that once you have mastered Julia, you are also able to understand (and if needed, adapt) the code for all predefined Julia functions and for the Julia packages. This is very different from languages such as Python or R, large parts of which are written in C or C++.

You should be able to follow the arguments in this book without even switching on your computer. Nevertheless, I recommend that you do have a look at Julia and maybe even try your hand at the code for the simulations that I am making available. According to the mathematician Gregory Chaitin,

> the computer changes epistemology, it changes the meaning of "to understand." To me, you understand something only if you can program it. (You, not someone else!) Otherwise you don't really understand it, you only think you understand it. (Chaitin, 2006, p. xiii)

I could not agree more. Because Julia is so new, I am assuming that most readers will not be familiar with it. I therefore start with a brief introduction to the language, focusing on the elements of the language that are most essential to the simulations.

If you have heard about Julia previously, then probably you have heard about its speed, which is generally seen as one of Julia's main selling points. In many applications, one should not be overly concerned about one's code not being optimally fast: as computer scientists like to say, your own time is more valuable than the computer's time. Indeed, that performance is not always important is especially true now that we can run computations in the cloud for little money. In my experience, though, sometimes speed really does matter. Especially in the exploratory stage of one's research, when the code is still being developed and bugs easily slip in, it can be helpful to see the outcome of a computation *quickly*. It may just take too much of one's patience if it takes hours to get some first results that might indicate whether the ideas one is pursuing have merit or not. Even when one is happy with one's code and deems it ready to go—if it takes a week to run in the cloud, then that will cost you more than just a little money. Depending on the software one uses, a week may be a realistic expectation for the simulations that were reported in this book. When Sylvia Wenmackers and I were working on our joint 2017 paper, we wrote the simulations mentioned in chapter 7 in *Mathematica*. That code could then be run only on a supercomputer to which we had institutional access at the time. Even on that computer, the simulations took four days to complete. Running the same simulations, now written in Julia, on my own desktop machine (which, admittedly, is quite a beefy machine, with twelve physical and twenty-four logical cores) takes under half an hour.[1] The Julia website (https://julialang.org/) exhibits interesting benchmarks, comparing Julia with a host of other languages, including *Mathematica*, R, and Python, and showing that, for some common computational operations, Julia is at least one hundred times faster than each of those other widely used languages.

What makes Julia a fantastic language to work with is not simply its speed but rather the combination of speed and user-friendliness. Languages like C, C++, FORTRAN, and some others are fast as well, sometimes even still a bit faster than Julia. But all these languages have a steep learning curve, and even professional programmers working in them on a daily basis tend to spend a fair amount of their time debugging code, simply because it is so easy to make mistakes in these languages. Julia code is very different in this respect. While Julia code can run as fast, or nearly as fast, as code written in the aforementioned languages, it looks almost like pseudocode, and users of *Mathematica*,

1. In fairness, it should be mentioned that Sylvia Wenmackers and I used *Mathematica 8*, the most recent version being *Mathematica 12*. I have limited experience with the latter, but it is clearly more performant than the older version that we used.

Python, or R should have little difficulty familiarizing themselves with Julia and can become productive in the language within a day or two.

I started using Julia in mid-2017, when it was still in beta; to be precise, I started using version 0.5.2. Julia 1.0 was released in August 2018, the current version (November 2020) being 1.5.3. Over those years, I have not encountered any downside of using Julia. To be sure, precisely because the language is so new, its "ecosystem" is still not quite at the level of Python's or R's, both of which have been around for decades. For instance, there are not nearly as many packages for Julia as exist for Python or R, and many of the Julia packages are still in beta. But, first, it is just a matter of time for Julia to catch up, and second, there are easy ways to call Python and R from within Julia, which allows one to use any Python or R package also in Julia, subsequently explained. The available documentation for Julia, in print or online, is also rather limited still. That situation, too, will soon improve. Besides, the community of Julia users is very friendly, and I have received valuable help by asking questions on the main online forum for Julia (https://discourse.julialang.org/).

There are free services that allow you to run Julia in the cloud (e.g., Google's Colab, and NextJournal), but if you are serious about giving Julia a try, then I recommend that you install it on your own computer. Details about how to install Julia (which are different for different operating systems) can be found on the download page from the Julia language website: https://julialang.org/downloads/. There are various ways to interact with Julia. It is possible to run Julia from within the terminal, but I have mostly been using either Microsoft's Visual Studio Code editor or JupyterLab, both of which are free and open-source and easy to set up (although details depend on the operating system). Recently, I have also come to enjoy using Pluto, which is a new, also free and open-source, technology that was especially developed for Julia and which can be installed via Julia's package manager (see below).

Language Basics

Two key elements of virtually any computer language are what one may generally refer to as *containers* and *control flow elements*. The former are where you store, at least for a while, your data or the outcomes of your computations; the latter are elements that let you automate repetitive tasks, which is the kind of operation that many people most immediately associate with computers.

The most common type of container in Julia, and in many other languages, is the so-called array. In Julia, arrays can have any number of dimensions. Perhaps it is easiest to think of a two-dimensional array as the most basic case. Such an array is characterized by its number of rows and its number of columns. But if we have a series of such two-dimensional arrays, we may want to store them together in one container. Provided that they have the same number of rows and columns, that is possible, namely, by putting them in a *three*-dimensional array, the first two dimensions still being rows and columns, and the third dimension being an enumeration of the two-dimensional arrays that we brought together. And, of course, if we want to store a number of three-dimensional arrays together, we can do that as well (given that they have the same dimensions), namely, in a four-dimensional array (arrays with more than two dimensions are also referred to as "tensors"). And so on. Julia recognizes `Vector` as shorthand for a one-dimensional array and `Matrix` as shorthand for a two-dimensional array.

To give an example, the code `rand(3, 4)` creates a 3×4 (i.e., three rows by four columns) array of random floating point numbers (or floats) between 0 and 1 inclusive:[2]

```
julia> rand(3, 4)
3×4 Array{Float64,2}:
 0.481211  0.205386  0.968688  0.239485
 0.385465  0.691358  0.295478  0.215576
 0.911691  0.224254  0.984399  0.541366
```

We can similarly create a $3 \times 4 \times 2$ array of random floats between 0 and 1 inclusive:

```
julia> rand(3, 4, 2)
3×4×2 Array{Float64,3}:
[:, :, 1] =
 0.190611   0.656259  0.556281  0.858714
 0.0814072  0.164408  0.954222  0.92906
```

2. Although modern technology still will not let *you* run a Julia session within this book, it will let *me* do so. Well, sort of. I can run Julia from within the document that will eventually be typeset into this book, thanks to Geoffrey Poore's wonderful `pythontex` package (Poore, 2015). What *looks* like a Julia prompt actually *is* a Julia prompt, and everything printed in sans serif immediately below a prompt is actual Julia output.

```
 0.83987    0.974812  0.474558  0.0514526
[:, :, 2] =
 0.765418  0.750336  0.0654037  0.476914
 0.834557  0.95696   0.0157077  0.491732
 0.39551   0.831768  0.428758   0.279625
```

Think of this as having created two two-dimensional arrays (of size 3×4) that are then stored together in one container.

From the output above, it can already be seen that Julia is a *typed* language. For instance, the line immediately below `rand(3, 4, 2)` indicates that this code outputs an array containing 64-bits floats and that this array has three dimensions. You do not have to pay any attention to types while working in Julia. However, doing so often pays off in terms of performance. One reason why Julia is fast is that it has a clever way of reserving chunks of your computer's memory to store whatever values are going to be computed when your code is executed. Generally speaking, code runs much faster if the data it has to retrieve is stored orderly in a contiguous chunk of memory. To know in advance how much memory must be reserved, the compiler needs to know the type of the data that are going to be stored in it. For instance, if it has to store a thousand 64-bit floats it needs more space than if it has to store a thousand 32-bit integers.

In Julia, we access specific entries in an array, or slices of it, by indexing, much in the way in which this is done in R, Python, and many other languages.[3] Suppose that we have a two-dimensional array

```
julia> ar = rand(3, 4)
3×4 Array{Float64,2}:
 0.160707  0.569687  0.797983  0.760631
 0.922007  0.57953   0.358656  0.0114223
 0.286155  0.796654  0.456668  0.698085
```

Then

```
julia> ar[2, 3]
0.3586560380995427
```

returns the entry in the second row, third column, and

3. However, Python users should keep in mind that Julia is not 0-indexed but (like R) 1-indexed.

```
julia> ar[:, 4]
3-element Array{Float64,1}:
 0.7606313907554907
 0.011422317555960682
 0.6980853689654172
```

returns the whole fourth column of the array, and

```
julia> ar[1, :]
4-element Array{Float64,1}:
 0.1607070692312922
 0.569687258606496
 0.7979833103099228
 0.7606313907554907
```

returns the first row in its entirety. Similarly, we can get chunks of `ar` by using, for example,

```
julia> ar[2:3, 1:3]
2×3 Array{Float64,2}:
 0.922007  0.57953   0.358656
 0.286155  0.796654  0.456668
```

which returns from rows 2 and 3 the numbers that are in columns 1 to 3. The same principles apply no matter the dimensionality of the array. To work with arrays, you have to know a bit about how to permute their dimensions (e.g., via the `permutedims` function), how to reshape them more generally (e.g., via the `reshape` function), how to apply functions to specific dimensions (row-wise or column-wise, e.g., via the `mapslices` function), how to concatenate two or more arrays (e.g., using `hcat` or `vcat`), how to find or select specific values or values meeting some criterion, in an array (e.g., using `findall` or `findfirst` or `filter`), and so on. I will not cover these issues here. In general, with the help of the internet, you should have no difficulty figuring out how to perform whichever operation you would like to perform on an array.

Computers can repeat an operation over and over again, thousands of times, hundred thousands of times, millions of times, and more, and they can do so *fast*, or in any event much, much faster than we ever could. In most simulations, and certainly in the ones that we reported, that is exactly what a computer is needed for. There is no step in any of those simulations that we could not have calculated by hand. But if we had had to do that for each step, this book would never have come into existence. We could not have calculated all the steps of our simulations in our life time, no matter the effort we would have been willing to make. To define such repetition procedures, we need control flow elements.

An especially important automated repetition procedure, often used in the simulations, is the **`for`** loop. In such a loop, we go through a range of values, and for each value, run whichever procedure in which we are interested. There are variants, such as **`while`** loops, that repeat a procedure while some prespecified condition holds. If you come from R, you may be inclined to think that **`for`** loops are to be avoided for being relatively slow, and that instead one should use so-called vectorized code, where, roughly, a function is mapped over a vector, or an array, or another structure. That is not true in Julia, however: **`for`** loops are in general fast, and faster than vectorized code (although almost anything in Julia is fast, including vectorized code; but **`for`** loops are normally still a bit faster).

In the following code, we first pre-allocate a vector (i.e., a one-dimensional array) that can contain 1,000 integers and then use a **`for`** loop to fill it with the squares of the integers from 1 to 1,000. We time the code via the `@time` macro.

```
julia> v = Vector{Int64}(undef, 1000);
julia> @time for i in 1:1000; v[i] = i^2; end
  0.000025 seconds (1.47 k allocations: 22.922 KiB)
```

Using vectorized code, we can create the same vector as follows:

```
julia> @time v = [ i^2 for i in 1:1000 ];
  0.019676 seconds (41.84 k allocations: 2.221 MiB)
```

But this is clearly slower and also allocates more memory.

If one needs very precise measurements, one should use the `@btime` macro, or for more informative output the `@benchmark` macro, from the package `BenchmarkTools.jl`.[4] (See the subsequent explanation on how to install and load packages.) To see this in action, we show an even faster way to square the

4. Using the `@time` macro (but not the `@btime` or `@benchmark` macro), one can verify that the first call to a function always takes a bit longer than subsequent calls. The reason is that although Julia *feels* like an interactive language (you type some code, e.g., in a Julia session running in the terminal, or in a Jupyter notebook, and it will immediately execute), it really is a compiled language, like C, C++, FORTRAN, and many others. However, whereas these

first thousand positive integers. This defines a vector containing just these integers and then squares each of them in place, meaning that the vector gets replaced by another vector containing the squares of the first thousand positive integers. That does not require any moving around of data in memory, which makes this a very fast operation:

```
julia> @benchmark v = (1:1000).^2
BenchmarkTools.Trial:
  memory estimate:  7.94 KiB
  allocs estimate:  1
  --------------
  minimum time:     738.293 ns (0.00% GC)
  median time:      810.732 ns (0.00% GC)
  mean time:        1.055 μs (14.74% GC)
  maximum time:     7.697 μs (87.11% GC)
  --------------
  samples:          10000
  evals/sample:     164
```

We see that, on my computer, this takes just about one microsecond, on average. That makes this way of creating v about eighteen times faster than the **for** loop that we used above.[5] (For details on how to use this package, see the information at https://github.com/JuliaCI/BenchmarkTools.jl.)

Here is an example of a **while** loop:

```
julia> i = 1; while i < 5; println(i); global i += 1; end
1
2
3
4
```

We first set *i* equal to 1 outside the loop and then increment it by 1 each time we go through the loop. In each iteration, we want the value of *i* in that iteration to be printed.

---

more traditional languages require the user first to compile an executable of his or her code, Julia does what is called "just-in-time compilation," meaning that, for instance, a function gets compiled just before it is needed for the first time. All further calls to the function make use of the compiled code that was created on the first call and that also got stored then.

5. This was actually determined using the `@btime` macro, not via the `@time` macro reported in the text above.

Another important class of control flow elements is conditional statements. Suppose we want a program to output the value of some variable x if that value is greater than 0.95 and else want it to output π. One way that we can get this done in Julia uses the so-called ternary operator:

```
julia> x = rand()
0.5020104104255427
julia> x > 0.95 ? x : π
π = 3.1415926535897...
```

(Note, incidentally, that Julia understands Unicode!) Another way to achieve the same result uses the `ifelse` operator:

```
julia> ifelse(x > 0.95, x, π)
π = 3.1415926535897...
```

There are still further conditional operators in Julia; see subsequent discussion and the Supplementary Materials.

At least in the simulations, what we want the control flow elements to help us do is perform mathematical operations on synthetic data. Julia's notation of mathematical operations is fairly standard, for instance:

```
julia> 2 + 3
5
julia> 2 * 3
6
julia> 2^3
8
```

Boolean evaluations and set-theoretic operations are also fairly standard:[6]

```
julia> a = 12 < 3
false
julia> b = 3 + 4 ≤ 7
true
julia> a && b  # conjunction operator
false
```

6. Julia interprets anything following the # symbol as a comment and does not try to evaluate it.

```
julia> a || b  # disjunction operator
true
julia> !a      # negation operator
true
julia> [2, 3] ∩ [2, 5, 6]
1-element Array{Int64,1}:
 2
julia> [2, 3] ∪ [2, 5, 6]
4-element Array{Int64,1}:
 2
 3
 5
 6
julia> [2, 3] ⊆ [2, 5, 6]
false
julia> 3 ∈ [2, 5, 6]
false
julia> 3 ∉ [2, 5, 6]
true
```

You can also define your own functions:

```
julia> f(x) = sin(x) + sqrt(x)
f (generic function with 1 method)
julia> f(4)
1.2431975046920718
```

Multiline function definitions must be prepended with **function** and must end with **end**. For instance, the following function defines Bayes's rule specifically for the kind of data that we use in our simulations:

```
function bayes_rule(probs, toss)
    likelihood_heads = 0:1/(length(probs) - 1):1
    likelihood_tails = reverse(likelihood_heads)
    if toss == 1
        (probs .* likelihood_heads) ./ dot(probs, likelihood_heads)
    else
        (probs .* likelihood_tails) ./ dot(probs, likelihood_tails)
    end
end
```

Suppose that probs is your current probabilities for a set of jointly exhaustive and mutually exclusive hypotheses about the bias of a given coin and that toss is the outcome of a given toss. Then the function looks at whether the outcome was heads (coded as 1) or tails (coded as 0 in our simulations, though any value other than 1 would have done) and then, if the former, first multiplies your probabilities with the likelihoods bestowed on heads by the various hypotheses and then renormalizes them (by dividing each by the dot product of your probabilities and the likelihoods for heads) or, if the latter, does the same but now with likelihoods for tails replacing likelihoods for heads.[7]

Note the use of the period before the multiplication and division operators: that ensures that those operators are *broadcasted*, meaning that they apply element-wise (e.g., your probability for the hypothesis that the coin has a bias of x gets multiplied with the likelihood of the outcome on *that* bias hypothesis) and do not try to multiply, respectively divide, two whole *vectors* (which in the first case would throw an error and in the second would give the wrong outcome).

Broadcasting is also very convenient for selecting and replacing elements that have certain properties. For example, the following shows where in the array ar defined previously we find the elements greater than .5:

```
julia> ar .> .5
3×4 BitArray{2}:
 0  1  1  1
 1  1  0  0
 0  1  0  1
```

We can select those,

```
julia> ar[ar .> .5]
7-element Array{Float64,1}:
```

7. To make the dot product available, we must load the LinearAlgebra.jl package. It is also to be noted that the definition of Bayes's rule in the Jupyter notebooks contained in the Supplementary Materials looks a bit more complicated, but that is just because, thus defined, it is more convenient for how we set up the simulations. (The definition in the notebooks is also type annotated, as are all other definitions used for the simulations. Sometimes, using type annotations makes your code more performant, but more often, it has no influence on performance. Even then, I like to use type annotations, simply because they help me remember what kind of input or inputs a function expects.)

```
 0.9220074819252688
 0.569687258606496
 0.5795303164581178
 0.796653818790771
 0.7979833103099228
 0.7606313907554907
 0.6980853689654172
```

which we may for instance want to do if we need to add them up:

```
julia> sum(ar[ar .> .5])
5.124578945811484
```

Or perhaps we want to replace them all by some other value:

```
julia> ar[ar .> .5] .= 5;

julia> ar
3×4 Array{Float64,2}:
 0.160707  5.0  5.0       5.0
 5.0       5.0  0.358656  0.0114223
 0.286155  5.0  0.456668  5.0
```

And broadcasting together with the negation operator makes it easy to omit particular elements of a vector, or to omit rows or columns of an array. For instance,

```
julia> ar[1:end .!=2, 1:end .!=3]
2×3 Array{Float64,2}:
 0.160707  5.0  5.0
 0.286155  5.0  5.0
```

tells Julia to return `ar`, but with the second row and the third column removed.

Using the feature that Julia is a typed language, we can overload functions:

```
julia> g(x::Int64) = x^3
g (generic function with 1 method)
julia> g(x::Float64) = sqrt(x)
g (generic function with 2 methods)
julia> g(3)
```

```
27
julia> g(16.0)
4.0
```

In Julia jargon, there are now two *methods* associated with g. Which method is called depends on the type of the input argument.

In Julia, you can even define your own types, which are then on a par with Julia's built-in types. Suppose that we define

```
julia> struct my_type; a::String; end
```

which introduces a new type, my_type, which has one field that can contain a string.[8] Then we could add a further method to the ones already defined for g, for instance,

```
julia> g(x::my_type) = "Hello, " * x.a
g (generic function with 3 methods)
```

Then if

```
julia> s = my_type("Alice and Bob");
```

the function g allows us to greet our friends Alice and Bob:

```
julia> g(s)
"Hello, Alice and Bob"
```

(The type and method serve merely illustrative purposes here: I cannot think of any reason why one would want to introduce either.) This feature also enables multiple dispatch, which we briefly discussed in chapter 1.

It was already mentioned that Julia uses packages (or libraries). These contain functions that serve special purposes. Some of those packages are preinstalled, but most are not and must be installed via the package manager. For instance, to install the `Distributions.jl` package, press the] key and enter (and execute) `add Distributions`. (To leave the package manager, press the backspace key.) To use a package in a session, you must explicitly load it. For instance, to load the `Distributions.jl` package, enter **`using`** `Distributions`.

8. While type annotations are often optional in function definitions, as mentioned in footnote 7 in this appendix, in type definitions it is strongly recommended to annotate the field or fields, as this will help the Julia compiler to optimize the assembly code.

To repeat, at the time of this writing, a number of packages are still in beta, which means that later versions may introduce breaking changes. Often, package developers then include deprecation warnings, telling you how to update your old code—often, but not always, which can lead to frustrating experiences, as can (at the moment) dependency issues and some other quirks.[9]

Two packages that deserve special mention are `RCall.jl` and `PyCall.jl`. The first allows you to call R from within Julia, the second, from within Python. (They require R and, respectively, Python to be installed on your computer.) For instance, the following lines of code perform an ANOVA test using R's `aov` function:

```
julia> using RCall
julia> u = rand(100);
julia> v = rand(100) .+ 0.1;
julia> w = rand(100) .- 0.1;
julia> x = repeat(1:3, inner = 100);
julia> @rput u; @rput v; @rput w; @rput x;
julia> R"summary(aov(c(u, v, w) ~ as.factor(x)))"
RCall.RObject{RCall.VecSxp}
              Df Sum Sq Mean Sq F value   Pr(>F)
as.factor(x)   2  2.369  1.1844    15.9 2.76e-07 ***
Residuals    297 22.127  0.0745
---
Signif. codes:  0 '***' 0.001 '**' 0.01 '*' 0.05 '.' 0.1 ' ' 1
```

Here, the calls to `rand(100)` create vectors of one hundred random floats between 0 and 1 inclusive. To create `v`, we add 0.1 to *each* random value in the generated vector, by broadcasting the + operator (if we used + instead of `.+`, we would be trying to add 0.1 to a vector, which makes no sense and accordingly would throw an error), and similarly for `w`. The vector `x` is to serve as our predictor. Then the `@rput` macro is used to make `u`, `v`, `w`, and `x` available to R. The use of `PyCall.jl` is similar, for instance,

```
julia> using PyCall
julia> math = pyimport("math");
```

9. You can ignore warnings because they will not prevent your code from running. Errors *cannot* be ignored; they have to be fixed, or your code will not run.

```
julia> math.sqrt(2)
1.4142135623730951
```

To use R or Python packages from within Julia, these must first be installed via the package managers for those languages (so, using `install.packages` for R and `pip` or `conda` for Python).

As a final general comment, I should mention how easy parallel computing is in Julia. Parallel computing leverages that nowadays virtually all computers have more than one core. Even computers with twenty-four or more cores have become relatively affordable. For the simulations we ran, having multiple cores available helped to speed up computation tremendously. In these simulations, various procedures are repeated over and over again, each time starting from a somewhat different random configuration, the results from the various runs then typically being averaged in some way at the end. Julia makes it very easy to divide those runs over different cores, which can then run the requisite computations in parallel, so that, for instance, if your computer has twenty-four cores, running twenty-four simulations takes about as long as running just one simulation (the former will take a *bit* longer because of the overhead associated with parallel computation). There are different ways to set up parallel computations in Julia, and which is best depends on details of the computations to be executed. I refer here to the highly readable part on parallel computing in the Julia manual, at https://docs.julialang.org/en/v1/manual/parallel-computing/, which also contains all the information about the macros and functions for parallel computing used in the notebooks in the Supplementary Materials.

Statistics in Julia

R is a computer language specifically for statistical computing. Whatever type of advanced statistical analysis you may want to conduct, there will be at least one dedicated R package for it. Julia is still comparatively limited in this respect. However, for all the standard types of analysis, there are excellent packages available, including `Statistics.jl`, `StatsBase.jl`, `Bootstrap.jl`, `HypothesisTests.jl`, `GLM.jl`, `Lathe.jl`, and the previously mentioned package `Distributions.jl`. In fact, Douglas Bates, the main developer of the `lme4` package for R, has switched to Julia precisely because it is so much faster than R and has written the excellent `MixedModels.jl` package, which was used for the analyses reported in section 3.4, which readers can try out in the corresponding notebook in the Supplementary Materials. `Turing.jl` is a probabilistic programming package currently under active development but already excellently suited for running a rich variety of Bayesian statistical analyses. For any kind of analysis that is not yet available in Julia, you can call the requisite R package via `RCall.jl`, as just explained. See Nazarathy and Klok (2021) for more on statistics in Julia.

Optimization

There are a number of packages for optimization. Probably the best-developed and best-maintained of these is `JuMP.jl`. This package is an interface to a variety of different solvers, both open-source and proprietary ones. It does not provide those solvers itself, so they have to be installed independently. Which solver is best to use depends on the optimization problem at hand. For instance, the optimization problem considered in section 5.3 was nonlinear, for which one can use the Ipopt optimizer, a free and open-source optimizer that is automatically installed when one installs the `Ipopt.jl` package in Julia.

The `JuMP.jl` package offers a highly intuitive language for defining optimization problems: one defines the variables, the objective function to be optimized, constraints (if there are any), and then simply solves the resulting model, indicating the type of optimization (minimization, in our case) one is after. We illustrate the use of this package by showing how we found the minimal VS rule penalty for David's update on evidence E in section 5.3:

```
using JuMP
using Ipopt

model = Model(Ipopt.Optimizer)
@variable(model, 0 ≤ x[1:3] ≤ 1)
@NLobjective(
    model,
    Min,
    .01(.1(1 - x[1])^2 + .3x[2]^2 + .6x[3]^2) +
    .25(.45x[1]^2 + .1(1 - x[2])^2 + .45x[3]^2) +
    .1(.6x[1]^2 + .3x[2]^2 + .1(1 - x[3])^2)
)
@constraint(model, sum(x) == 1)
optimize!(model)
```

This actually gives a lot of output, which is why it is suppressed here. You may want to retrieve the minimum of the objective function specifically by entering `objective_value(model)`, and similarly for the values of the variables for which that minimum is obtained, which can be accessed by entering `value.(x)` and so on.

To ascertain that Bayes's rule does minimize expected RPS penalty (as we saw in the same section), replace the specification of the objective function in the preceding by

```
@NLobjective(
    model,
    Min,
    .01(((x[1] - 1)^2 +
        (x[1] + x[2] - 1)^2 +
        (x[1] + x[2] + x[3] - 1)^2)/2) +
    .25(((x[1] - 0)^2 +
        (x[1] + x[2] - 1)^2 +
        (x[1] + x[2] + x[3] - 1)^2)/2) +
     .1(((x[1] - 0)^2 +
        (x[1] + x[2] - 0)^2 +
        (x[1] + x[2] + x[3] - 1)^2)/2)
)
```

One section earlier in the same chapter it was shown that Bayesian updates are not necessarily more accurate than EXPL updates if accuracy is measured by a VS rule. This involved finding weights that, given David's probabilities, made his VS score come out lower if he updated by EXPL than if he updated by Bayes's rule. Interested readers can reproduce the result by using the code below. We first define variables containing the post-update probabilities for the two rules, for which we load the `LinearAlgebra.jl` package:

```
using LinearAlgebra

bayes = normalize!([.01, .25, .1], 1)
expl = normalize!([.11, .25, .1], 1)
```

Next we define the corresponding scoring functions:

```
function bayes_scr(w1, w2, w3, w4, w5)
    x = .01(w1*(1 - bayes[1])^2 + w2*bayes[2]^2 + w3*bayes[3]^2)
    y = .25(w5*bayes[1]^2 + w4*(1 - bayes[2])^2 + w5*bayes[3]^2)
    z = .10(w3*bayes[1]^2 + w2*bayes[2]^2 + w1*(1 - bayes[3])^2)
    return x + y + z
end

function expl_scr(w1, w2, w3, w4, w5)
    x = .01(w1*(1 - expl[1])^2 + w2*expl[2]^2 + w3*expl[3]^2)
    y = .25(w5*expl[1]^2 + w4*(1 - expl[2])^2 + w5*expl[3]^2)
    z = .10(w3*expl[1]^2 + w2*expl[2]^2 + w1*(1 - expl[3])^2)
    return x + y + z
end
```

And finally we define a function to loop through combinations of weights to find ones (if there are any, and perhaps if we are lucky) that make David incur a lower VS penalty for his EXPL update than for his Bayesian update:

```
function find_weights(iter)
    i = 1
    res = ()
    weights = Vector{Float64}(undef, 5)
    while i ≤ iter
        v = sort(rand(Dirichlet([1, 1, 1])))
        w = sort(rand(Dirichlet([1, 1])))
        weights = vcat(v, w)
        res = bayes_scr(weights...), expl_scr(weights...)
        if first(res) > last(res)
            break
        end
        i += 1
    end
    if first(res) > last(res)
        return res, weights
    else
        println("No weights found. Try again!")
    end
end
```

In my experiments, running `find_weights(1000)` was typically enough to yield weights of the sought-after kind. In other words, such weights exist, and apparently it is not difficult to find them.

If you are curious by *how much* EXPL can outperform Bayes's rule, in principle, we can again use the `JuMP.jl` package to find the weights that will maximize the difference between the EXPL-update score and the Bayesian-update score, with the former being lower than the latter. Specifically, the following chunk of code defines a function that takes weights as input and then outputs David's EXPL-update VS score (given the input weights) minus his Bayesian-update VS score. It then lets the computer look for the weights that minimize that function:

```
model = Model(Ipopt.Optimizer)
@variables(model, begin
    0 ≤ w[1:3] ≤ 1
    0 ≤ v[1:2] ≤ 1
end)
@NLobjective(
    model,
    Min,
    (.01(w[1]*(1 - expl[1])^2 +
         w[2]*expl[2]^2 +
         w[3]*expl[3]^2) +
     .25(v[1]*expl[1]^2 +
         v[2]*(1 - expl[2])^2 +
         v[1]*expl[3]^2) +
      .1(w[3]*expl[1]^2 +
         w[2]*expl[2]^2 +
         w[1]*(1 - expl[3])^2)) -
    (.01(w[1]*(1 - bayes[1])^2 +
         w[2]*bayes[2]^2 +
         w[3]*bayes[3]^2) +
     .25(v[1]*bayes[1]^2 +
         v[2]*(1 - bayes[2])^2 +
         v[1]*bayes[3]^2) +
      .1(w[3]*bayes[1]^2 +
         w[2]*bayes[2]^2 +
         w[1]*(1 - bayes[3])^2))
)
@constraints(model, begin
    sum(w) == 1
    sum(v) == 1
    w[1] ≤ w[2]
    w[2] ≤ w[3]
    v[2] ≤ v[1]
end)
optimize!(model)
```

Running `objective_value(model)` on this model shows that the EXPL-update VS score can at most be 0.001 lower than the Bayesian-update VS score; again, run `value.(x)` to obtain the weights that realize that difference in favor of EXPL.

Agent-Based Modeling

The simulations reported in chapter 7 relied heavily on agent-based models. Such models have become increasingly popular in philosophy and, to some extent, in cognitive science. Most colleagues working in this field use NetLogo (Wilensky & Rand, 2015), a language originally designed to introduce children to computer programming. In this language, the programmer basically tells turtles how to move and behave. Like Julia, NetLogo is free and open-source. The language has some other virtues as well (e.g., it is easy to learn, and it offers an attractive user interface). Nevertheless, I strongly recommend the use of Julia for agent-based modeling.

That is not just because you might be embarrassed to tell your friends that you program turtles for a living and would much prefer to say that you use this cool new language for high-performance computing that has been adopted by Google, IBM, NASA, Pfizer, and other high-end users. Most importantly, NetLogo is *extremely* slow. (What else can one expect from turtles?)

Users of NetLogo may tell you that the lack of speed is not really a problem, given that one can always move performance-critical parts of the code to C++, or FORTRAN, or some other compiled language. While that is true, that one has to take such an extra step is exactly the problem, known as "the two-language problem," that motivated the Julia developers. Julia can be used both for prototyping—designing your program—and for running the simulations. To repeat, it does so by offering a language that is easy to work with—as easy as NetLogo and Python—and yet is about as fast as C++.

Getting started with agent-based modeling in Julia has become particularly easy with the advent of the `Agents.jl` package (Vahdati, 2019), which requires from the user no more than a specification of agent and model properties and then takes care of all the rest (in particular, the running of the simulations and the gathering of relevant data). The appendix of Douven and Hegselmann (2021) contains links to various tutorials that help readers to get started with the package. It is to be noted, however, that the package is geared primarily toward ease of use and generality, and only then toward performance. And while it still allows one to write fast code, I have found it easier to write highly performant code by starting from scratch, without using the package. In particular, the code used for chapter 7 does not rely on `Agents.jl`. Still, the package offers an excellent starting point for new users.

Graphics

There is a variety of high-quality graphics packages for Julia. The figures in this book have all been produced using `Gadfly.jl`, but I have also had good experiences with `PlotlyJS.jl`, and the standard package `Plots.jl` yields publication-quality output as well. Nevertheless, I know of users who came to Julia from R and prefer to use the `ggplot` package via `RCall.jl`, as explained previously, and also of users who came from Python and have stuck with `pyplot`, which can be used in Julia thanks to `PyCall.jl`. And users with previous experience with `gnuplot` may want to have a look at the `Gaston.jl` and `Gnuplot.jl` packages. Probably, the most powerful visualization package for Julia is `Makie.jl`, which is under rapid development at the time of this writing but which at least for now offers a less than intuitive user interface (in my admittedly limited experience with the package).

Running the Simulations

I am making the code for the simulations available, because, first, as mentioned, I am a strong believer in a learning-by-doing approach, which in the present case means that I believe experimenting with the code to be the best and fastest way to obtain a good understanding of the simulations that were reported. In fact, because Julia code is relatively easy to understand, just reading it will deepen one's understanding of the simulations. To run the code, and later possibly alter it, first download it from the following public GitHub repository: https://github.com/IgorDouven/ABDUCTION.git. There are four Jupyter notebooks in the repository, which correspond to different parts of the book (as indicated in the titles of the notebooks), as well as the data from Douven and Schupbach (2015a), which are necessary to run the analysis reported in section 3.4. For instructions on how to work with Jupyter notebooks, see the documentation on this website: https://jupyter.org. Since very recently, you can also open, and work with, Jupyter notebooks in Visual Studio Code, a combination that I have found to offer a wonderful user experience.

References

Ackerman, R., Douven, I., Elqayam, S., & Teodorescu, K. (2020). Satisficing, meta-reasoning, and the rationality of further deliberation. In S. Elqayam, I. Douven, N. Cruz, & J. St. B. T. Evans (eds.), *Logic and uncertainty in the human mind* (pp. 10–26). London, UK: Routledge.

Ackerman, R. & Thompson, V. A. (2015). Meta-reasoning: What we can learn from meta-memory. In A. Feeney & V. A. Thompson (eds.), *Reasoning as memory* (pp. 164–178). Hove, UK: Psychology Press.

Ackerman, R. & Thompson, V. A. (2017). Meta-reasoning: Monitoring and control of thinking and reasoning. *Trends in Cognitive Sciences* 21: 607–617.

Ackerman, R. & Thompson, V. A. (2018). Meta-reasoning: Shedding meta-cognitive light on reasoning research. In L. Ball & V. A. Thompson (eds.), *The Routledge international handbook of thinking and reasoning* (pp. 1–15). New York, NY: Routledge.

Aczél, J. & Wagner, C. (1980). A characterization of weighted arithmetic means. *SIAM Journal on Algebraic and Discrete Methods* 1: 259–260.

Adler, J. (1994). Testimony, trust, knowing. *Journal of Philosophy* 91: 264–275.

Agresti, A. (2007). *An introduction to categorical data analysis*. Hoboken, NJ: Wiley.

Akaike, H. (1973). Information theory as an extension of the maximum likelihood principle. In B. Petrov & F. Csaki (eds.), *Second international symposium on information theory* (pp. 267–281). Budapest, Hungary: Akademiai Kiado.

Ali, N., Chater, N., & Oaksford, M. (2011). The mental representation of causal conditional reasoning: Mental models or causal models. *Cognition* 119: 403–418.

Ali, N., Schlottmann, A., Shaw, A., Chater, N., & Oaksford, M. (2010). Causal discounting and conditional reasoning in children. In M. Oaksford & N. Chater (eds.), *Cognition and conditionals: Probability and logic in human thinking* (pp. 117–134). Oxford, UK: Oxford University Press.

Aliseda, A. (2006). *Abductive reasoning: Logical investigations into discovery and explanation*. Berlin, Germany: Springer.

Anderson, J. R. (1990). *The adaptive character of thought*. Hillsdale, NJ: Lawrence Erlbaum Associates.

Anderson, J. R. (1991). Is human cognition adaptive? *Behavioral and Brain Sciences* 14: 471–517.

Arkes, H. R., Gigerenzer, G., & Hertwig, R. (2016). How bad is incoherence? *Decision* 3: 20–39.

Atkinson, D. (2003). Experiments and thought experiments in natural science. In M. C. Galavotti (ed.), *Observation and experimentation in the natural and social sciences* (pp. 209–225). Dordrecht, Netherlands: Kluwer.

Baayen, R. H. (2008). *Analyzing linguistic data*. Cambridge, UK: Cambridge University Press.

Bach, K. & Harnish, R. (1979). *Linguistic communication and speech acts*. Cambridge, MA: MIT Press.

Bacharach, M. (1979). Normal Bayesian dialogues. *Journal of the American Statistical Association* 74: 837–846.

Bäck, T. (1996). *Evolutionary algorithms in theory and practice*. Oxford, UK: Oxford University Press.

Baillargeon, R. & DeJong, G. F. (2017). Explanation-based learning in infancy. *Psychonomic Bulletin & Review* 24: 1511–1526.

Baker, J. E. (1985). Adaptive selection methods for genetic algorithms. In J. J. Grefenstette (ed.), *Proceedings of the First International Conference on Genetic Algorithms and Their Applications* (pp. 100–111). Hillsdale, NJ: Lawrence Erlbaum Associates.

Ballantyne, N. & Coffman, E. J. (2011). Uniqueness, evidence, and rationality. *Philosophers Imprint* 11: 1–13.

Ballantyne, N. & Coffman, E. J. (2012). Conciliationism and uniqueness. *Australasian Journal of Philosophy* 90: 657–670.

Baratgin, J. & Politzer, G. (2007). The psychology of dynamic probability judgment: Order effect, normative theories, and experimental methodology. *Mind & Society* 6: 53–66.

Barbati, M., Bruno, G., & Genovese, A. (2012). Applications of agent-based models for optimization problems: A literature review. *Expert Systems with Applications* 39: 6020–6028.

Bar-Hillel, M. (1980). The base-rate fallacy in probability judgments. *Acta Psychologica* 44: 211–233.

Beach, L. R. & Swenson, R. G. (1966). Intuitive estimation of means. *Psychonomic Science* 5: 161–162.

Berg, N., Biele, G., & Gigerenzer, G. (2016). Consistent Bayesians are no more accurate than non-Bayesians: Economists surveyed about PSA. *Review of Behavioral Economics* 3: 189–219.

Bernardo, J. M. (1979). Expected information as expected utility. *Annals of Statistics* 7: 686–690.

Bernardo, J. M. & Smith, A. F. M. (2000). *Bayesian theory*. New York, NY: Wiley.

Bes, B., Sloman, S., Lucas, C. G., & Raufaste, E. (2012). Non-Bayesian inference: Causal structure trumps correlation. *Cognitive Science* 36: 1178–1203.

Bezanson, J., Edelman, A., Karpinski, S., & Shah, V. B. (2017). Julia: A fresh approach to numerical computing. *SIAM Review* 59: 65–98.

Bickel, J. E. (2007). Some comparisons between quadratic, spherical, and logarithmic scoring rules. *Decision Analysis* 4: 49–65.

Bickel, J. E. (2010). Scoring rules and decision analysis education. *Decision Analysis* 7: 346–357.

Biggs, S. & Wilson, J. (2017). The a priority of abduction. *Philosophical Studies* 174: 735–758.

Bird, A. (1998). *Philosophy of science*. Montreal, Canada: McGill–Queen's University Press.

Bird, A. (2010). Eliminative abduction: Examples from medicine. *Studies in the History and Philosophy of Science* 41: 345–352.

BonJour, L. (1985). *The structure of empirical knowledge*. Cambridge, MA: Harvard University Press.

BonJour, L. & Sosa, E. (2003). *Epistemic justification*. Malden, MA: Blackwell.

Borsboom, D. & Haig, B. D. (2013). How to practice Bayesian statistics outside the Bayesian church: What philosophy for Bayesian modeling? *British Journal of Mathematical and Statistical Psychology* 66: 39–44.

Bovens, L. & Hartmann, S. (2003). *Bayesian epistemology*. Oxford, UK: Oxford University Press.

Boyd, R. N. (1984). On the current status of scientific realism. *Erkenntnis* 19: 45–90.

Boyd, R. N. (1985). Lex orandi est lex credendi. In P. Churchland & C. Hooker (eds.), *Images of science* (pp. 3–34). Chicago, IL: University of Chicago Press.

Boyd, R. N. (1990). Realism, approximate truth, and philosophical method. In C. Wade Savage (ed.), *Scientific theories* (pp. 355–391). Minneapolis, MN: University of Minnesota Press.

Boyd, R. & Richerson, P. J. (1985). *Culture and the evolutionary process*. Chicago, IL: University of Chicago Press.

Boyd, R., Richerson, P. J., & Henrich, J. (2011). The cultural niche: Why social learning is essential for human adaptation. *Proceedings of the National Academy of Sciences of the United States of America* 108: 10918–10925.

Brewer, M. B. (2004). Taking the origins of human nature seriously: Toward a more imperialist social psychology. *Personality and Social Psychology Review* 8: 107–113.

Brier, G. W. (1950). Verification of forecasts expressed in terms of probability. *Monthly Weather Review* 78: 1–3.

Briggs, R. A. & Pettigrew, R. (2020). An accuracy-dominance argument for conditionalization. *Noûs* 54: 162–181.

Brössel, P. (2013). The problem of measure sensitivity redux. *Philosophy of Science* 80: 378–397.

Brössel, P. (2015). On the role of explanatory and systematic power in scientific reasoning. *Synthese* 192: 3877–3913.

Brouwer, L. E. J. (1911). Über Abbildungen von Mannigfaltigkeiten. *Mathematische Annalen* 71: 97–115.

Bulger, P. M. J., Hiles, D. R., & Lowe, G. (1969). Presentation time and the intuitive estimation of means. *Psychonomic Science* 15: 191–192.

Bunt, H. C. & Black, W. J. (2000). *Abduction, belief, and context in dialogue: Studies in computational pragmatics*. Cambridge, MA: MIT Press.

Burnham, K. P. & Anderson, D. R. (2002). *Model selection and multi-model inference: A practical information-theoretic approach* (2nd ed.). Berlin, Germany: Springer.

Byerly, T. R. (2013). Explanationism and justified beliefs about the future. *Erkenntnis* 78: 229–243.

Byrne, A. (2004). How hard are the sceptical paradoxes? *Noûs* 38: 299–325.

Cabrera, F. (2017). Can there be a Bayesian explanationism? On the prospects of a productive parternship. *Synthese* 194: 1245–1272.

Callebaut, W. (1993). *Taking the naturalistic turn*. Chicago, IL: University of Chicago Press.

Campos, D. (2011). On the distinction between Peirce's abduction and Lipton's inference to the best explanation. *Synthese* 180: 419–442.

Caporael, L R. (2001). Evolutionary psychology: Toward a unifying theory and a hybrid science. *Annual Review of Psychology* 52: 607–628.

Carnap, R. (1945). On inductive logic. *Philosophy of Science* 12: 72–97.

Carnap, R. (1947a). *Meaning and necessity*. Chicago, IL: University of Chicago Press.

Carnap, R. (1947b). On the application of inductive logic. *Philosophy and Phenomenological Research* 8: 133–148.

Carnap, R. (1950). *Logical foundations of probability*. Chicago, IL: University of Chicago Press.

Carnap, R. (1952). Meaning postulates. *Philosophical Studies* 3: 65–73.

Carnap, R. (1980). A basic system of inductive logic II. In R. C. Jeffrey (ed.), *Studies in inductive logic and probability* (vol. 2, pp. 7–155). Berkeley, CA: University of California Press.

Cesa-Bianchi, N. & Lugosi, G. (2006). *Prediction, learning, and games*. Cambridge, UK: Cambridge University Press.

Cevolani, G., Festa, R., & Kuipers, T. A. F. (2013). Verisimilitude and belief change for nomic conjunctive theories. *Synthese* 190: 3307–3324.

Chabris, C. F. & Simons, D. J. (2010). *The invisible gorilla*. New York, NY: Crown.

Chaitin, G. (2006). *Meta math! The quest for omega*. New York, NY: Vintage Books.

Chandler, J. (2013). Acceptance, aggregation and scoring rules. *Erkenntnis* 78: 201–217.

Chater, N. & Oaksford, M. (1999). The probability heuristics model of syllogistic reasoning. *Cognitive Psychology* 38: 191–258.

Chi, M. T. H., de Leeuw, N., Chiu, M.-H., & LaVancher, C. (1994). Eliciting self-explanations improves understanding. *Cognitive Science* 18: 439–477.

Chien, A. (2008). Scalar implicature and contrastive explanation. *Synthese* 161: 47–66.

Christensen, D. (1983). Glymour on evidential relevance. *Philosophy of Science* 50: 471–481.

Christensen, D. (1994). Conservatism in epistemology. *Noûs* 28: 69–89.

Christensen, D. (1997). What is relative confirmation? *Noûs* 31: 370–384.

Christensen, D. (2007). Epistemology of disagreement: The good news. *Philosophical Review* 116: 187–218.

Christensen, D. (2011). Disagreement, question-begging and epistemic self-criticism. *Philosophers Imprint* 11: 1–22.

Christensen, D. (2016). Uniqueness and rational toxicity. *Noûs* 50: 584–603.

Climenhaga, N. (2017a). How explanation guides confirmation. *Philosophy of Science* 84: 359–368.

Climenhaga, N. (2017b). Inference to the best explanation made incoherent. *Journal of Philosophy* 114: 251–273.

Coello Coello, C. A., Lamont, G. B., & Van Veldhuizen, D. A. (2007). *Evolutionary algorithms for solving multi-objective problems* (2nd ed.). New York, NY: Springer.

Cohen, M. P. (2016). On three measures of explanatory power with axiomatic representations. *British Journal for the Philosophy of Science* 67: 1077–1089.

Colombo, M. (2017). Experimental philosophy of explanation rising: The case for a plurality of concepts of explanation. *Cognitive Science* 41: 503–517.

Colombo, M., Bucher, L., & Sprenger, J. (2017). Determinants of judgments of explanatory power: Credibility, generality, and statistical relevance. *Frontiers in Psychology* 8, https://doi:10.3389/fpsyg.2017.01430.

Colombo, M., Elkin, L., & Hartmann, S. (2021). Being realist about Bayes, and the predictive processing theory of mind. *British Journal for the Philosophy of Science* 72: 185–220.

Conee, E. & Feldman, R. (2004). *Evidentialism: Essays in epistemology*. Oxford, UK: Oxford University Press.

Conee, E. & Feldman, R. (2008). Evidence. In Q. Smith (ed.), *Epistemology: New essays* (pp. 83–104). Oxford, UK: Oxford University Press.

Cooke, R. M. (1991). *Experts in uncertainty*. Oxford, UK: Oxford University Press.

Costello, F. & Watts, P. (2016). People's conditional probability judgments follow probability theory (plus noise). *Cognitive Psychology* 89: 106–133.

Costello, F. & Watts, P. (2018). Invariants in probabilistic reasoning. *Cognitive Psychology* 100: 1–16.

Crosscombe, M. & Lawry, J. (2016). A model of multi-agent consensus for vague and uncertain beliefs. *Adaptive Behavior* 24: 249–260.

Crupi, V. & Tentori, K. (2012). A second look at the logic of explanatory power (with two novel representation theorems). *Philosophy of Science* 79: 365–385.

Crupi, V., Tentori, K., & Gonzalez, M. (2007). On Bayesian measures of evidential support: Theoretical and empirical issues. *Philosophy of Science* 74: 229–252.

Czarnowski, I., Jędrzejowicz, P., & Kacprzyk, J. (2013). *Agent-based optimization*. Berlin, Germany: Springer.

Darwin, C. (1859). *On the origin of species by means of natural selection*. London, UK: John Murray.

Darwin, C. (1876). *On the origin of species by means of natural selection* (6th ed.). London, UK: John Murray.

Dascal, M. (1979). Conversational relevance. In A. Margalit (ed.), *Meaning and use* (pp. 153–174). Dordrecht, Netherlands: Reidel.

Davidson, D. (1996). The folly of trying to define truth. *Journal of Philosophy* 93: 263–278.

Day, T. & Kincaid, H. (1994). Putting inference to the best explanation in its place. *Synthese* 98: 271–295.

Deb, K. (2001). *Multi-objective optimization using evolutionary algorithms*. Chichester, UK: Wiley.

Decock, L. & Douven, I. (2012). Putnam's internal realism: A radical restatement. *Topoi* 31: 111–120.

Decock, L., Douven, I., & Sznajder, M. (2016). A geometric principle of indifference. *Journal of Applied Logic* 19: 54–70.

De Cruz, H. (2017). Religious disagreement: An empirical study among academic philosophers. *Episteme* 14: 71–87.

De Cruz, H. & De Smedt, J. (2013). The value of epistemic disagreement in scientific practice: The case of Homo floresiensis. *Studies in History and Philosophy of Science* 44: 169–177.

De Cruz, H. & De Smedt, J. (2015). *A natural history of natural theology: The cognitive science of theology and philosophy of religion*. Cambridge, MA: MIT Press.

Deffuant, G., Neau, D., Amblard, F., & Weisbuch, G. (2000). Mixing beliefs among interacting agents. *Advances in Complex Systems* 3: 87–98.

de Finetti, B. (1937). Foresight: Its logical laws, its subjective sources. In H. Kyburg & H. Smokler (eds.), *Studies in subjective probability* (1980, 2nd ed., pp. 53–118). New York, NY: Krieger.

de Finetti, B. (1962). Does it make sense to speak of "good probability appraisers"? In I. J. Good (ed.), *The scientist speculates: An anthology of partly-baked ideas* (pp. 357–364). New York, NY: Basic Books.

De Langhe, R. (2013). Peer disagreement under multiple epistemic constraints. *Synthese* 190: 2547–2556.

De Langhe, R. & Greiff, M. (2010). Standards and the distribution of cognitive labour: A model of the dynamics of scientific activity. *Logic Journal of the IGPL* 18: 278–294.

Dellsén, F. (2017). Reactionary responses to the bad lot objection. *Studies in History and Philosophy of Science* 61: 32–40.

Dellsén, F. (2018). The heuristic conception of inference to the best explanation. *Philosophical Studies* 175: 1745–1766.

Dellsén, F. (2021). Explanatory consolidation: From "best" to "good enough." *Philosophy and Phenomenological Research* 103: 157–177.

de Regt, H. W. (2017). *Understanding scientific understanding*. Oxford, UK: Oxford University Press.

de Regt, H. W. & Dieks, D. (2005). A contextual approach to scientific understanding. *Synthese* 144: 137–170.

de Ridder, J. (2014). Epistemic dependence and collective scientific knowledge. *Synthese* 191: 37–53.

Devitt, M. (1991). *Realism and truth* (2nd ed.). Oxford, UK: Blackwell.

Dhanalakshmi, S., Kanna, S., Mahadevan, K., & Baskar, S. (2011). Application of modified NSGA-II algorithm to combined economic and emission dispatch problem. *International Journal of Electrical Power & Energy Systems* 33: 992–1002.

Dietz, R. (2010). On generalizing Kolmogorov. *Notre Dame Journal of Formal Logic* 51: 323–335.

Dittmer, J. C. (2001). Consensus formation under bounded confidence. *Nonlinear Analysis* 7: 4615–4621.

Djulbegovic, B. & Elqayam, S. (2017). Many faces of rationality: Implications of the great rationality debate for clinical decision-making. *Journal of Evaluation in Clinical Practice* 23: 915–922.

Djulbegovic, B., Elqayam, S., Reljic, T., Hozo, I., Miladinovic, B., Tsalatsanis, A., Kumar, A., Beckstead, J., Taylor, S., & Cannon-Bowers, J. (2014). How do physicians decide to treat: An empirical evaluation of the threshold model. *BMC Medical Informatics and Decision Making* 14: 47, http://www.biomedcentral.com/1472-6947/14/47.

Douven, I. (1995). Boyd's miraculous no miracle argument. In P. Cortois (ed.), *The many problems of realism* (pp. 89–116). Tilburg, Netherlands: Tilburg University Press.

Douven, I. (1999). Inference to the best explanation made coherent. *Philosophy of Science* 66: S424–S435.

Douven, I. (2000). The antirealist argument for underdetermination. *Philosophical Quarterly* 50: 371–375.

Douven, I. (2002a). Testing inference to the best explanation. *Synthese* 130: 355–377.

Douven, I. (2002b). A new solution to the paradoxes of rational acceptability. *British Journal for the Philosophy of Science* 53: 391–410.

Douven, I. (2005). Evidence, explanation, and the empirical status of scientific realism. *Erkenntnis* 63: 253–291.

Douven, I. (2006). Assertion, knowledge, and rational credibility. *Philosophical Review* 115: 449–485.

Douven, I. (2008). Underdetermination. In S. Psillos & M. Curd (eds.), *The Routledge companion to philosophy of science* (pp. 292–301). London, UK: Routledge.

Douven, I. (2009). Assertion, Moore, and Bayes. *Philosophical Studies* 144: 361–375.

Douven, I. (2010). Simulating peer disagreements. *Studies in History and Philosophy of Science* 41: 148–157.

Douven, I. (2012a). The lottery paradox and the pragmatics of belief. *Dialectica* 66: 351–373.

Douven, I. (2012b). Learning conditional information. *Mind & Language* 27: 239–263.

Douven, I. (2013). Inference to the best explanation, Dutch books, and inaccuracy minimisation. *Philosophical Quarterly* 69: 428–444.

Douven, I. (2016a). *The epistemology of indicative conditionals*. Cambridge, UK: Cambridge University Press.

Douven, I. (2016b). Explanation, updating, and accuracy. *Journal of Cognitive Psychology* 28: 1004–1012.

Douven, I. (2017a). Abduction. In E. N. Zalta (ed.), *The Stanford encyclopedia of philosophy*, https://plato.stanford.edu/archives/sum2017/entries/abduction/.

Douven, I. (2017b). Inference to the best explanation: What is it? And why should we care? In K. McCain & T. Poston (eds.), *Best explanations: New essays on inference to the best explanation* (pp. 4–22). Oxford, UK: Oxford University Press.

Douven, I. (2018). A Bayesian perspective on Likert scales and central tendency. *Psychonomic Bulletin & Review* 25: 1203–1211.

Douven, I. (2019a). Optimizing group learning: An evolutionary computing approach. *Artificial Intelligence* 275: 235–251.

Douven, I. (2019b). Putting prototypes in place. *Cognition* 193: 104007, https://doi.org/10.1016/j.cognition.2019.104007.

Douven, I. (2019c). Computational models in social epistemology. In M. Fricker, P. Graham, D. Henderson, N. J. Pedersen, & J. Wyatt (eds.), *The Routledge handbook of social epistemology* (pp. 457–465). London, UK: Routledge.

Douven, I. (2020a). Can the evidence for explanatory reasoning be explained away? *IEEE Transactions on Cognitive and Developmental Systems* 12: 12–17.

Douven, I. (2020b). Scoring in context. *Synthese* 197: 1565–1580.

Douven, I. (2020c). The ecological rationality of explanatory reasoning. *Studies in History and Philosophy of Science* 79: 1–14.

Douven, I. (2020d). The rationality of vagueness. In R. Dietz (ed.), *Vagueness and rationality in language use and cognition* (pp. 115–134). Cham, Switzerland: Springer.

Douven, I. (2021). Tracking confirmation. *Philosophy of Science* 88: 398–414.

Douven, I. (in press). Explaining the success of induction. *British Journal for the Philosophy of Science*.

Douven, I. & Dietz, R. (2011). A puzzle about Stalnaker's hypothesis. *Topoi* 30: 31–37.

Douven, I. & Elqayam, S. (2021). Rethinking the lottery paradox: A dual processing perspective. In I. Douven (ed.), *Lotteries, knowledge, and rational belief: Essays on the lottery paradox* (pp. 110–127). Cambridge, UK: Cambridge University Press.

Douven, I., Elqayam, S., & Mirabile, P. (in press). Inference strength predicts the probability of conditionals better than conditional probability does. *Journal of Memory and Language*, in press.

Douven, I., Elqayam, S., Singmann, H., & van Wijnbergen-Huitink, J. (2018). Conditionals and inferential connections: A hypothetical inferential theory. *Cognitive Psychology* 101: 50–81.

Douven, I., Elqayam, S., Singmann, H., & van Wijnbergen-Huitink, J. (2020). Conditionals and inferential connections: Toward a new semantics. *Thinking & Reasoning* 26: 311–351.

Douven, I. & Gärdenfors, P. (2020). What are natural concepts? A design perspective. *Mind & Language* 35: 313–334.

Douven, I. & Hegselmann, R. (2021). Mis- and disinformation in a bounded confidence model. *Artificial Intelligence* 291: 103415, https://doi.org/10.1016/j.artint.2020.103415.

Douven, I. & Horsten, L. (1998). Earman on underdetermination and empirical indistinguishability. *Erkenntnis* 49: 303–320.

Douven, I. & Kelp, C. (2011). Truth approximation, social epistemology, and opinion dynamics. *Erkenntnis* 75: 271–283.

Douven, I. & Kelp, C. (2013). Proper bootstrapping. *Synthese* 190: 171–185.

Douven, I. & Krzyżanowska, K. H. (2019). The semantics–pragmatics divide: An empirical perspective. In A. Capone, M. Carapezza, & F. Lo Piparo (eds.), *Further advances in pragmatics and philosophy: Theories and applications* (pp. 81–100). Dordrecht, Netherlands: Springer.

Douven, I. & Meijs, W. (2006). Bootstrap confirmation made quantitative. *Synthese* 49: 97–132.

Douven, I. & Meijs, W. (2007). Measuring coherence. *Synthese* 156: 405–425.

Douven, I. & Mirabile, P. (2018). Best, second-best, and good-enough explanations: How they matter to reasoning. *Journal of Experimental Psychology: Language, Memory, and Cognition* 44: 1792–1813.

Douven, I. & Riegler, A. (2010). Extending the Hegselmann–Krause model I. *Logic Journal of the IGPL* 18: 323–335.

Douven, I. & Rott, H. (2018). From probabilities to categorical beliefs: Going beyond toy models. *Journal of Logic and Computation* 28: 1099–1124.

Douven, I. & Schupbach, J. N. (2015a). The role of explanatory considerations in updating. *Cognition* 142: 299–311.

Douven, I. & Schupbach, J. N. (2015b). Probabilistic alternatives to Bayesianism: The case of explanationism. *Frontiers in Psychology* 6, https://doi.org/10.3389/fpsyg.2015.00459.

Douven, I. & Uffink, J. B. M. (2003). The preface paradox revisited. *Erkenntnis* 59: 389–420.

Douven, I. & Wenmackers, S. (2017). Inference to the best explanation versus Bayes' rule in a social setting. *British Journal for the Philosophy of Science* 68: 535–570.

Douven, I. & Williamson, T. (2006). Generalizing the lottery paradox. *British Journal for the Philosophy of Science* 57: 755–779.

Dragulinescu, S. (2016). Inference to the best explanation and mechanisms in medicine. *Theoretical Medicine and Bioethics* 37: 211–232.

Dyke, M. M. (in press). Could our epistemic reasons be *collective* practical reasons? *Noûs*.

Earman, J. (1992). *Bayes or bust?* Cambridge, MA: MIT Press.

Earman, J. (1993). Underdetermination, realism, and reason. In P. French, T. Uehling, Jr., & H. Wettstein (eds.), *Midwest Studies in Philosophy* (vol. 18, pp. 19–38). Notre Dame, IL: University of Notre Dame Press.

Edgington, D. (1995). On conditionals. *Mind* 104: 235–329.

Edwards, B. J., Williams, J. J., Gentner, D. & Lombrozo, T. (2019). Explanation recruits comparison in a category-learning task. *Cognition* 185: 21–38.

Edwards, W. (1968). Conservatism in human information processing. In B. Kleinmuntz (ed.), *Formal representation of human judgment* (pp. 17–52). New York, NY: Wiley.

Eells, E. & Fitelson, B. (2002). Symmetries and asymmetries in evidential support. *Philosophical Studies* 107: 129–142.

Einstein, A. (1915). Erklärung der Perihelbewegung des Merkur aus der allgemeinen Relativitätstheorie. *Preussische Akademie der Wissenschaften, Sitzungsberichte* 2: 831–839.

Einstein, A. (1950). *Out of my later years*. New York, NY: Philosophical Library.

Elqayam, S. (2011). Grounded rationality: A relativist framework for normative rationality. In K. I. Manktelow, D. E. Over, & S. Elqayam (eds.), *The science of reason* (pp. 397–420). Hove, UK: Psychology Press.

Elqayam, S. (2012). Grounded rationality: Descriptivism in epistemic context. *Synthese* 189: 39–49.

Elqayam, S. (2018). New psychology of reasoning. In L. J. Ball & V. A. Thompson (eds.), *The Routledge international handbook of thinking and reasoning* (pp. 130–150). New York, NY: Routledge.

Elqayam, S. & Evans, J. St. B. T. (2011). Subtracting "ought" from "is": Descriptivism versus normativism in the study of human thinking. *Behavioral and Brain Sciences* 34: 233–248.

Elqayam, S. & Evans, J. St. B. T. (2013). Rationality in the new paradigm: Strict versus soft Bayesian approaches. *Thinking & Reasoning* 19: 453–470.

Elqayam, S. & Over, D. E. (2012). Probabilities, beliefs, and dual processing: The paradigm shift in the psychology of reasoning. *Mind & Society* 11: 27–40.

Epstein, E. S. (1969). A scoring system for probability forecasts of ranked categories. *Journal of Applied Meteorology* 8: 985–987.

Eva, B., Hartmann, S., & Rafiee Rad, S. (2020). Learning from conditionals. *Mind* 129: 461–508.

Evans, J. St. B. T. (2007a). *Hypothetical thinking: Dual processes in reasoning and judgement.* Hove, UK: Psychology Press.

Evans, J. St. B. T. (2007b). On the resolution of conflict in dual process theories of reasoning. *Thinking & Reasoning* 13: 321–339.

Evans, J. St. B. T. (2008). Dual-processing accounts of reasoning, judgment, and social cognition. *Annual Review of Psychology* 59: 255–278.

Evans, J. St. B. T. (2010a). *Thinking twice: Two minds in one brain.* Oxford, UK: Oxford University Press.

Evans, J. St. B. T. (2010b). Intuition and reasoning: A dual-process perspective. *Psychological Inquiry* 21: 313–326.

Evans, J. St. B. T., Barston, J. L., & Pollard, P. (1983). On the conflict between logic and belief in syllogistic reasoning. *Memory & Cognition* 11: 295–306.

Evans, J. St. B. T. & Over, D. E. (1996). *Rationality and reasoning.* Hove, UK: Psychology Press.

Fairchild, M. D. (2013). *Color appearance models.* Chichester, UK: Wiley.

Faisal, A. A., Selen, L. P. J., & Wolpert, D. M. (2008). Noise in the nervous system. *Nature Reviews Neuroscience* 9: 292–303.

Fann, K. T. (1970). *Peirce's theory of abduction.* The Hague, Netherlands: Martinus Nijhoff.

Fantl, J. & McGrath, M. (2009). *Knowledge in an uncertain world.* Oxford, UK: Oxford University Press.

Fernbach, P. M. & Erb, C. D. (2013). A quantitative causal model theory of conditional reasoning. *Journal of Experimental Psychology: Learning, Memory, and Cognition* 39: 1327–1343.

Fine, A. I. (1984). The natural ontological attitude. In J. Leplin (ed.), *Scientific realism* (pp. 261–277). Berkeley, CA: University of California Press.

Fischhoff, B. & Lichtenstein, S. (1978). Don't attribute this to Reverend Bayes. *Psychological Bulletin* 85: 239–243.

Fitelson, B. (1999). The plurality of Bayesian measures of confirmation and the problem of measure sensitivity. *Philosophy of Science* 66: S362–S378.

Fitelson, B. (2001). *Studies in Bayesian confirmation theory.* Doctoral dissertation. University of Wisconsin, Madison.

Fitelson, B. (2003). A probabilistic theory of coherence. *Analysis* 63: 194–199.

Fitelson, B. & Eells, E. (2000). Measuring confirmation and evidence. *Journal of Philosophy* 97: 663–672.

Fodor, J. A. (2001). *The mind doesn't work that way.* Cambridge, MA: MIT Press.

Fodor, J. A. & Lepore, E. (1992). *Holism: A shopper's guide.* Cambridge, MA: Blackwell.

Foley, R. (1987). *The theory of epistemic rationality.* Cambridge, MA: Harvard University Press.

Foley, R. (1992). The epistemology of belief and the epistemology of degrees of belief. *American Philosophical Quarterly* 29: 111–124.

Foley, R. (1993). *Working without a net.* Oxford, UK: Oxford University Press.

Frankfurt, H. (1958). Peirce's notion of abduction. *Journal of Philosophy* 55: 593–596.

Fricker, E. (1994). Against gullibility. In B. K. Matilal & A. Chakrabarti (eds.), *Knowing from words* (pp. 125–161). Dordrecht, Netherlands: Kluwer.

Fricker, E. (2017). Inference to the best explanation and the receipt of testimony: Testimonial reductionism vindicated. In K. McCain & T. Poston (eds.), *Best explanations: New essays on inference to the best explanation* (pp. 262–294). Oxford, UK: Oxford University Press.

Friedman, M. (1974). Explanation and scientific understanding. *Journal of Philosophy* 71: 5–19.

Fumerton, R. A. (1980). Induction and reasoning to the best explanation. *Philosophy of Science* 47: 589–600.

Fumerton, R. A. (2017). Reasoning to the best explanation. In K. McCain & T. Poston (eds.), *Best explanations: New essays on inference to the best explanation* (pp. 65–79). Oxford, UK: Oxford University Press.

Gabbay, D. M. & Woods, J. (2005). *The reach of abduction: Insight and trial.* Amsterdam, Netherlands: North-Holland.

Gaifman, H. (1986). A theory of higher order probabilities. In J. Halpern (ed.), *Theoretical aspects of reasoning about knowledge: Proceedings of the 1986 conference* (pp. 275–292). Los Altos, CA: Morgan Kaufmann.

Gaifman, H. & Snir, M. (1982). Probabilities over rich languages. *Journal of Symbolic Logic* 47: 495–548.

Galison, P. (1997). *Image and logic: A material culture of microphysics.* Chicago, IL: University of Chicago Press.

Gärdenfors, P. (2000). *Conceptual spaces: The geometry of thought.* Cambridge, MA: MIT Press.

Gärdenfors, P. (2014). *Geometry of meaning: Semantics based on conceptual spaces.* Cambridge, MA: MIT Press.

Gelman, A., Carlin, J. B., Stern, H. S., Dunson, D. B., Vehtari, A., & Rubin, D. B. (2013). *Bayesian data analysis* (3rd ed.). Boca Raton, FL: CRC Press.

Gelman, A. & Hill, J. (2009). *Data analysis using regression and multilevel/hierarchical models.* Cambridge, UK: Cambridge University Press.

Gelman, A. & Shalizi, C. R. (2012). Philosophy and the practice of Bayesian statistics in the social sciences. In H. Kincaid (ed.), *The Oxford handbook of philosophy of social science* (pp. 259–273). Oxford, UK: Oxford University Press.

Gelman, A. & Shalizi, C. R. (2013). Philosophy and the practice of Bayesian statistics. *British Journal of Mathematical and Statistical Psychology* 66: 8–38.

Genest, C. (1984). A characterization theorem for externally Bayesian groups. *Annals of Statistics* 12: 1100–1105.

Genest, C. & Zidek, J. V. (1986). Combining probability distributions: A critique and an annotated bibliography. *Statistical Science* 1: 114–148.

Giere, R. N. (1988). *Explaining science: A cognitive approach*. Chicago, IL: University of Chicago Press.

Gigerenzer, G. (2000). *Adaptive thinking: Rationality in the real world*. New York, NY: Oxford University Press.

Gigerenzer, G. (2001). The adaptive toolbox. In G. Gigerenzer & R. Selten (eds.), *Bounded rationality: The adaptive toolbox* (pp. 37–50). Cambridge, MA: MIT Press.

Gigerenzer, G. & Goldstein, D. G. (1996). Reasoning the fast and frugal way: Models of bounded rationality. *Psychological Review* 103: 650–669.

Gigerenzer, G., Hertwig, R., & Pachur, T. (2011). *Heuristics: The foundations of adaptive behavior*. New York, NY: Oxford University Press.

Gigerenzer, G., Todd, P. M., & ABC Research Group (1999). *Simple heuristics that make us smart*. Oxford, UK: Oxford University Press.

Gillies, D. (2000). *Philosophical theories of probability*. London, UK: Routledge.

Glass, D. H. (2007). Coherence measures and inference to the best explanation. *Synthese* 157: 275–296.

Glass, D. H. (2012). Inference to the best explanation: Does it track truth? *Synthese* 185: 411–427.

Glass, D. H. (2018a). An evaluation of probabilistic approaches to inference to the best explanation. *International Journal of Approximate Reasoning* 103: 184–194.

Glass, D. H. (2021a). Coherence, explanation, and hypothesis selection. *British Journal for the Philosophy of Science* 72: 1–26.

Glass, D. H. (2021b). Competing hypotheses and abductive inference. *Annals of Mathematics and Artificial Intelligence* 89: 161–178.

Glass, C. A. & Glass, D. H. (2021). Opinion dynamics of social learning with a conflicting source. *Physica A: Statistical Mechanics and its Applications* 563: 125480.

Glymour, C. (1980). *Theory and evidence*. Princeton, NJ: Princeton University Press.

Goldberg, D. (1991). What every computer scientist should know about floating-point arithmetic. *ACM Computing Surveys* 23: 5–48.

Goldberg, S. C. (2015). *Assertion: On the philosophical significance of assertoric speech*. Oxford, UK: Oxford University Press.

Goldman, A. H. (1988). *Empirical knowledge*. Berkeley, CA: University of California Press.

Goldman, A. I. (1999). *Knowledge in a social world*. Oxford, UK: Oxford University Press.

Goldman, A. I. (2010). Epistemic relativism and reasonable disagreement. In R. Feldman & T. Warfield (eds.), *Disagreement* (pp. 187–215). Oxford, UK: Oxford University Press.

Goldstein, D. G. & Gigerenzer, G. (2002). Models of ecological rationality: The recognition heuristic. *Psychological Review* 109: 75–90.

Good, I. J. (1952). Rational decisions. *Journal of the Royal Statistical Society* B14: 107–114.

Good, I. J. (1960). Weight of evidence, corroboration, explanatory power, information and the utility of experiment. *Journal of the Royal Statistical Society* B22: 319–331.

Good, I. J. (1967). On the principle of total evidence. *British Journal for the Philosophy of Science* 17: 319–321.

Goodfellow, I., Bengio, Y., & Courville, A. (2016). *Deep learning*. Cambridge, MA: MIT Press.

Gopnik, A. & Tenenbaum, J. B. (2007). Bayesian networks, Bayesian learning and cognitive development. *Developmental Science* 10: 281–287.

Gottifredi, S., Tamargo, L. H., García, A. J., & Simari, G. R. (2018). Arguing about informant credibility in open multi-agent systems. *Artificial Intelligence* 259: 91–109.

Greaves, H. & Wallace, D. (2006). Justifying conditionalization: Conditionalization maximizes expected epistemic utility. *Mind* 115: 607–632.

Greco, J. (2015). Inference to the best explanation. In R. Audi (ed.), *The Cambridge dictionary of philosophy* (3rd ed.). Cambridge, UK: Cambridge University Press.

Gribbin, J. (2002). *Science: A history*. London, UK: Penguin.

Grice, H. P. (1989). *Studies in the way of words*. Cambridge, MA: Harvard University Press.

Griffiths, T. L., Lieder, F., & Goodman, N. D. (2015). Rational use of cognitive resources: Levels of analysis between the computational and the algorithmic. *Topics in Cognitive Science* 7: 217–229.

Griffiths, T. L. & Tenenbaum, J. B. (2006). Optimal predictions in everyday cognition. *Psychological Science* 17: 767–773.

Günther, M. (2018). Learning conditional information by Jeffrey imaging on Stalnaker conditionals. *Journal of Philosophical Logic* 47: 851–876.

Haack, S. (1976). The justification of deduction. *Mind* 85: 112–119.

Hacking, I. (1967). Slightly more realistic personal probability. *Philosophy of Science* 34: 311–325.

Hacking, I. (1981). Do we see through a microscope? *Pacific Philosophical Quarterly* 62: 305–322. (Reprinted in P. Churchland & C. Hooker (eds.), *Images of science* (pp. 132–152). Chicago, IL: University of Chicago Press, 1985; the page reference is to the reprint.)

Hacking, I. (1984). Trial by number. *Science* 84: 69–70.

Hahn, U., Hansen, J. U., & Olsson, E. J. (2020). Truth tracking performance of social networks: How connectivity and clustering can make groups less competent. *Synthese* 197: 1511–1541.

Haig, B. D. (2005a). An abductive theory of scientific method. *Psychological Methods* 10: 371–388.

Haig, B. D. (2005b). Exploratory factor analysis, theory generation, and scientific method. *Multivariate Behavioral Research* 40: 303–329.

Harman, G. H. (1965). The inference to the best explanation. *Philosophical Review* 74: 88–95.

Harman, G. H. (1968). Enumerative induction as inference to the best explanation. *Journal of Philosophy* 65: 529–533.

Harman, G. H. (1973). *Thought*. Princeton, NJ: Princeton University Press.

Harré, R. (1986). *Varieties of realism*. Oxford, UK: Blackwell.

Harré, R. (1988). Realism and ontology. *Philosophia Naturalis* 25: 386–398.

Hartmann, S. (in press). Bayes nets and rationality. In M. Knauff & W. Spohn (eds.), *Handbook of rationality*. Cambridge, MA: MIT Press.

Hastie, T., Tibshirani, R., & Friedman, J. (2009). *The elements of statistical learning* (2nd ed.). New York, NY: Springer.

Hawley, K. (2006). Science as a guide to metaphysics? *Synthese* 149: 451–470.

Hegselmann, R. & Krause, U. (2002). Opinion dynamics and bounded confidence: Models, analysis, and simulations. *Journal of Artificial Societies and Social Simulation* 5, http://jasss.soc.surrey.ac.uk/5/3/2.html.

Hegselmann, R. & Krause, U. (2005). Opinion dynamics driven by various ways of averaging. *Computational Economics* 25: 381–405.

Hegselmann, R. & Krause, U. (2006). Truth and cognitive division of labor: First steps towards a computer aided social epistemology. *Journal of Artificial Societies and Social Simulation* 9, http://jasss.soc.surrey.ac.uk/9/3/10.html.

Hegselmann, R. & Krause, U. (2009). Deliberative exchange, truth, and cognitive division of labour: A low-resolution modeling approach. *Episteme* 6: 130–144.

Hegselmann, R. & Krause, U. (2015). Opinion dynamics under the influence of radical groups, charismatic leaders, and other constant signals: A simple unifying model. *Networks and Heterogeneous Media* 10: 477–509.

Hempel, C. G. (1965). *Aspects of scientific explanation and other essays in the philosophy of science*. New York, NY: Free Press.

Heris, S. M. K. & Khaloozadeh, H. (2011). Open- and closed-loop multiobjective optimal strategies for HIV therapy using NSGA-II. *IEEE Transactions on Biomedical Engineering* 58: 1678–1685.

Hildebrand, T. & Metcalf, T. (in press). The nomological argument for the existence of God. *Noûs*, https://doi.org/10.1111/nous.12364.

Hill, A. & Paris, J. B. (2013). The counterpart principle of analogical support by structural similarity. *Erkenntnis* 79: 1–16.

Hjortland, O. (2019). What counts as evidence for a logical theory? *Australasian Journal of Logic* 16: 250–282.

Hobbs, J. R. (2004). Abduction in natural language understanding. In L. Horn & G. Ward (eds.), *Handbook of pragmatics* (pp. 724–741). Oxford, UK: Blackwell.

Hoefer, C. & Martí, G. (2020). Realism, reference and perspective. *European Journal for Philosophy of Science* 10: 38.

Hohwy, J. (2013). *The predictive mind*. Oxford, UK: Oxford University Press.

Holland, J. (1975). *Adaptation in natural and artificial systems*. Cambridge, MA: MIT Press.

Horry, R. & Brewer, N. (2016). How target–lure similarity shapes confidence judgments in multiple-alternative decision tasks. *Journal of Experimental Psychology: General* 145: 1615–1634.

Horsten, L. (2011). *The Tarskian turn*. Cambridge, MA: MIT Press.

Horwich, P. (1982). *Probability and evidence*. Cambridge, UK: Cambridge University Press.

Horwich, P. (1990). *Truth*. Oxford, UK: Oxford University Press.

Howson, C. (1984). Bayesianism and support by novel facts. *British Journal for the Philosophy of Science* 35: 245–251.

Howson, C. & Oddie, G. (1979). Miller's so-called paradox of information. *British Journal for the Philosophy of Science* 30: 253–261.

Howson, C. & Urbach, P. (1993). *Scientific reasoning* (2nd ed.). La Salle, IL: Open Court.

Huang, Y. (1994). *The syntax and pragmatics of anaphora*. Cambridge, UK: Cambridge University Press.

Hume, D. (1748/2006). *An inquiry concerning human understanding*. Fairford, UK: Echo Library.

Inhelder, B. & Piaget, J. (1958). *The growth of logical thinking from childhood to adolescence*. New York, NY: Basic Books.

Jackson, F. (1979). On assertion and indicative conditionals. *Philosophical Review* 88: 565–589.

Jeffrey, R. C. (1983). *The logic of decision* (2nd ed.). Chicago, IL: University of Chicago Press.

Jeffrey, R. C. (1992). *Probability and the art of judgment*. Cambridge, UK: Cambridge University Press.

Jern, A., Derrow-Pinion, A., & Piergiovanni, A. J. (2021). A computational framework for understanding the roles of simplicity and rational support in people's behavior explanations. *Cognition* 210: 104606, https://doi.org/10.1016/j.cognition.2021.104606.

Johnson, S. G. B., Rajeev-Kumar, G., & Keil, F. C. (2016). Sense-making under ignorance. *Cognitive Psychology* 89: 39–70.

Johnson, S. G. B., Valenti, J. J., & Keil, F. C. (2019). Simplicity and complexity preferences in causal explanation: An opponent heuristic account. *Cognitive Psychology* 113: 101222.

Johnston, A. M., Johnson, S. G. B., Koven, M. L., & Keil, F. C. (2017). Little Bayesians or little Einsteins? Probability and explanatory virtue in children's inferences. *Developmental Science* 20: e12483, https://doi.org/10.1111/desc.12483.

Josephson, J. R. & Josephson, S. G. (1994). *Abductive inference*. Cambridge, UK: Cambridge University Press.

Joyce, J. (1998). A nonpragmatic vindication of probabilism. *Philosophy of Science* 65: 575–603.

Joyce, J. (1999). *The foundations of causal decision theory*. Cambridge, UK: Cambridge University Press.

Joyce, J. (2009). Accuracy and coherence: Prospects for an alethic epistemology of partial belief. In F. Huber & C. Schmidt-Petri (eds.), *Degrees of belief* (pp. 263–297). Berlin, Germany: Springer.

Judd, C. M., Westfall, J., & Kenny, D. A. (2012). Treating stimuli as a random factor in social psychology: A new and comprehensive solution to a pervasive but largely ignored problem. *Journal of Personality and Social Psychology* 103: 54–69.

Kagehiro, D. & Stanton, W. (1985). Legal vs. quantified definitions of standards of proof. *Law and Human Behavior* 9: 159–178.

Kahneman, D., Slovic, P., & Tversky, A. (1982). *Judgment under uncertainty: Heuristics and biases*. Cambridge, UK: Cambridge University Press.

Kaplan, M. (1981). A Bayesian theory of rational acceptance. *Journal of Philosophy* 78: 305–330.

Kasirzadeh, A. (in press). A new role for mathematics in empirical sciences. *Philosophy of Science*.

Keeney, R. L. & Raiffa, H. (1976). *Decisions with multiple objectives: Preferences and value tradeoffs*. New York, NY: John Wiley.

Keil, F. C. (2006). Explanation and understanding. *Annual Review of Psychology* 57: 227–254.

Kelp, C. (2016). Assertion: A function first account. *Noûs* 50: 411–442.

Keynes, J. M. (1921) *A treatise of probability*. London, UK: Macmillan.

Khalifa, K., Doble, G., & Millson, J. (2020). Counterfactuals and explanatory pluralism. *British Journal for the Philosophy of Science* 71: 1439–1460.

Khemlani, S. S., Sussman, A. B., & Oppenheimer, D. M. (2011). Harry Potter and the sorcerer's scope: Latent scope biases in explanatory reasoning. *Memory & Cognition* 39: 527–535.

Kitcher, P. (1981). Explanatory unification. *Philosophy of Science* 48: 507–531.

Kitcher, P. (1993). *The advancement of science*. Oxford, UK: Oxford University Press.

Kitcher, P. (2001). Real realism: The Galilean strategy. *Philosophical Review* 110: 151–197.

Klein, P. (1985). The virtues of inconsistency. *Monist* 68: 105–135.

Kneer, M. (2018). The norm of assertion: Empirical data. *Cognition* 177: 165–171.

Kneer, M. (2021). Norms of assertion in the US, Germany, and Japan: A cross-cultural exploration of what we expect of each other in linguistic communication. *Proceedings of the National Academy of Sciences of the United States of America* 118: e2105365118, https://doi.org/10.1073/pnas.2105365118.

Kochenderfer, M. J. & Wheeler, T. A. (2019). *Algorithms for optimization*. Cambridge, MA: MIT Press.

Koehler, D. J. (1991). Explanation, imagination, and confidence in judgment. *Psychological Bulletin* 110: 499–519.

Konek, J. (2016). Probabilistic knowledge and cognitive ability. *Philosophical Review* 125: 509–587.

Korb, K. (1992). The collapse of collective defeat: Lessons from the lottery paradox. *Proceedings of the Biennial Meeting of the Philosophy of Science Association* I: 230–236.

Kornblith, H. (1993). *Inductive inference and its natural ground: An essay in naturalistic epistemology*. Cambridge, MA: MIT Press.

Koscholke, J. (2016). Evaluating test cases for probabilistic measures of coherence. *Erkenntnis* 81: 155–181.

Koscholke, J. & Jekel, M. (2017). Probabilistic coherence measures: A psychological study of coherence assessment. *Synthese* 194: 1303–1322.

Koscholke, J. & Schippers, M. (2019). Coherence and common causes: Against relevance-sensitive measures of coherence. *British Journal for the Philosophy of Science* 70: 771–785.

Koscholke, J., Schippers, M., & Stegmann, A. (2019). New hope for relative overlap measures of coherence. *Mind* 128: 1261–1284.

Koza, J. (1992). *Genetic programming: On the programming of computers by means of natural selection*. Cambridge, MA: MIT Press.

Kreisel, G. (1980). Kurt Gödel. *Biographical Memoirs of the Fellows of the Royal Society* 26: 149–224.

Kruschke, J. K. (2011). *Doing Bayesian data analysis*. Burlington, MA: Academic Press.

Kruschke, J. K. (2013a). Bayesian estimation supersedes the *t* test. *Journal of Experimental Psychology: General* 142: 573–603.

Kruschke, J. K. (2013b). Posterior predictive checks can and should be Bayesian. *British Journal of Mathematical and Statistical Psychology* 66: 45–56.

Krzyżanowska, K. H. (2015). *Between "if" and "then."* Doctoral dissertation. University of Groningen, Netherlands.

Kuhn, T. (1962). *The structure of scientific revolutions.* Chicago, IL: University of Chicago Press.

Kuipers, T. A. F. (1984). Approaching the truth with the rule of success. *Philosophia Naturalis* 21: 244–253.

Kuipers, T. A. F. (1992). Naive and refined truth approximation. *Synthese* 93: 299–341.

Kuipers, T. A. F. (2000). *From instrumentalism to constructive realism.* Dordrecht, Netherlands: Kluwer.

Kuipers, T. A. F. (2001). *Structures in science.* Dordrecht, Netherlands: Kluwer.

Kuipers, T. A. F. (2014). Empirical progress and nomic truth approximation revisited. *Studies in History and Philosophy of Science* 46: 64–72.

Kuipers, T. A. F. (2019). *Nomic truth approximation revisited.* Basel, Switzerland: Springer.

Kuipers, T. A. F. (2020). Refined nomic truth approximation by revising models and postulates. *Synthese* 197: 1601–1625.

Kummerfeld, E. & Zollman, K. J. S. (2016). Conservatism and the scientific state of nature. *British Journal for the Philosophy of Science* 82: 956–968.

Kyburg, H. E. (1961). *Probability and the logic of rational belief.* Middletown, CT: Wesleyan University Press.

Kyburg, H. E. (1990). *Science and reason.* Oxford, UK: Oxford University Press.

Lange, M. (2000). Is Jeffrey conditionalization defective by virtue of being non-commutative? Remarks on the sameness of sensory experiences. *Synthese* 123: 393–403.

Lange, M. (2016). *Because without cause: Non-causal explanation in science and mathematics.* Oxford, UK: Oxford University Press.

Lange, M. (2017). The evidential relevance of explanatoriness. *Analysis* 77: 303–312.

Latus, A. (2000). Our epistemic goal. *Australasian Journal of Philosophy* 78: 28–39.

Laudan, L. (1981). A confutation of convergent realism. *Philosophy of Science* 48: 19–49.

Laudan, L. & Leplin, J. (1991). Empirical equivalence and underdetermination. *Journal of Philosophy* 88: 449–472.

Lawry, J. & Lee, C. (2020). Probability pooling for dependent agents in collective learning. *Artificial Intelligence* 288: 103371.

LeCun, Y., Bengio, Y., & Hinton, G. (2015). Deep learning. *Nature* 521: 436.

Lee, C. (2020). *Decision making and consensus in multi-agent systems*. Doctoral dissertation. Bristol Robotics Laboratory, University of Bristol, Bristol, UK.

Leeds, S. (1994). Constructive empiricism. *Synthese* 101: 187–221.

Legare, C. H. & Lombrozo, T. (2014). Selective effects of explanation on learning during early childhood. *Journal of Experimental Child Psychology* 126: 198–212.

Lehrer, K. (1974). *Knowledge*. Oxford, UK: Oxford University Press.

Lehrer, K. & Wagner, C. (1981). *Rational consensus in science and society*. Dordrecht, Netherlands: Reidel.

Leitgeb, H. (2017). *The stability of belief*. Oxford, UK: Oxford University Press.

Leitgeb, H. & Pettigrew, R. (2010a). An objective justification of Bayesianism I: Measuring inaccuracy. *Philosophy of Science* 77: 201–235.

Leitgeb, H. & Pettigrew, R. (2010b). An objective justification of Bayesianism II: The consequences of minimizing inaccuracy. *Philosophy of Science* 77: 236–272.

Leplin, J. (1997). *A novel defense of scientific realism*. Oxford, UK: Oxford University Press.

Levenson, T. (2015). *The hunt for Vulcan*. New York, NY: Random House.

Levinstein, B. A. (2012). Leitgeb and Pettigrew on accuracy and updating. *Philosophy of Science* 79: 413–424.

Levinstein, B. A. (2017). A pragmatist's guide to epistemic utility. *Philosophy of Science* 84: 613–638.

Lewis, D. K. (1973). *Counterfactuals*. Oxford, UK: Blackwell.

Lewis, D. K. (1980). A subjectivist's guide to objective chance. In R. C. Jeffrey (ed.), *Studies in inductive logic and probability* (vol. 2, pp. 263–293). Berkeley, CA: University of California Press.

Lewis, D. K. (1986). *On the plurality of worlds*. Oxford, UK: Blackwell.

Lewis, D. K. (1999). Why conditionalize? In his *Papers in metaphysics and epistemology* (pp. 403–407). Cambridge, UK: Cambridge University Press.

Lindenbaum-Hosiasson, J. (1941). Induction et analogie: Comparaison de leur fondement. *Mind* 50: 351–365.

Lindley, D. V. (1972). *Bayesian statistics, a review*. Philadelphia, PA: Society for Industrial and Applied Mathematics.

Lindley, D. V. (1976). Bayesian statistics. In W. Harper & C. Hooker (eds.), *Foundations of probability theory, statistical inference and statistical theories of science* (Vol. 2, pp. 353–362). Dordrecht, Netherlands: Reidel.

Lindskog, M., Winman, A., & Juslin, P. (2013a). Naïve point estimation. *Journal of Experimental Psychology: Learning, Memory, and Cognition* 39: 782–800.

Lindskog, M., Winman, A., & Juslin, P. (2013b). Calculate or wait: Is man an eager or a lazy intuitive statistician? *Journal of Cognitive Psychology* 25: 994–1014.

Lipton, P. (1993). Is the best good enough? *Proceedings of the Aristotelian Society* 93: 89–104.

Lipton, P. (1998). The epistemology of testimony. *Studies in History and Philosophy of Science* 29: 1–31.

Lipton, P. (2004). *Inference to the best explanation* (2nd ed.). London, UK: Routledge.

Lombrozo, T. (2007). Simplicity and probability in causal explanation. *Cognitive Psychology* 55: 232–257.

Lombrozo, T. (2020). "Learning by thinking" in science and in everyday life. In P. Godfrey-Smith & A. Levy (eds.), *The scientific imagination: Philosophical and psychological perspectives* (pp. 230–249). Oxford, UK: Oxford University Press.

Lombrozo, T. & Gwynne, N. Z. (2014). Explanation and inference: Mechanistic and functional explanations guide property generalization. *Frontiers in Human Neuroscience* 8, https://doi.org/10.3389/fnhum.2014.00700.

Longino, H. (1995). Gender, politics and theoretical virtues. *Synthese* 104: 383–397.

Lorenz, J. (2003). *Mehrdimensionale Meinungsdynamik bei wechselndem Vertrauen*. Diploma thesis. University of Bremen, Germany, available at http://nbn-resolving.de/urn:nbn:de:gbv:46-dipl000000564.

Lorenz, J. (2008). Fostering consensus in multidimensional continuous opinion dynamics under bounded confidence. In D. Helbing (ed.), *Managing complexity: Insights, concepts, applications* (pp. 321–334). Berlin, Germany: Springer.

Lorenz, J., Rauhut, H., Schweitzer, F., & Helbing, D. (2011). How social influence can undermine the wisdom of crowd effect. *Proceedings of the National Academy of Sciences of the United States of America* 108: 9020–9025.

Lovie, P. & Lovie, A. D. (1976). Teaching intuitive statistics I: Estimating means and variances. *International Journal of Mathematical, Educational, and Scientific Technology* 7: 29–39.

Lycan, W. G. (1988). *Judgement and justification*. Cambridge, UK: Cambridge University Press.

Lycan, W. G. (2002). Explanation and epistemology. In P. Moser (ed.), *The Oxford handbook of epistemology* (pp. 408–433). Oxford, UK: Oxford University Press.

Machery, E., Mallon, R., Nichols, S., & Stich, S. P. (2004). Semantics, cross-cultural style. *Cognition* 92: B1–12.

Maher, P. (1992). Diachronic rationality. *Philosophy of Science* 59: 120–141.

Maher, P. (1993). *Betting on theories*. Cambridge, UK: Cambridge University Press.

Malmi, R. A. & Samson, D. J. (1983). Intuitive averaging of categorized numerical stimuli. *Journal of Verbal Learning and Verbal Behavior* 22: 547–559.

Mancosu, P. (2015). Explanation in mathematics. In E. N. Zalta (ed.), *The Stanford encyclopedia of philosophy*, https://plato.stanford.edu/archives/sum2015/entries/mathematics-explanation/.

Marks, D. F. & Clarkson, J. K. (1972). An explanation of conservatism in the bookbag-and-pokerchips situation. *Acta Psychologica* 36: 145–160.

Marsili, N. & Wiegmann, A. (2021). Should I say that? An experimental investigation of the norm of assertion. *Cognition* 212: 104657, https://doi.org/10.1016/j.cognition.2021.104657.

Mason, W. A. & Watts, D. J. (2012). Collaborative learning in networks. *Proceedings of the National Academy of Sciences of the United States of America* 109: 764–769.

McAuliffe, W. (2015). How did abduction get confused with inference to the best explanation? *Transactions of the Charles S. Peirce Society* 51: 300–319.

McCain, K. & Poston, T. (2014). Why explanatoriness is evidentially relevant. *Thought* 3: 145–153.

McCain, K. & Poston, T. (2019). Dispelling the disjunction objection to explanatory inference. *Philosophers' Imprint* 19: 36, http://hdl.handle.net/2027/spo.3521354.0019.036.

McCarthy, J. (1956). Measures of the value of information. *Proceedings of the National Academy of Sciences of the United States of America* 42: 654–655.

McCutcheon, R. G. (2019). In favor of logarithmic scoring. *Philosophy of Science* 86: 286–303.

McElreath, R. (2020). *Statistical rethinking*. Boca Raton, FL: CRC Press.

McGrew, T. (2003). Confirmation, heuristics, and explanatory reasoning. *British Journal for the Philosophy of Science* 54: 553–567.

McMullin, E. (1985). Galilean idealization. *Studies in History and Philosophy of Science* 16: 247–273.

McMullin, E. (1992). *The inference that makes science*. Milwaukee, WI: Marquette University Press.

McMullin E. (1996). Epistemic virtue and theory appraisal. In I. Douven & L. Horsten (eds.), *Realism in the sciences* (pp. 13–34). Leuven, Belgium: Leuven University Press.

Meheus, J. & Batens, D. (2006). A formal logic for abductive reasoning. *Logic Journal of the IGPL* 14: 221–236.

Meijs, W. (2005). *Probabilistic measures of coherence*. Doctoral dissertation. Erasmus University Rotterdam, Netherlands.

Meijs, W. & Douven, I. (2007). On the alleged impossibility of coherence. *Synthese* 157: 347–360.

Mellers, B., Stone, E., Murray, T., Minster, A., Rohrbaugh, N., Bishop, M., Chen, E., Baker, J., Hou, Y., Horowitz, M., Ungar, L., & Tetlock, P. (2015). Identifying and cultivating superforecasters as a method of improving probabilistic predictions. *Perspectives on Psychological Science* 10: 267–281.

Merton, R. K. (1973). *The sociology of science*. Chicago, IL: University of Chicago Press.

Michelon, C. (2019). The inference to the best legal explanation. *Oxford Journal of Legal Studies* 39: 878–900.

Miller, D. (1966). A paradox of information. *British Journal for the Philosophy of Science* 17: 59–61.

Mirabile, P. (2020). *How we reason about explanations: Philosophy and psychology of explanatory reasoning*. Doctoral dissertation. Sorbonne University, Paris.

Mirabile, P. & Douven, I. (2020). Abductive conditionals as a test case for inferentialism. *Cognition* 200: 104232, https://doi.org/10.1016/j.cognition.2020.104232.

Mirabile, P. & Horne, Z. (2019). Explanatory virtues and belief in conspiracy theories. *Proceedings of the 41st annual meeting of the Cognitive Science Society*, https://cogsci.mindmodeling.org/2019/papers/0410/index.html.

Moore, G. E. (1962). Proof of an external world. In his *Philosophical Papers* (pp. 126–149). New York, NY: Collier Books.

Morgenstern, Y., Murray, R. F., & Harris, L. R. (2011). The human visual system's assumption that light comes from above is weak. *Proceedings of the National Academy of Sciences of the United States of America* 108: 12551–12553.

Moser, P. K. (1990). Rationality in action. In P. Moser (ed.), *Rationality in action* (pp. 1–16). Cambridge, UK: Cambridge University Press.

Moser, P. K. & Tlumak, J. (1985). Two paradoxes of rational acceptance. *Erkenntnis* 23: 127–141.

Moss, S. (2011). Scoring rules and epistemic compromise. *Mind* 120: 1053–1069.

Murphy, A. H. (1969). On the "ranked probability score." *Journal of Applied Meteorology* 8: 988–989.

Musgrave, A. (1988). The ultimate argument for scientific realism. In R. Nola (ed.), *Relativism and realism in science* (pp. 229–252). Dordrecht, Netherlands: Kluwer.

Nagel, J. (2014). *Knowledge: A very short introduction*. Oxford, UK: Oxford University Press.

Nagel, J. (2021). The psychological dimension of the lottery paradox. In I. Douven (ed.), *Lotteries, knowledge, and rational belief: Essays on the lottery paradox* (pp. 48–73). Cambridge, UK: Cambridge University Press.

Nazarathy, Y. & Klok, H. (2021). *Statistics with Julia*. Cham, Switzerland: Springer.

Nickerson, R. S., Butler, S. F., & Barch, D. H. (in press). Looking behind: Turning cards in the selection task. *Quarterly Journal of Experimental Psychology*, https://doi.org/10.1177/17470218211001293.

Niiniluoto, I. (1984). *Is science progressive?* Dordrecht, Netherlands: Reidel.

Niiniluoto, I. (1987). *Truthlikeness*. Dordrecht, Netherlands: Reidel.

Niiniluoto, I. (1998). Verisimilitude: The third period. *British Journal for the Philosophy of Science* 49: 1–29.

Niiniluoto, I. (1999a). Defending abduction. *Philosophy of Science* 66: 436–451.

Niiniluoto, I. (1999b). *Critical scientific realism*. Oxford, UK: Oxford University Press.

Niiniluoto, I. (2018). *Truth-seeking by abduction*. Cham, Switzerland: Springer.

Niiniluoto, I. (2020). Truthlikeness: Old and new debates. *Synthese* 197: 1581–1599

Nowak, L. (1980). *The structure of idealization*. Dordrecht, Netherlands: Reidel.

Oaksford, M. & Chater, N. (1994). A rational analysis of the selection task as optimal data selection. *Psychological Review* 101: 608–631.

Oaksford, M. & Chater, N. (1996). Rational explanation of the selection task. *Psychological Review* 103: 381–391.

Oaksford, M. & Chater, N. (2001). The probabilistic approach to human reasoning. *Trends in Cognitive Sciences* 5: 349–357.

Oaksford, M. & Chater, N. (2003). Conditional probability and the cognitive science of conditional reasoning. *Mind & Language* 18: 359–379.

Oaksford, M. & Chater, N. (2007). *Bayesian rationality*. Oxford, UK: Oxford University Press.

Oaksford, M. & Chater, N. (2013). Dynamic inference and everyday conditional reasoning in the new paradigm. *Thinking & Reasoning* 19: 346–379.

Oaksford, M. & Chater, N. (2014). Probabilistic single function dual process theory and logic programming as approaches to non-monotonicity in human vs. artificial reasoning. *Thinking & Reasoning* 20: 269–295.

Oaksford, M. & Chater, N. (2017). Causal models and conditional reasoning. In M. R. Waldmann (ed.), *The Oxford handbook of causal reasoning* (pp. 327–346). Oxford, UK: Oxford University Press.

Oaksford, M. & Chater, N. (2020a). Integrating causal Bayes nets and inferentialism in conditional inference. In S. Elqayam, I. Douven, J. St. B. T. Evans, & N. Cruz (eds.), *Logic and uncertainty in the human mind* (pp. 116–132). London, UK: Routledge.

Oaksford, M. & Chater, N. (2020b). New paradigms in the psychology of reasoning. *Annual Review of Psychology* 71: 1–26.

Oaksford, M. & Hall, S. (2016). On the source of human irrationality. *Trends in Cognitive Sciences* 20: 336–344.

O'Hagan, A., Buck, C. E., Daneshkhah, A., Eiser, J. R., Garthwaite, P. H., Jenkinson, D. J., Oakley, J. E., & Rakow, T. (2006). *Uncertain judgements: Eliciting experts' probabilities*. Hoboken, NJ: Wiley.

Okasha, S. (2000). Van Fraassen's critique of inference to the best explanation. *Studies in History and Philosophy of Science* 31: 691–710.

Olin, D. (2003). *Paradox*. Chesham, UK: Acumen.

Olsson, E. J. (2002). What is the problem of coherence and truth? *Journal of Philosophy* 99: 246–272.

Olsson, E. J. (2005). *Against coherence: Truth, probability, and justification*. Oxford, UK: Oxford University Press.

Olsson, E. J. (2008). Knowledge, truth, and bullshit: Reflections on Frankfurt. *Midwest Studies in Philosophy* 32: 94–110.

Over, D. E. (2009). New paradigm psychology of reasoning. *Thinking & Reasoning* 15: 431–438.

Overton, M. L. (2001). *Numerical computing with IEEE floating point arithmetic*. Philadelphia, PA: SIAM.

Page, S. E. (2007). *The difference: How the power of diversity creates better groups, firms, schools, and societies*. Princeton, NJ: Princeton University Press.

Papineau, D. (1997). *The philosophy of science*. Oxford, UK: Oxford University Press.

Paris, J. B. & Vencovská, A. (2017). Combining analogical support in pure inductive logic. *Erkenntnis* 82: 401–419.

Patel, V. L. & Groen, G. J. (1986). Knowledge-based solution strategies in medical reasoning. *Cognitive Science* 10: 91–116.

Patel, V. L. & Groen, G. J. (1991). Developmental accounts of the transition from medical student to doctor: Some problems and suggestions. *Medical Education* 25: 527– 535.

Patel, V. L., Groen, G. J., & Arocha, J. F. (1990). Medical expertise as a function of task difficulty. *Memory & Cognition* 18: 394–406.

Pauker, S. G. & Kassirer, J. P. (1980). The threshold approach to clinical decision making. *New England Journal of Medicine* 302: 1109–1117.

Pearl, J. (1988). *Probabilistic reasoning in intelligent systems: Networks of plausible inference*. San Francisco, CA: Morgan Kaufmann.

Pearl, J. (2000). *Causality: Models, reasoning, and inference*. Cambridge, UK: Cambridge University Press.

Peirce, C. S. (1934). *The collected papers of Charles S. Peirce, vol. V: Pragmatism and pragmaticism* (edited by C. Hartshorne & P. Weiss). Cambridge, MA: Harvard University Press.

Pennington, N. & Hastie, R. (1988). Explanation-based decision making: The effects of memory structure on judgment. *Journal of Experimental Psychology: Learning, Memory, and Cognition* 14: 521–533.

Pennington, N. & Hastie, R. (1992). Explaining the evidence: Tests of the story-model for juror decision making. *Journal of Personality and Social Psychology* 62: 189–206.

Pennington, N. & Hastie, R. (1993). Reasoning in explanation-based decision making. *Cognition* 49: 123–163.

Pettigrew, R. (2016). *Accuracy and the laws of credence*. Oxford, UK: Oxford University Press.

Phillips, L. D. & Edwards, W. (1966). Conservatism in a simple probability inference task. *Journal of Experimental Psychology* 72: 346–354.

Pluchino, A., Latora, V., & Rapisarda, A. (2006). Compromise and synchronization in opinion dynamics. *European Physical Journal B* 50: 169–176.

Pollock, J. (1995). *Cognitive carpentry*. Cambridge, MA: MIT Press.

Poore, G. (2015). PythonTEX: Reproducible documents with LATEX, Python, and more. *Computational Science & Discovery* 8: 014010.

Popper, K. R. (1959). *The logic of scientific discovery*. London, UK: Hutchinson.

Popper, K. R. (1963). *Conjectures and refutations*. London, UK: Routledge and Kegan Paul.

Poston, T. (2014). *Reason and explanation*. New York, NY: Palgrave Macmillan.

Prasetya, Y. (in press). Which models of scientific explanation are (in)compatible with IBE? *British Journal for the Philosophy of Science*.

Priest, G. (2005). *Doubt truth to be a liar*. Oxford, UK: Oxford University Press.

Pryor, J. (2000). The skeptic and the dogmatist. *Noûs* 34: 517–549.

Pryor, J. (2004). What's wrong with Moore's argument? *Philosophical Issues* 14: 349–378.

Psillos, S. (1996). On van Fraassen's critique of abductive reasoning. *Philosophical Quarterly* 46: 31–47.

Psillos, S. (1999). *Scientific realism: How science tracks truth*. London, UK: Routledge.

Psillos, S. (2000). Abduction: Between conceptual richness and computational complexity. In A. K. Kakas & P. Flach (eds.), *Abduction and induction: Essays on their relation and integration* (pp. 59–74). Dordrecht, Netherlands: Kluwer.

Psillos, S. (2004). Inference to the best explanation and Bayesianism. In F. Stadler (ed.), *Induction and deduction in the sciences* (pp. 83–91). Dordrecht, Netherlands: Kluwer.

Psillos, S. (2007). The fine structure of inference to the best explanation. *Philosophy and Phenomenological Research* 74: 441–448.

Putnam, H. (1981). *Reason, truth and history*. Cambridge, UK: Cambridge University Press.

Putnam, H. (1994). Sense, nonsense, and the senses: An inquiry into the powers of the human mind. *Journal of Philosophy* 91: 445–517.

Qian, C., Yang, Y, & Zhou, Z.-H. (2013). An analysis on recombination in multi-objective evolutionary optimization. *Artificial Intelligence* 204: 99–119.

Quine, W. V. O. (1992). *Pursuit of truth*. Cambridge, MA: Harvard University Press.

Rakshit, P. & Konar, A. (2015). Differential evolution for noisy multiobjective optimization. *Artificial Intelligence* 227: 165–189.

Ramsey, F. P. (1926). Truth and probability. In his *The foundations of mathematics* (pp. 156–198). London, UK: Routledge, 1931.

Ramsey, F. P. (1929). General propositions and causality. In his *The foundations of mathematics* (pp. 237–255). London, UK: Routledge, 1931.

Rastogi, V., Puri, N., Arora, S., Kaur, G., Yadav, L., & Sharma, R. (2013). Artefacts: A diagnostic dilemma. *Journal of Clinical and Diagnostic Research* 7: 2408–2413.

Reichenbach, H. (1938). *Experience and prediction*. Chicago, IL: University of Chicago Press.

Rellihan, M. J. (2009). Fodor's riddle of abduction. *Philosophical Studies* 144: 313–338.

Reuter, K. & Brössel, P. (2018). No knowledge required. *Episteme* 16: 303–321.

Reutlinger, A., Colyvan, M., & Krzyżanowska, K. H. (in press). The prospects for a monist theory of non-causal explanation in science and mathematics. *Erkenntnis*.

Riegler, A. & Douven, I. (2009). Extending the Hegselmann–Krause model III: From single beliefs to complex belief states. *Episteme* 6: 145–163.

Rinard, S. (2017). No exception for belief. *Philosophy and Phenomenological Research* 94: 121–143.

Rittle-Johnson, B. & Loehr, A. M. (2017). Eliciting explanations: Constraints on when self-explanation aids learning. *Psychonomic Bulletin & Review* 24: 1501–1510.

Robbins Landon, H. C. (1999). *1791: Mozart's last year*. New York, NY: Schirmer Books.

Roche, W. & Shogenji, T. (2018). Information and inaccuracy. *British Journal for the Philosophy of Science* 69: 577–604.

Roche, W. & Sober, E. (2013). Explanatoriness is evidentially irrelevant, or inference to the best explanation meets Bayesian confirmation theory. *Analysis* 73: 659–668.

Rock, I. (1983). *The logic of perception*. Cambridge, MA: MIT Press.

Rock, I. & Ebenholtz, S. (1962). Stroboscopic movement based on change of phenomenal rather than retinal location. *American Journal of Psychology* 75: 193–207.

Rock, I. & Gilchrist, A. (1975). The conditions for the perception of the covering and uncovering of a line. *American Journal of Psychology* 88: 571–582.

Rock, I. & Halper, F. (1969). Form perception without a retinal image. *American Journal of Psychology* 82: 425–440.

Rock, I. & Smith, D. (1981). Alternative solutions to kinetic stimulus transformations. *Journal of Experimental Psychology: Human Perception and Performance* 7: 19–29.

Rosenkrantz, R. D. (1981). *Foundations and applications of inductive probability*. Atascadero, CA: Ridgeview.

Rosenkrantz, R. D. (1992). The justification of induction. *Philosophy of Science* 59: 527–539.

Rosenstock, S., Bruner, J., & O'Connor, C. (2017). In epistemic networks, is less really more? *Philosophy of Science* 84: 234–252.

Rostworowski, W., Pietrulewicz, N., & Będkowski, M. (in press). Conditionals and specific links—an experimental study. *Synthese*, https://doi.org/10.1007/s11229-021-03119-2.

Ruggeri, A., Xu, F., & Lombrozo, T. (2019). Effects of explanation on children's question asking. *Cognition* 186: 82–94.

Ruse, M. (1975). Darwin's debt to philosophy: An examination of the influence of the philosophical ideas of John F. W. Herschel and William Whewell on the development of Charles Darwin's theory of evolution. *Studies in the History and Philosophy of Science* 6: 159–181.

Ruse, M. (1995). *Evolutionary naturalism*. London, UK: Routledge.

Russell, B. (1912). *The problems of philosophy*. Oxford, UK: Oxford University Press.

Russell, G. (2015). The justification of the basic laws of logic. *Journal of Philosophical Logic* 44: 793–803.

Russell, G. (2019). Deviance and vice: Strength as a theoretical virtue in the epistemology of logic. *Philosophy and Phenomenological Research* 99: 548–563.

Ryan, S. (1996). The epistemic virtues of consistency. *Synthese* 109: 121–141.

Saatsi, J. (in press). Non-causal explanations in physics. In E. Knox & A. Wilson (eds.), *The Routledge companion to the philosophy of physics*. London, UK: Routledge.

Salmon, W. C. (1963). On vindicating induction. *Philosophy of Science* 30: 252–261.

Salmon, W. C. (1984). *Scientific explanation and the causal structure of the world*. Princeton, NJ: Princeton University Press.

Sarkar, D. & Modak, J. M. (2005). Pareto-optimal solutions for multi-objective optimization of fed-batch bioreactors using nondominated sorting genetic algorithm. *Chemical Engineering Science* 60: 481–492.

Schippers, M. (2014). Probabilistic measures of coherence: From adequacy constraints towards pluralism. *Synthese* 191: 3821–3845.

Schippers, M. & Schurz, G. (2017). Genuine coherence as mutual confirmation between content elements. *Studia Logica* 105: 299–329.

Schmitt, F. (1995). *Truth: A primer*. Boulder, CO: Westview.

Schum, D. A. & Martin, A. W. (1982). Formal and empirical research on cascaded inference in jurisprudence. *Law and Society Review* 17: 105–151.

Schupbach, J. N. (2005). Paley's inductive inference to design. *Philosophia Christi* 7: 491–502.

Schupbach, J. N. (2011a). Comparing probabilistic measures of explanatory power. *Philosophy of Science* 78: 813–829.

Schupbach, J. N. (2011b). New hope for Shogenji's coherence measure. *British Journal for the Philosophy of Science* 62: 125–142.

Schupbach, J. N. (2014). Is the bad lot objection just misguided? *Erkenntnis* 79: 55–64.

Schupbach, J. N. (2017). Inference to the best explanation, cleaned up and made respectable. In T. Poston & K. McCain (eds.), *Best explanations: New essays on inference to the best explanation* (pp. 39–61). Oxford, UK: Oxford University Press.

Schupbach, J. N. & Glass, D. H. (2017). Hypothesis competition beyond mutual exclusivity. *Philosophy of Science* 84: 810–824.

Schupbach, J. N. & Sprenger, J. (2011). The logic of explanatory power. *Philosophy of Science* 78: 105–127.

Schurz, G. (1987). A new definition of verisimilitude and its applications. In P. Weingartner & G. Schurz (eds.), *Logic, Philosophy of Science and Epistemology* (pp. 177–184). Vienna, Austria: Hölder-Pichler-Tempsky.

Schurz, G. (1991). Relevant deduction. *Erkenntnis* 35: 391–437.

Schurz, G. (2008a). Patterns of abduction. *Synthese* 164: 201–234.

Schurz, G. (2008b). The meta-inductivist's winning strategy in the prediction game: A new approach to Hume's problem. *Philosophy of Science* 75: 278–305.

Schurz, G. (2009). Meta-induction and social epistemology. *Episteme* 6: 200–220.

Schurz, G. (2011). Verisimilitude and belief revision. *Erkenntnis* 75: 203–221.

Schurz, G. (2012). Meta-induction in epistemic networks and the social spread of knowledge. *Episteme* 9: 151–170.

Schurz, G. (2014). *Philosophy of science: A unified approach*. New York, NY: Routledge.

Schurz, G. (2017). No free lunch theorem, inductive skepticism, and the optimality of meta-induction. *Philosophy of Science* 84: 825–839.

Schurz, G. (2019). *Hume's problem solved: The optimality of meta-induction*. Cambridge, MA: MIT Press.

Schurz, G. (2021). Abduction as a method of inductive metaphysics. *Grazer Philosophische Studien* 98: 50–74.

Schurz, G. (in press). Meta-induction over unboundedly many prediction methods: A reply to Arnold and Sterkenburg. *Philosophy of Science*.

Schurz, G. & Hertwig, R. (2019). Cognitive success: A consequentialist account of rationality and cognition. *Topics in Cognitive Science* 11: 7–36.

Schurz, G. & Thorn, P. D. (2016). The revenge of ecological rationality: Strategy-selection by meta-induction within changing environments. *Minds and Machines* 26: 31–59.

Schwarz, G. (1978). Estimating the dimension of a model. *Annals of Statistics* 6: 461–465.

Schwitzgebel, E. (2008). The unreliability of naive introspection. *Philosophical Review* 117: 245–273.

Selten, R. (1998). Axiomatic characterization of the quadratic scoring rule. *Experimental Economics* 1: 43–62.

Semeshenko, V., Gordon, M. B., & Nadal, J.-P. (2008). Collective states in social systems with interacting learning agents. *Physica A: Statistical Mechanics and its Applications* 387: 4903–4916.

Shafer, G. (1985). Conditional probability. *International Statistical Review* 85: 261–277.

Shepard, R. N. (1987). Toward a universal law for psychological science. *Science* 237: 1317–1323.

Sheredos, B., Burnston, D., Abrahamsen, A., & Bechtel, W. (2013). Why do biologists use so many diagrams? *Philosophy of Science* 80: 931–944.

Shimojo, A., Miwa, K., & Terai, H. (2020). How does explanatory virtue determine probability estimation?—Empirical discussion on effect of instruction. *Frontiers in Psychology* 11, https://doi.org/10.3389/fpsyg.2020.575746.

Shogenji, T. (1999). Is coherence truth-conducive? *Analysis* 59: 338–345.

Shoots-Reinhard, B. L., Rucker, D. D., Petty, R. E., & Shakarchi, R. (2014). Not all contrast effects are created equal: Extent of processing affects contrast strength. *Journal of Applied Social Psychology* 44: 523–535.

Sider, T. (2009). Ontological antirealism. In D. Chalmers, D. Manley, & R. Wasserman (eds.), *Metametaphysics: New essays on the foundations of ontology* (pp. 384–421). Oxford, UK: Oxford University Press.

Sidney, P. G., Hattikudur, S., & Alibali, M. W. (2015). How do contrasting cases and self-explanation promote learning? Evidence from fraction division. *Learning and Instruction* 40: 29–38.

Siebel, M. & Wolff, W. (2008). Equivalent testimonies as a touchstone of coherence measures. *Synthese* 161: 167–182.

Sigman, E. & Rock, I. (1974). Stroboscopic movement based on perceptual intelligence. *Perception* 3: 9–28.

Simon, H. A. (1982). *Models of bounded rationality* (vol. 1). Cambridge, MA: MIT Press.

Singmann, H. & Kellen, D. (2020). An introduction to mixed models for experimental psychology. In D. Spiler & E. Schumacher (eds.), *New methods in cognitive psychology* (pp. 4–31). New York, NY: Routledge.

Skyrms, B. (1987). Coherence. In N. Rescher (ed.), *Scientific inquiry in philosophical perspective* (pp. 225–241). Lanham, MD: University Press of America.

Skyrms, B. (1993a). Analogy by similarity in hyper-carnapian inductive logic. In J. Earman, A. I. Janis, J. Massey, & N. Rescher (eds.), *Philosophical problems of the internal and external worlds* (pp. 273–282). Pittsburgh, PA: University of Pittsburgh Press.

Skyrms, B. (1993b). A mistake in dynamic coherence arguments? *Philosophy of Science* 60: 320–328.

Sloman, S. A. (2005). *Causal models: How people think about the world and its alternatives*. Oxford, UK: Oxford University Press.

Sober, E. (2002). Instrumentalism, parsimony, and the Akaike framework. *Philosophy of Science* 69: S112–S123.

Sosa, E. (1997). Reflective knowledge in the best circles. *Journal of Philosophy* 94: 410–430.

Spencer, J. (1961). Estimating averages. *Ergonomics* 4: 317–328.

Spencer, J. (1963). A further study of estimating averages. *Ergonomics* 6: 255–265.

Sprenger, J. (2018). Foundations of a probabilistic theory of causal strength. *Philosophical Review* 127: 371–398.

Sprenger, J. (2020). Conditional degree of belief and Bayesian inference. *Philosophy of Science* 87: 319–335.

Sprenger, J. & Hartmann, S. (2019). *Bayesian philosophy of science*. Oxford, UK: Oxford University Press.

Stegenga, J. (2018). *Medical nihilism*. Oxford, UK: Oxford University Press.

Steglich-Petersen, A. & Skipper, M. (2019). An instrumentalist account of how to weigh epistemic and practical reasons for belief. *Mind* 129: 1071–1094.

Sterkenburg, T. F. (2019). The meta-inductive justification of induction: The pool of strategies. *Philosophy of Science* 86: 981–992.

Sterkenburg, T. F. (2020). The meta-inductive justification of induction. *Episteme* 17: 519–541.

Stevens, S. S. (1971). Issues in psychophysical measurement. *Psychological Review* 78: 426–450.

Stewart, A. J., Singmann, H., Haigh, M., Wood, J. S., & Douven, I. (2021). Tracking the eye of the beholder: Is explanation subjective? *Journal of Cognitive Psychology* 33: 199–206.

Stilgenbauer, J.-L. & Baratgin, J. (2018). Étude des stratégies de raisonnement causal dans l'estimation de la probabilité diagnostique à travers un paradigme expérimental de production de règle. *Canadian Journal of Experimental Psychology/Revue canadienne de psychologie expérimentale* 72: 58–70.

Stone, M. (1961). The opinion pool. *Annals of Mathematical Statistics* 32: 1339–1342.

Strevens, M. (2013). No understanding without explanation. *Studies in History and Philosophy of Science* 44: 510–515.

Surowiecki, J. (2004). *The wisdom of crowds.* New York, NY: Anchor Books.

Swami, V., Coles, R., Stieger, S., Pietschnig, J., Furnham, A., Rehim, S., & Voracek, M. (2011). Conspiracist ideation in Britain and Austria: Evidence of a monological belief system and associations between individual psychological differences and real-world and fictitious conspiracy theories. *British Journal of Psychology* 102: 443–463.

Tamargo, L. H., Garcia, A. J., Falappa, M. A., & Simari, G. R. (2014). On the revision of informant credibility orders. *Artificial Intelligence* 212: 36–58.

Teller, P. (1973). Conditionalization and observation. *Synthese* 26: 218–258.

Tenenbaum, J. B., Griffiths, T. L., & Kemp, C. (2006). Theory-based Bayesian models of inductive learning and reasoning. *Trends in Cognitive Science* 10: 304–318.

Tenenbaum, J. B., Kemp, C., Griffiths, T. L., & Goodman, N. D. (2011). How to grow a mind: Statistics, structure, and abstraction. *Science* 331: 1279–1285.

Tenney, E. R., Cleary, H. M. D., & Spellman, B. A. (2009). Unpacking the doubt in "beyond a reasonable doubt": Plausible alternative stories increase not guilty verdicts. *Basic and Applied Social Psychology* 31: 1–8.

Tentori, K., Crupi, V., Bonini, N., & Osherson, D. (2007). Comparison of confirmation measures. *Cognition* 103: 107–119.

Tentori, K., Crupi, V., & Russo, S. (2013). On the determinants of the conjunction fallacy: Probability versus inductive confirmation. *Journal of Experimental Psychology: General* 142: 235–255.

Tetlock, P. & Gardner, D. (2015). *Superforecasting: The art and science of prediction*. London, UK: Penguin Random House.

Thagard, P. R. (1977). Discussion: Darwin and Whewell. *Studies in the History and Philosophy of Science* 8: 353–356.

Thagard, P. R. (1978). The best explanation: Criteria for theory choice. *Journal of Philosophy* 75: 76–92.

Thagard, P. R. (1989). Explanatory coherence. *Behavioral and Brain Sciences* 12: 435–502.

Thagard, P. R. (2000). *Coherence in thought and action*. Cambridge, MA: MIT Press.

Thompson, V. A., Prowse Turner, J. A., & Pennycook, G. (2011). Intuition, reason, and metacognition. *Cognitive Psychology* 63: 107–140.

Thomson, J. J. (1897). Cathode rays. *Philosophical Magazine* 44: 303–326.

't Hooft, G. (1994). Questioning the answers or stumbling upon good and bad theories of everything. In J. Hilgevoord (ed.), *Physics and our view of the world* (pp. 16–37). Cambridge, UK: Cambridge University Press.

Thorn, P. & Schurz, G. (2019). Meta-inductive prediction based on attractivity weighting: Mathematical and empirical performance evaluation. *Journal of Mathematical Psychology* 89: 13–30.

Tichý, P. (1974). On Popper's definition of verisimilitude. *British Journal for the Philosophy of Science* 25: 155–160.

Todd, P. M. & Brighton, H. (2016). Building the theory of ecological rationality. *Minds and Machines* 26: 9–30.

Todd, P. M. & Gigerenzer, G. (2012). *Ecological rationality: Intelligence in the world*. Oxford, UK: Oxford University Press.

Tregear, M. (2004). Utilising explanatory factors in induction? *British Journal for the Philosophy of Science* 55: 505–519.

Trpin, B. & Pellert, M. (2019). Inference to the best explanation in uncertain evidential situations. *British Journal for the Philosophy of Science* 70: 977–1001.

Tversky, A. & Kahneman, D. (1983). Extensional versus intuitive reasoning: The conjunction fallacy in probability judgment. *Psychological Review* 90: 293–315.

Uffink, J. B. M. (1995). Can the maximum entropy principle be explained as a consistency requirement? *Studies in History and Philosophy of Modern Physics* 26: 223–261.

Vahdati, A. R. (2019). `Agents.jl`: Agent-based modeling framework in Julia. *Journal of Open Source Software* 4: 1611, https://doi.org/10.21105/joss.01611.

Vahid, H. (2004). Varieties of epistemic conservatism. *Synthese* 141: 97–122.

Vallinder, A. & Olsson, E. J. (2013). Do computer simulations support the argument from disagreement? *Synthese* 190: 1437–1454.

van Elswyk, P. (2019). Propositional anaphors. *Philosophical Studies* 176: 1055–1075.

van Fraassen, B. C. (1977). The only necessity is verbal necessity. *Journal of Philosophy* 74: 71–85.

van Fraassen, B. C. (1980). *The scientific image*. Oxford, UK: Oxford University Press.

van Fraassen, B. C. (1983). Theory comparison and relevant evidence. In J. Earman (ed.), *Testing scientific theories* (pp. 27–42). Minneapolis, MN: University of Minnesota Press.

van Fraassen, B. C. (1985). Empiricism in the philosophy of science. In P. Churchland & C. Hooker (eds.), *Images of science* (pp. 245–308). Chicago, IL: University of Chicago Press.

van Fraassen, B. C. (1989). *Laws and symmetry*. Oxford, UK: Oxford University Press.

Vasilyeva, N. & Lombrozo, T. (2020). Structural thinking about social categories: Evidence from formal explanations, generics, and generalization. *Cognition* 204: 104383.

Vassend, O. B. (2019). New semantics for Bayesian inference: The interpretive problem and its solutions. *Philosophy of Science* 86: 696–718.

Vassend, O. B. (in press). Justifying the norms of inductive inference. *British Journal for the Philosophy of Science*.

Vehtari, A., Gelman, A., & Gabry, J. (2017). Practical Bayesian model evaluation using leave-one-out cross-validation and WAIC. *Statistics and Computing* 27: 1413–1432

Vineberg, S. (2016). Dutch book arguments. In E. N. Zalta (ed.), *The Stanford encyclopedia of philosophy*, https://plato.stanford.edu/archives/spr2016/entries/dutch-book/.

Vogel, J. (1990). Cartesian skepticism and inference to the best explanation. *Journal of Philosophy* 87: 658–666.

Vogel, J. (2005). The refutation of skepticism. In M. Steup & E. Sosa (eds.), *Contemporary debates in epistemology* (pp. 72–84). Malden, MA: Blackwell.

Vogel, J. (2008). Epistemic bootstrapping. *Journal of Philosophy* 105: 518–539.

von Helmholtz, H. (1867). *Handbuch der physiologischen Optik*. Leipzig, Germany: Leopold Voss.

Walker, C. M. & Lombrozo, T. (2017). Explaining the moral of the story. *Cognition* 167: 266–281.

Walker, C. M., Lombrozo, T., Williams, J. J., Rafferty, A., & Gopnik, A. (2017). Explaining constrains causal learning in childhood. *Child Development* 88: 229–246.

Walton, D. N. (2005). *Abductive reasoning*. Tuscaloosa, AL: University of Alabama Press.

Wang, H. (1996). *A logical journey: From Gödel to philosophy*. Cambridge, MA: MIT Press.

Wason, P. C. (1968). Reasoning about a rule. *Quarterly Journal of Experimental Psychology* 20: 273–281.

Weatherson, B. (2003). From classical to intuitionistic probability. *Notre Dame Journal of Formal Logic* 44: 111–123.

Weinberg, J. M., Gonnerman, C., Buckner, C., & Alexander, J. (2010). Are philosophers expert intuiters? *Philosophical Psychology* 23: 331–355.

Weinstein, J. (2008). A postgenomic visual icon. *Science* 319: 1772–1773.

Weintraub, R. (2013). Induction and inference to the best explanation. *Philosophical Studies* 166: 203–216.

Weintraub, R. (2017). Scepticism about inference to the best explanation. In K. McCain & T. Poston (eds.), *Best explanations: New essays on inference to the best explanation* (pp. 188–202). Oxford, UK: Oxford University Press.

Weisberg, J. (2009). Locating IBE in the Bayesian framework. *Synthese* 167: 125–143.

Weisberg, J. (2011). Varieties of Bayesianism. In D. M. Gabbay, S. Hartmann, & J. Woods (eds.), *Handbook of the history of logic, vol. 10: Inductive logic* (pp. 477–551). Amsterdam, Netherlands: North-Holland.

Weisberg, J. (2020). Belief in psyontology. *Philosophers' Imprint* 20: 1–27.

Weisbuch, G., Deffuant, G., Amblard, F., & Nadal, J.-P. (2002). Meet, discuss and segregate! *Complexity* 7: 55–63.

Wenmackers, S. (2013). Ultralarge lotteries: Analyzing the lottery paradox using non-standard analysis. *Journal of Applied Logic* 11: 452–467.

Wenmackers, S., Vanpoucke, D., & Douven, I. (2012). Probability of inconsistencies in theory revision: A multi-agent model for updating logically interconnected beliefs under bounded confidence. *European Physical Journal B* 85, https://doi.org/10.1140/epjb/e2011-20617-8.

Wenmackers, S., Vanpoucke, D., & Douven, I. (2014). Rationality: A social-epistemology perspective. *Frontiers in Psychology* 5: 581, https://doi.org/10.3389/fpsyg.2014.00581.

Westfall, R. S. (1980). *Never at rest*. Cambridge, UK: Cambridge University Press.

Whewell, W. (1847). *The philosophy of the inductive sciences, founded upon their history* (2 vols., 2nd ed.). London, UK: John W. Parker.

Wilenski, U. & Rand, W. (2015). *An introduction to agent-based modeling: Modeling natural, social, and engineered complex systems with NetLogo*. Cambridge, MA: MIT Press.

Williams, J. J. & Lombrozo, T. (2010). The role of explanation in discovery and generalization: Evidence from category learning. *Cognitive Science* 34: 776–806.

Williams, J. J. & Lombrozo, T. (2013). Explanation and prior knowledge interact to guide learning. *Cognitive Psychology* 66: 55–84.

Williams, J. R. G. (2012). Generalized probabilism: Dutch books and accuracy domination. *Journal of Philosophical Logic*, *41*, 811–840.

Williamson, T. (1998). Conditionalizing on knowledge. *British Journal for the Philosophy of Science* 49: 89–121.

Williamson, T. (2000). *Knowledge and its limits*. Oxford, UK: Oxford University Press.

Williamson, T. (2007). *The philosophy of philosophy*. Oxford, UK: Blackwell.

Williamson, T. (2016). Modal science. *Canadian Journal of Philosophy* 46: 453–492.

Williamson, T. (2017). Semantic paradoxes and abductive methodology. In B. Armour-Garb (ed.), *Reflections on the liar* (pp. 325–346). Oxford, UK: Oxford University Press

Williamson, T. (2018). *Doing philosophy*. Oxford, UK: Oxford University Press.

Wilson, E. O. (2009). Foreword. In M. Ruse & R. J. Richards (eds.), *The Cambridge companion to the "Origin of Species"* (pp. xv–xvi). Cambridge, UK: Cambridge University Press.

Wilson, N. L. (1959). Substances without substrata. *Review of Metaphysics* 12: 521–539.

Winkler, R. L. (1969). Scoring rules and the evaluation of probability assessors. *Journal of the American Statistical Association* 64: 1073–1078.

Winkler, R. L. (1996). Scoring rules and the evaluation of probabilities. *Test* 5: 1–60.

Winkler, R. L. & Murphy, A. H. (1968). "Good" probability assessors. *Journal of Applied Meteorology* 7: 751–758.

Woody, A. I. (2004). More telltale signs: What attention to representation reveals about scientific explanation. *Philosophy of Science* 71: 780–793.

Woody, A. I. (2015). Re-orienting discussions of scientific explanation: A functional perspective. *Studies in History and Philosophy of Science* 52: 79–87.

Wouters, A. (1995). Viability explanation. *Biology & Philosophy* 10: 435–457.

Wouters, A. (2007). Design explanation: Determining the constraints on what can be alive. *Erkenntnis* 67: 65–80.

Wright, C. (1985). Facts and certainty. *Proceedings of the British Academy* 71: 429–472.

Wright, C. (1991). Scepticism and dreaming: Imploding the demon. *Mind* 100: 87–116.

Yu, S. & Zenker, F. (2018). Peirce knew why abduction isn't IBE—A scheme and critical questions for abductive argument. *Argumentation* 32: 569–587.

Yu, Y., Yao, X., & Zhou, Z.-H. (2012). On the approximation ability of evolutionary optimization with application to minimum set cover. *Artificial Intelligence* 180–181: 20–33.

Zahar, E. (1973). Why did Einstein's programme supersede Lorentz's? (II). *British Journal for the Philosophy of Science* 24: 223–262.

Zegers, R. H. C., Weigl, A., & Steptoe, A. (2009). The death of Wolfgang Amadeus Mozart: An epidemiological perspective. *Annals of Internal Medicine* 151: 274–278.

Zhao, J., Shah, A., & Osherson, D. (2009). On the provenance of judgments of conditional probability. *Cognition* 113: 26–36.

Zollman, K. J. S. (2007). The communication structure of epistemic communities. *Philosophy of Science* 74: 574–587.

Zuur, A. F., Ieno, E. N., Walker, N. J., Savaliev, A. A., & Smith, G. M. (2009). *Mixed effects models and extensions in ecology with* R. New York, NY: Springer.

Zwart, S. D. (2001). *Refined verisimilitude.* Dordrecht, Netherlands: Kluwer.

Index